PPG INDUSTRIES
COATINGS & RESINS DIV.
RESEARCH LIBRARY
P.O. BOX 1009/ROSANNA DR.
ALLISON PARK, PA. 15101

AF737497

Graft Copolymerization of Lignocellulosic Fibers

Graft Copolymerization of Lignocellulosic Fibers

David N.-S. Hon, Editor

Virginia Polytechnic Institute and State University

Based on a symposium sponsored by the ACS Division of Cellulose, Paper, and Textile Chemistry at the 182nd Meeting of the American Chemical Society, New York, New York, August 23–28, 1981.

Pub. Std. Order 7-17-82

678.013
G737

ACS SYMPOSIUM SERIES **187**

AMERICAN CHEMICAL SOCIETY

WASHINGTON, D. C. 1982

PPG INDUSTRIES
COATINGS & RESINS DIV.
RESEARCH LIBRARY
P.O. BOX 1009, ROSANNA DR.
ALLISON PARK, PA. 15101

Library of Congress Cataloging in Publication Data

Graft copolymerization of lignocellulosic fibers.
(ACS symposium series, ISSN 0097-6156; 187)

"Based on a symposium sponsored by the ACS Division of Cellulose, Paper, and Textile Chemistry at the 182nd Meeting of the American Chemical Society, New York, New York, August 23-28, 1981."

Includes bibliographies and index.

1. Graft copolymers—Congresses. 2. Cellulose—Congresses.
I. Hon, David N.-S. II. American Chemical Society. Cellulose, Paper, and Textile Division. III. Series.

QD382.G7G73 547.7'82 82-6717
ISBN 0-8412-0721-6 AACR2 ACSMC8 189 1-382
1982

Copyright © 1982

American Chemical Society

All Rights Reserved. The appearance of the code at the bottom of the first page of each article in this volume indicates the copyright owner's consent that reprographic copies of the article may be made for personal or internal use or for the personal or internal use of specific clients. This consent is given on the condition, however, that the copier pay the stated per copy fee through the Copyright Clearance Center, Inc. for copying beyond that permitted by Sections 107 or 108 of the U.S. Copyright Law. This consent does not extend to copying or transmission by any means—graphic or electronic—for any other purpose, such as for general distribution, for advertising or promotional purposes, for creating new collective work, for resale, or for information storage and retrieval systems.

The citation of trade names and/or names of manufacturers in this publication is not to be construed as an endorsement or as approval by ACS of the commercial products or services referenced herein; nor should the mere reference herein to any drawing, specification, chemical process, or other data be regarded as a license or as a conveyance of any right or permission, to the holder, reader, or any other person or corporation, to manufacture, reproduce, use, or sell any patented invention or copyrighted work that may in any way be related thereto.

PRINTED IN THE UNITED STATES OF AMERICA

ACS Symposium Series

M. Joan Comstock, *Series Editor*

Advisory Board

David L. Allara
Robert Baker
Donald D. Dollberg
Robert E. Feeney
Brian M. Harney
W. Jeffrey Howe
James D. Idol, Jr.
Herbert D. Kaesz
Marvin Margoshes
Robert Ory
Leon Petrakis
Theodore Provder
Charles N. Satterfield
Dennis Schuetzle
Davis L. Temple, Jr.
Gunter Zweig

FOREWORD

The ACS SYMPOSIUM SERIES was founded in 1974 to provide a medium for publishing symposia quickly in book form. The format of the Series parallels that of the continuing ADVANCES IN CHEMISTRY SERIES except that in order to save time the papers are not typeset but are reproduced as they are submitted by the authors in camera-ready form. Papers are reviewed under the supervision of the Editors with the assistance of the Series Advisory Board and are selected to maintain the integrity of the symposia; however, verbatim reproductions of previously published papers are not accepted. Both reviews and reports of research are acceptable since symposia may embrace both types of presentation.

CONTENTS

PREFACE

The shortage of oil and natural gas has been reflected in shortages and spiraling prices for polymers based on petrochemical resources. Providentially, such polymers may be in part replaced by lignocellulosic materials that are the most abundant and most economical organic renewable resources available. In their natural state as wood and plant fibers, and as the principal constituent used in the manufacture of paper, textile fibers, and many other industrial products, lignocellulosic materials will continue to be fundamental to human welfare.

Lignocellulosic materials are eternally renewable raw materials that can be used in their polymeric form without degradation. However, their properties and performance must be improved before they can compete with petrochemically derived synthetic polymers. Many new techniques for improving the competitive position of lignocellulosic materials are available. The grafting of vinyl monomers onto lignocellulosic fibers is one of these techniques. However, this technique has need of amendment to obtain economically feasible grafted lignocellulosic copolymers of superior quality.

This volume is a collection of topics discussed at a symposium organized to provide an international forum for discussion of problems and challenges related to grafting reactions of lignocellulosic materials. Participating scientists were invited from Australia, Bangladesh, Canada, Egypt, France, India, Japan, Romania, Sweden, Taiwan, and the United States.

This volume contains the latest information on grafting reactions important to the polymer, fiber, and wood products industries. Chapters of this volume are divided into four general categories dealing with grafting reactions of (1) cellulose and cellulose derivatives, (2) starch and bamboo, (3) lignocellulosic fibers, and (4) chemically modified lignocellulosic fibers. These chapters provide only a small sampling of the activity in the field and the high level of interest in grafting research is continuing. One can thus look forward to a vigorous activity in this area. It is hoped that this volume will be timely and helpful and will serve as a source of information and a base from which further development of the science and technology of grafted lignocellulosics can emerge.

Finally, I would like to thank all of the patient and enthusiastic contributors who prepared the manuscripts, participated as speakers at the symposium, and reviewed the papers of this volume. Special thanks also go to V. Stannett (North Carolina State University) and Roger Rowell (USDA Forest Products Laboratory) for chairing sessions at the symposium.

DAVID N.-S. HON
Virginia Polytechnic Institute
and State University
Department of Forest Products
Blacksburg, Virginia 24061

December 1, 1981

GRAFT COPOLYMERIZATION OF CELLULOSE AND CELLULOSE DERIVATIVES

1

Some Challenges in Grafting to Cellulose and Cellulose Derivatives

V. STANNETT

North Carolina State University, Department of Chemical Engineering, Raleigh, NC 27650

Grafting reactions to cellulose and its derivatives began in the early 1950's. There has been intense research and development activity in the field since those early days and many hundreds of papers and patents have been published. In spite of all this effort there has been comparatively little commercialization of the grafting process. The turning around of this situation represents the greatest challenge to those of us who are active in the field. The reasons for the lack of industrialization on a large scale have been partly economic and partly the ability of researchers to find alternative means to achieve comparable properties. The author feels, however, that the situation will change once certain technical challenges are met. A number of these are discussed together with some still promising fields of application. Among the technical problems which still remain to a considerable extent are the concurrent formation of homopolymer in most cases, and the lack of reproducibility in these largely heterogeneous reactions. In addition, the grafted side chains are often too few and too long, leading to a limited involvement also of the cellulose molecules themselves. Although there is often some concurrent degradation of the cellulose itself, this is not believed to be normally an important consideration.

Conscious grafting to cellulose was first reported in 1953 (1), although relevant and somewhat earlier work in the U.S.S.R. (2) and in England (3) certainly preceded it. This work was variously described as copolymerization, actually accompanied by extensive crosslinking (2), and as polymer deposition (3). The author was privileged to follow on the 1953 work that same year and has continued to work in the field since that time.

0097-6156/82/0187-0003$6.00/0
© 1982 American Chemical Society

The possibility of grafting synthetic polymers to cellulose immediately attracted worldwide attention as a new and exciting way to modify cellulose and extend its uses against the rapidly growing competition from synthetic polymers themselves. The well known phrase attributed to Theodore Roosevelt, "if you can't beat them, join them", if somewhat trite, seems particularly appropriate at this point. Research in the field blossomed quickly and is still an extremely active subject of study. For example, in a very recent (and the first) book on the subject by Hebeish and Guthrie (4) more than one thousand references are quoted.

In spite of these enormous efforts, there is still no large-scale commercial application of cellulose graft copolymers. The reasons for this situation and the challenge it represents to cellulose and polymer scientists and engineers will be the subject of this introductory paper. It is convenient to break down such a discussion into the following areas, synthesis, characterization, properties and, finally, applications. The discussion will be mainly devoted to cellulose itself, although grafting to cellulose derivatives has also been actively pursued.

Synthesis of Cellulose Graft Copolymers

This area of research has almost totally dominated the field and continues to do so, partly because of the patent opportunities which arise. There are now a considerable number of methods available, each with its own particular advantages and disadvantages. Many of the methods are discussed in this symposium series. Among the older methods, studies of redox systems, direct oxidation and high energy radiation are presented. The comparatively new method, but actually developed by the Scott Paper Co. as early as 1964, is also discussed (5). There seems to be a growing interest in the use of uv for surface grafting to cellulose and considerable work by Arthur (6), Garnett (7), Hon (8) and others has been reported in recent years. Plasma has also been investigated, but to a lesser extent, by Yasuda, Simionescu and their colleagues, among others, again for surface modifications. Extensive reviews of work up to 1965 and 1970 have been presented by Krassig and Stannett (9) and by Arthur (10), respectively. The whole field, including quite recent development, has been described by Hebeish and Guthrie (4). Until recently all the synthetic approaches have been heterogeneous reactions, i.e., to cellulose fibers, film and powder. This is clearly more convenient, especially in the absence of practical solvent systems for cellulose itself. Recently, however, Shiraishi *et al.* (11,12) have carried out grafting to cellulose in some of the newer cellulose solvents. In principle, this type of grafting would become practical if it could be followed by direct spinning or film casting. Alternatively, powdered, fibrous or granular products could be obtained by precipitation. It is interesting and important that Shiraishi *et al.* (12) did use the thin layer technique

discussed later in this paper to characterize their products. The consequences of the heterogeneous nature of the grafting reaction are quite far reaching, as will be described later.

There are some formidable challenges facing those engaged in the synthetic aspects of cellulose grafting if viable commercial processes are to be developed. The most important of these may be enumerated as follows:

1. The elimination, or at least minimization, of concurrent homopolymer formation.
2. The involvement of all or most of the cellulose molecules in the grafting process.
3. Better control of the molecular weights and molecular weight distribution of the grafted side chains.
4. Better reproducibility of the grafting yields, properties and other features of the graft copolymers.

There are, no doubt, other considerations which have been overlooked. The present situation with regard to these significant problems appears to be the following.

Elimination or Reduction of Homopolymer

It is very difficult to experimentally determine the percent of homopolymer in the case of grafting to cellulose itself. Often the grafted fiber or film is simply extracted to constant weight with a suitable solvent for the homopolymer. It is known that this is inadequate. A better extraction may be made by swelling the graft in a cellulose solvent followed by, or simultaneously, shaking with the homopolymer solvent. Recently Inagaki *et al.* (13) have hydrolyzed away the cellulose from the grafted product after first extracting with a solvent for the homopolymer. The residue is then quantitatively separated by thin layer chromatography into the parts which contain sugar residues and true grafts, and those which do not. If cellulose derivatives are used as substrates, much better separations can often be achieved.

Most cellulose grafting methods are free radical in nature, involving various vinyl monomers and monomer mixtures. If the initiation reaction can be confined to cellulose macroradicals, there should be little or no homopolymer. This can, in principle, be achieved by preirradiation or with various redox systems. In practice, however, there always seems to be some homopolymer, but sometimes as low as 5%, i.e., 95% grafting efficiency. The homopolymer is produced in these cases by radical fragments from the cellulose or, more likely, by various chain transfer mechanisms, e.g., to the monomer itself, swelling agents and so on. With most radical grafting reactions both macroradicals and small molecule radicals are produced, resulting often in as much as 50% of homopolymer. Coupled with transfer reactions, up to 80-90% of homopolymer can be formed. The grafting reaction can also result in direct initiation of the vinyl monomer, as is the case with the mutual radiation and the peroxide method, with ceric ion initiation with many monomers and with other systems. It is clear that each initiator monomer system will have its own particular yield of homopolymer. One very practical method of minimizing homopolymer formation is by the addition of very small amounts of a difunctional monomer, such as ethylene glycol dimethacrylate. This effectively ties any homopolymers into the grafted chains increasing the grafting efficiency. A good example of this is in the grafting of acrylic acid to cotton to increase water sorbency and other properties (14). The literature contains many other examples of methods of increasing the grafting efficiency with various methods.

There remains the question of the seriousness, from a practical standpoint, of homopolymer formation. Again, this clearly depends on the particular synthetic method and the application envisaged for the graft copolymer. Even if only small quantities of homopolymer are produced in the reaction medium, usually liquid, or after washing the grafted material, this could be sufficient cause to abandon the process. Apart from the cost of such wasted polymer, there is the question of disposing of it and the attendant pollution and other problems. On the other hand, if the homopolymer is firmly sealed into the substrate film or fiber, it could be quite harmless and even as beneficial as if actually grafted. This would be true, for example, in the case of flame retardant polymer grafting. This is probably not true, however, when changes in the mechanical behavior and some other properties of the fibers or films is the objective. In the case of surface grafting, homopolymer would certainly be undesirable, although easily removed. Again, each situation would have to be investigated and judged on its own merits.

Greater Involvement of the Cellulose Molecules

The problem of increased efficiency of the grafting process from the point of view of the minimizing of concurrent homopolymer formation has been discussed in the previous section. The related problem of maximizing the participation of the cellulose substrate is also of great importance. In most of the methods of grafting, including ceric ion and radiation, only a comparatively small proportion of the cellulose is usually grafted. The homopolymer can be removed, the cellulose backbone hydrolyzed away, and the molecular weights of the side chains determined in many instances. It is then easy to estimate the number of anhydro glucose units per grafted side chain. If the degree of polymerization of the cellulose starting material is known, the proportion of chains actually grafted can be estimated. If some cellulose degradation occurs during the reaction, the measured values will be somewhat low. On the other hand, it is difficult to remove all of the homopolymer which will lead to higher values than the actual ones. In general, it can be assumed that the measured values are, if anything, rather too high.

Some typical results of such studies are presented for the radiation, ceric ion, manganic pyrophosphate and xanthate grafting methods in Tables I-IV, respectively. It can be seen that results vary, not only from method to method, but also with varying conditions within a given method. It is clear, however, that the only method giving consistently and comparatively full involvement of the cellulose chains is the xanthate method. In a sense, this is natural since the xanthate groups are first introduced by classical cellulose chemistry or, as in the case of rayon, left as residuals after the regeneration process. Since these groups are part of the initiating system, their participation, although by no means complete, could be expected perhaps to be more involved. This is probably an oversimplification, however, since radiation and peroxidation, including ozonolysis, should also be efficient. No doubt the competing reactions of radical losses by recombination, for example, and chain growth present serious problems. The diffusion of the monomer and the reaction of the radicals with themselves are both dependent on swelling. A delicate balance is set up, therefore, in any practical heterogeneous grafting system. As with the problem of homopolymer, it is not clear whether the presence of considerable amounts of ungrafted cellulose is, although disquieting, always deleterious. It undoubtedly depends on the application and each situation would have to be directly evaluated. The difficulty of not being able to involve each cellulose molecule in the grafting process, or at least control this feature of the reaction, does, however, present a negative feature of cellulose grafting.

Concurrent degradation. A problem which attends most radical reactions with cellulose is some concurrent chain scission. This has not been well studied but Nakamura *et al.* (19) showed that with ceric ion treatment under conventional grafting conditions the degree of polymerization of rayon fibers decreased from 210 to 170. Radiation in air to 6 Mrads. reduced a similar sample to 120 (20). Radiation in vacuum also reduces the degree of polymerization (21) and the mechanical strength of rayon (see reference 22 for example). It is not believed, however, that under most practical grafting conditions concurrent degradation will be a limiting factor in adopting the various methods for industrial use.

Molecular Weight Control

If one examines the data contained in Tables I-IV, it becomes clear that the molecular weights are all extremely high and do not vary enormously from method to method. The reasons are very obvious -- these are all heterogeneous reactions with substantial gel effects and growth of occluded free radicals. To the best knowledge of the writer, essentially all the various other methods involving free radical propagation give few and very high molecular weight grafted side chains.

Table I

Molecular Weights and Average Number of Cellulose Molecules per Grafted Side Chain. Radiation Grafting of Styrene to Cellophane and Cotton (Taken from Krässig) Ref. 15.

Cellulose	Method	Percent Graft	M.Wt.x10^{-5} of Grafted Side Chains	Cellulose Chains per Grafted Side Chain
Cellophane	Mutual	76.0	9.0	24.4
"	Preirradiation	96.0	13.0	27.8
Cotton	Mutual	47.5	3.1	2.8
"	Preirradiation	83.4	22.6	11.9

Dose ~3.0 Mrads.

Table II

Molecular Weights and Average Number of Cellulose Molecules per Grafted Side Chain. Ceric Ion Grafting of Styrene to Wood Pulp (Taken from Huang and Chandramouli) Ref. 16.

Percent "Graft"	Mwx10^{-5}	Mnx10^{-5}	$\frac{Mw}{Mn}$	Cellulose Chains per Grafted Side Chain
29.29	4.26	0.51	8.32	1.96
34.33	25.87	4.53	5.71	13.33
47.69	8.92	3.13	2.85	6.94
79.68	3.54	0.86	4.10	1.15
102.30	11.51	4.29	2.69	4.26

Table III

Molecular Weights and Grafting Frequency (Anhydrous glucose units (AGU) per Grafted Side Chain) Manganic Pyrophosphate Grafting of Acrylonitrile to Cotton Linters (taken from Rånby) Ref. 17.

Percent "Graft"	Grafting Efficiency	$\bar{Mn}$x10^{-5}	AGU per Grafted Side Chain
24.5	93.6	0.50	952
41.5	89.7	1.19	1,033
49.3	90.5	1.31	832

Table IV

Percent Extractable Homopolymer, Molecular Weights and Average Number of Cellulose Molecules per Grafted Cellulose. Xanthate Grafting of Various Monomers to Rayon (Taken from Krässig) Ref. 18.

Monomer (s)	Percent Add-on	Percent Extractable Homopolymer	MWtx10^{-5} of Grafted Side Chains	Cellulose Chains per Grafted Side Chain
Acrylonitrile	142.2	7.50	1.23	1.9
"	227.5	4.02	0.31	0.3
Styrene-Methacrylonitrile (1:12)	60.0	6.12	0.6	1.7
(1:12)	70.2	2.65	0.33	1.0

There are a few exceptions; for example, Morris et al. (23) found molecular weights as low as 33,000 with the mutual grafting of acrylonitrile to cotton. These were prepared in dimethyl formamide solution, a solvent for polyacrylonitrile. This could lead to a reduced gel effect. In any case, the degree of polymerization of the side chains was still as high as 620. There are, no doubt, other isolated examples in the literature. In a related study, but with styrene grafting to cellulose acetate using the mutual radiation method, the effect of swelling was clearly demonstrated (24). Some swelling was found necessary to allow more than surface grafting, but after a maximum the molecular weights and the yields steadily diminished. The effect of thickness demonstrated the diffusion-controlled nature of the reaction. Finally, in solution, quite low molecular weights and low yields were obtained. These results are described in detail in reference 24.

Another method of reducing the molecular weights of the grafted side chains is to introduce chain transfer agents such as carbon tetrachloride into the monomer solution. With radiation grafting, this appears attractive since the high radiation yields of radicals resulting from the radiolysis of carbon tetrachloride could lead to an increased rate of initiation in addition to lower molecular weights. In fact, this did appear to be the case with cellulose acetate films. In the case of cellulose, Huang (25) showed similar results with styrene. Some of his results are presented in Table V. Recently, however, the increased radiation yields have been questioned (see Table VI, for example). Rogovin has also presented a series of results without giving any details, although further references are given in the paper (26). These were conducted with styrene and with acrylonitrile and carbon tetrachloride and with methyl vinyl pyridine using triethylamine and dedecyl mercaptan as the chain transfer agents. Degrees of polymerization as low as 20 were obtained in one example. Recently, a much more detailed analysis has been presented by Min and Inagaki (13) on the mutual radiation grafting of styrene to viscose rayon fibers. By the use of thin layer chromatography they were able to differentiate between true graft copolymer and occluded homopolystyrene. Their conclusions were very informative. With most solutions, the molecular weights of the true grafted side chains were considerably larger than the inextractable homopolymer. With the styrene dioxane-water system, originally developed by Chapiro and Stannett (27), however, the molecular weights were similar. This was attributed to the ability of the solvent system to dissolve both the monomer and the polystyrene. Furthermore, the molecular weights were lower than the "non-solvent" systems. The differences with the latter systems were believed to be due to the growth of the attached side chains being essentially in a different "phase", with occluded chain ends continuing to grow. In the presence of carbon tetrachloride, however, the molecular weights were similar for both occluded and

Table V

Molecular Weights and Radiation Yields (G values) for Grafting Styrene to Rayon. Mutual Method with Preswelling. (Taken from Huang) Ref. 25.

Percent CCl_4 on Styrene	Percent "Grafting"	Mn of Side Chains	G (Side Chains)
0	284.1	910,000	0.19
1.0	127.1	296,000	0.26
2.0	117.9	220,000	0.32
3.0	100.9	133,000	0.45
4.0	92.9	89,000	0.63

Table VI

Effect of the Addition of Carbon Tetrachloride to the Styrene Grafting Solution, Preirradiation Method to Rayon Fibers. (Adapted from Min and Inagaki) Ref. 13.

CCl_4/Styrene Mole. Ratio	Percent Homopolymer	$Mn\times10^{-4}$ Polystyrene	True Percent Grafting	G (Side Chains) True	G (Side Chains) Apparent
0	15.1	40.2	36.9	2.8	3.6
0.02	15.3	20.2	17.9	2.6	4.5
0.05	13.9	13.6	10.8	2.2	4.8
0.10	17.7	8.4	2.7	1.0	6.8

Grafting Solution 5:10:1 Styrene, Methanol, Water. Total dose 0.35 Mrads.

truly grafted polystyrene, and were greatly reduced. The lowest reported value, however, was still 72,000. This was attributed to the ability of the primary radicals from the radiolysis of carbon tetrachloride to penetrate the grafting phase and terminate chain growth equally well as in the homopolymer phase. As with the experiments by Morris et al., the presence of a solvent for the polymer did give lower molecular weights. In the work of Min and Inagaki (13), these were lower than those obtained in the presence of carbon tetrachloride. Again, however, they were still in the 30,000 range. It is clear that the production of many short side chain grafts to cellulose in good yields and a minimum of homopolymer still remains an important challenge. This discussion has been restricted to the free radical grafting of vinyl monomers. It is possible that other techniques and non-vinyl monomers could overcome this problem. See the early work of Goldstein et al. (28) with β-propiolactone and reference 4, Chapter 5, for example.

Characterization and Properties

The characterization of grafted cellulosic materials is easier than with many grafting systems. The, often great, differences between the solubilities of the ungrafted cellulose and the homopolymer make separation procedures much easier. Furthermore, the cellulose backbone can be acid hydrolyzed, leading to determinations of the molecular weights and other details of the grafted side chains. The discussion of the results of the work of Min and Inagaki (13) presented in the previous section does, however, show that pitfalls do exist, and that the more refined TLC techniques give much more accurate information. A sound commercial process demands an accurate characterization of the product, so such considerations are important for the large-scale production of grafted cellulosic materials.

With regard to the properties of grafted cellulose, these have been studied in detail but to a much less extent than synthetic and kinetic aspects. The greatest amount of effort has been with regard to the mechanical properties of grafted fibers. In general, the results have been disappointing and only comparatively minor improvements have been achieved by grafting. An interesting paper by Blouin et al. (29) compares the mechanical properties of acrylonitrile grafted cotton fibers and fabrics prepared by different methods. The grafts were also characterized with respect to their degrees of substitution, side chain molecular weights and morphology. Their results are presented in Tables VII and VIII and can be summarized as follows.

In the case of the cotton fibers, the breaking strength was decreased in every case. In two cases, the elongation had significantly increased. The others were similar or slightly less. There was no apparent correlation between the location of the grafted polymers, their molecular weights, or the overall degree

Table VII

Effect of Grafting Acrylonitrile to Cotton Yarn Under Various Conditions on the Mechanical Properties. (Taken from Blouin et al.) Ref. 29.

Sample No.	Control	I	IV*	V	VI	VII
Percent Graft	0	37	35	37	37	28
Yarn No. tex.	252	404	398	360	362	352
Breaking Strength lb.	9.8	7.4	8.2	7.6	7.4	8.0
Percent Elong. at break	15.1	24.9	24.6	14.3	15.2	15.5
Breaking Toughens g/tex.	116	34	38	67	61	66
M.Wt. Side Chains x 10^{-5}	--	12.0	5.1	0.33	0.9	0.84
Distribution of Grafted Copolymer	--	Outer Layers	Uniform	Uniform	Uniform	--

*Original kidney shaped cross section had rounded.

Table VIII

Effect of Grafting Acrylonitrile to Cotton Fabric Under Various Conditions on the Mechanical Properties. (Taken from Blouin et al.) Ref. 29.

Sample No.	Control	II*	IV**	V	VI	VII
Percent Graft	0	25	27	25	26	26
Breaking Strength lb.	40.9	50.4	51.4	39.6	40.8	39.0
Percent Elong. at Break	22.0	25.9	33.6	20.5	17.8	21.7
Tear Resistance g.	573	507	650	447	393	413
Flex Abrasion Resistance $\frac{\text{Sample}}{\text{Control}}$	1.0	1.74	2.37	1.09	1.35	1.01
Flat Abrasion Resistance $\frac{\text{Sample}}{\text{Control}}$	1.0	6.05	6.32	2.04	3.36	2.29
Wrinkle Recovery-Conditioned (Warp + Fill) deg.	181	154	125	224	193	166

*MWt. of Side Chains 5.9×10^5
**Original kidney shaped cross section had rounded.

of substitution. The stiffness was reduced in every case by the grafting. With the fabrics there were modest increases in the breaking strength in two cases, accompanied with increases in elongation and abrasion resistance. Again, there was no obvious correlation with the location or molecular weight of the grafted side chains. The fabric sample with the rounded cross section, uniformly distributed but with high molecular weight side chains, had the best overall properties but not sufficiently improved over the control to be of obvious practical interest.

The principle reason for the rather modest changes in the mechanical properties of cellulose fibers, even with large amounts of grafting, is believed to be due to its semicrystalline nature. Since the grafting is carried out under heterogeneous conditions, the crystallinity remains essentially unchanged, the reaction taking place in the amorphous regions. The mechanical properties, on the other hand, are mainly governed by the crystalline morphology.

With very high degrees of grafting, above 100%, for example, the crystallinity begins to decrease and the mechanical properties start to change more. This is especially clear when elastomeric side chains, such as polyethyl acrylate, are grafted. This aspect has been studied in detail by Nakamura and Arthur et al. (19,20, 30) and by Williams and Stannett et al. (22,31). In the work of Nakamura et al. (20) it was interesting that the crystallinity of the cellulose rayon itself was only reduced from 40% to 37% with 719% graft, but the elongation at break increased from 32% to 680%. This was accomplished with 20 Mrads. of preirradiation when the degree of polymerization of the cellulose was reduced from 210 to 75. Electron microscopy showed that the cellulose fibers separated into a fibrillar structure forming a uniform network among which the amorphous polyethyl acrylate was uniformly distributed. Preswelling of the rayon with 57% zinc chloride solution followed by removing by washing out with water reduced the crystallinity of the rayon from 42% to 25%. Grafting onto such fibers further reduced the crystallinity to 22% to 800% grafting. The elongation only increased, however, from 110% to 160%. Presumably the structure of such heavily grafted fibers as described above means that the actual crystallinity of the cellulose network is not important within those limits. It should be emphasized that these results were obtained by ceric ion grafting whereas the previous results were related to preirradiation grafting. The preswelling in 56% zinc chloride solution did result, however, in a doubling of the grafting yield. The ceric ion method never gave such high elongations as the radiation method. Similar highly elastic rayon and cotton fibers were reported by Williams and Stannett et al. (32) in the same year, again the very high elasticities, more than 500% elongations were only obtained with the preirradiation grafting method. The same authors were able to achieve very high elongations at much lower degrees of grafting, 100% for example, by a decrystallization technique (33).

Realizing the role of the residual crystallinity in restricting the development of elasticity, the lower graft samples were decrystallized in concentrated zinc chloride solution or other cellulose solvents which was later removed by washing. The grafted polyethyl acrylate side chains being insoluble in the cellulose solvents held the fiber structure intact and also prevented full recrystallization and led to the development of the network structure at lower degrees of grafting. It was interesting that this technique also only worked well with radiation initiation. This is believed to be due to the more uniform deposition. Ceric ion initiation tends to promote grafting in the surface regions (22), presumably other hydrated inorganic ions have a similar tendency. The experiments of Arthur (34), however, indicated a uniform distribution with ceric ion, to cotton with acrylonitrile, under their conditions. It seems probable that similar high elasticities using the decrystallization technique could be achieved with chemically initiated grafts if the correct conditions were used. Paper and viscose films can also be rendered highly elastic by the above techniques.

In line with the above discussion, the properties of grafts must also depend on the location of the graft copolymer. Properties such as abrasion, enhanced adhesion, wetting and so on only need surface modification. Flame retardancy, water sorbency, and certain other properties, on the other hand, need essentially bulk grafting. These differences can be achieved rather easily in practice and the differences have been clearly demonstrated in a number of cases.

The sorption of water into cellulose is also restricted by the crystallinity and to develop so-called super-sorbency this must be overcome. If, say, acrylic acid is grafted followed by the "decrystallizing" procedure described above, super sorbency is immediately developed (35,36). The swelling in water is a mainly lateral phenomenon in fibers whereas elasticity is a longitudinal property. Almost any grafting procedure can, therefore, be used for sorbency provided it is applicable to grafting acrylic acid or other hydrophillic monomers. The sodium salts are much more effective than the free acids in developing very large, e.g., 3000%, water sorbency.

Numerous other properties, such as ion-exchange, mildew resistance, dimensional stability, coefficient of friction and electrical properties can also be imparted to cellulose by grafting. The presence of cellulose has also been shown to impart biodegradability to the cellulose-synthetic polymer grafted materials.

The properties of grafted cellulose fibers, films, powders, pulp and paper have been well described in a number of reviews and will not be discussed *per se* further in this paper.

Applications

The real challenge in the field of cellulose grafting is in

the development of viable, large scale industrial applications. It is significant that in the excellent and detailed book by Hebeish and Guthrie (4) only 16 out of 342 pages are devoted to this subject. Of these, 7 are on wood-plastic composites which are mainly in-situs polymerized monomers and only partially grafted. Nevertheless, they are closely related and with world-wide industrial applications. Details of these have been discussed not only in the above references but numerous other reviews and reports (37,38). These should be consulted for further details.

In the field of true cellulose grafting, there have been numerous attempts at commercialization in the cellulose fiber field. Some have reached pilot stage proportions, others have been full commercial processes such as the Deering-Millikan durable process now apparently in abeyance. Most of these developments have been in the field of textiles.

It seems clear that a number of grafted cellulosics are being manufactured in the Soviet Union. In particular, a graft copolymer of rayon-acrylonitrile named Mtylon B is used as a soft resilient fiber for carpet manufacture. They have also developed both cellulose and cellulose acetate grafted fibers with other useful properties such as flame retardancy, bactericidal properties, hemostatic agents for medical gauzes and water and oil repellency. Ion-exchange fibers have also been made for special uses such as the recovery of precious and heavy metals, for binding enzymes and for other applications. In general, it appears that the xanthate and redox methods of grafting are used.

The xanthate method has been used in the U.S. and Europe for a number of larger scale pilot plant studies. In the case of rayon, the technique has been explored for flameproofing, high water sorbency, ion exchange characteristics and bacteriostatic and fungistatic properties. Also nonwoven grafted fibers with excellent dispersibility for wet processing and improved binder affinities have been produced. It does not appear, however, that industrial exploitation of these technical successes will take place in the near future.

In Japan, pilot plant quantities of rayon and cotton grafted with styrene and other monomers have been prepared and evaluated. Improvements in the flex abrasion resistance and good water repellency were obtained. In addition, thermoplasticity was imparted plus an improved adhesion to rubber. Arthur *et al*. have also studied this type of grafted cellulose fibers. Again, the work has not led to any large scale developments. Research into grafting to textiles is continuing, but the prospects for any early successful industrialization seem remote.

Grafting to pulp and paper has also been studied but not to the same extent as textiles. A comprehensive review of the field was published in 1972 (39). Many of the useful properties imparted to pulp and paper are similar to those already discussed with textiles. These include the mechanical behavior, ion exchange properties, weather resistance, flame resistance and dimensional stability.

The strength properties of paper have a different origin in some respects from those of textiles. In particular, the fiber-fiber bonding plays a large role in the development of both dry and wet strength in addition to the strength of the fibers themselves. A number of investigations have, therefore, examined the grafting of hydrogen bonding side chains such as polyacrylic acid and polyacrylamide to increase the dry tensile and bursting strengths and other properties. Jayme and Hebbel (40) grafted acrylic acid and acrylonitrile to pulp and to performed paper and compared the results with polyacrylic acid which was added directly to the paper. The latter gave markedly better results than the grafting method. Polyacrylamide treated papers showed similar results. Grafting did result in substantial strength increases, however, with nonbeaten pulps with both polymers. With latices such as a polyethylacrylate-acrylonitrile copolymer, the best dry strengths were again improved to a much greater extent than the same percent polymer added by grafting (41). It was significant, however, that the grafted paper had much better wet strength. Thus, strong, tough, leathery papers with good wet and dry strength could be obtained by grafting. There is another advantage of grafting over simply adding polymers in the case of paper. Substantial grafting to paper can be achieved without greatly changing the porosity. In this way, filter papers can be made which have either cationic or anionic charges (42). Such materials might be useful, for example, as air filters. Paperboard grafted with polyacrylonitrile was shown to have improved stiffness and wet and dry strengths together with better abrasion grease and water resistance (43). In addition, the stiffness decreased markedly at about 135°C. This thermoplasticity could be a useful property in various converting operations. No commercialization, however, has resulted from these and numerous other studies.

The lack of any significant industrial production of grafted textiles and paper appears to be mainly economic. The costs of the process are too great to overcome the fact that similar or near-similar results can be obtained in both fields by simple well-established processes of adding preformed polymers. There are advantages, however, in grafting over simply adding as shown by the examples given above. In due time, it is felt, needs will arise which can only be brought about by grafting. The high production speeds now routinely achieved in both the textile and paper industries cannot be easily matched at this time by grafting. In the case of paper, if grafting could be combined with the pulping or beating operations, this would overcome this deterrent. It is known that such approaches have been made.

Perhaps the most promising applications of grafted cellulose are those related to ionic grafted side chains such as polyacrylic acid. These grafts can be synthesized either by direct grafting or by using acrylonitrile or acrylate esters followed by alkaline hydrolysis. A number of other acidic or quaternized basic systems are available. It is clear that such materials could have a number of interesting commercial possibilities. If the cellulose is in the form of film or parchmentized paper, reverse osmosis, hyperfiltration or ion exchange membranes can be prepared (44-46). Examples of super water sorbent materials and air filters have been presented earlier in this section together with other possible application.

Super water sorbent polymers which absorb more than thirty times their own weight have received much publicity (47,48). The first of such products was probably crosslinked carboxy methyl cellulose. Others have been developed by crosslinked acrylic acid derivatives and by graft copolymers of starch and cellulose. The latter are, of course, the interest for this review. Rayon, cotton and wood pulp have been used as substrates, although highly sorbent grafted viscose films have also been prepared. Probably industrial uses will be found for all these types of "super slurpers" as they have been called. The cellulosics are available in fibrous form whereas the others are prepared as powders. Both forms will find their own specific applications. Grafted wood pulps may well be the best compromise. In any case, it seems clear that this variety of grafted cellulose will reach commercialization in the not too distant future. These materials have found use in sanitary and similar sorbent products such as diapers for the removal of water from organic liquids and as rooting media for young plants. The chief challenge facing these products is to develop additional and, hopefully, large scale applications. Most of the information about these unique materials is contained in the patent and trade literature. However, useful details are given in the papers of Lepoutre *et al*. (49,50) and by Zahran *et al*. (36). Cationic grafted pulps have also been prepared by Lepoutre *et al*. using dimethylamino-ethyl acrylate. These were found to have less spectacular water retention properties, however, than their anionic counterparts (51).

Some of these products, when beaten in a Waring Blender, for example, change in character from fibrous structures containing a large amount of water, to gel-like viscous colloidal solutions (52). These show promise for applications as thickeners and dispersing agents.

Many of these grafted pulps also show promise as fibrous or colloidal beater additives to give added strength properties to paper such as those made from groundwood and high yield pulps (53).

Grafted cellulose powders such as Solka Floc have been grafted with polyacrylic acid and also rendered cationic by direct chemical reaction (54). Both of these modified cellulose products have been prepared on a large scale. They appear to show considerable promise for the removal of dyestuffs and other charged impurities from waste water such as textile effluents. They are also able to break emulsions of water and organics by removal of the surfactants (55).

Literature Cited

1. Waltcher, I.; Burroughs, R.; Jahn, E. C. Paper presented at the I.U.P.A.C. meeting. Stockholm. 1953, and Burroughs, R., Ph.D. Thesis, State University College of Forestry, Syracuse, N.Y. 1955.
2. Ushakov, S. N. Fiz-Mat. Nauk. 1946. 1, 35.
3. Landells, G.; Whewell, C. S. 1951. J. Soc. Dyers and Colourists 1951 67 338.
4. Hebeish, A.; Guthrie, J. T. "The Chemistry and Technology of Cellulosic Copolymers: Springer-Verlag. New York. 1981.
5. Kokta, B. V.; Lo, R. C.; Daneault, C. This volume, p.269.
6. Reine, A. H.; Arthur, J. C. Text. Res. J. 1972 42 155.
7. Ang, C. H., Garnett, J. L., Jankiewicz, S. V., Sangster, D. This volume, p.141.
8. Hon, D. N-S; Chan, H-C. This volume, p.101.
9. Kraessig, H. A.; Stannett, V. Adv. Polym. Sci., 1965 111.
10. Arthur, J. C. Adv. Macromol. Sci., 1970, 2 1.
11. Tsuziki, M.; Hagiwara, I.; Shiraishi, N.; Yokota, T. J. Appl. Polym. Sci., 1980 25 2721.
12. Tsuziki, M.; Hagiwara, I.; Shiraishi, N.; Yokota, T. J. Appl. Polym. Sci. 1980 25 2909.
13. Min, T. I.; Inagaki, H. Polymer. 1980 21 309.
14. Shiraishi, N.; Williams, J. L.; Stannett, V. Rad. Phys. and Chem. 1981 in press.
15. Kraessig, H. A. Tappi. 1963 46 654.
16. Huang, R. Y-M.; Chandramouli, P. J. Appl. Polym. Sci. 1968 12 2549.
17. Ranby, B. "Modified Cellulosics". Eds. R. M. Rowell and R. A. Young, Academic Press, 1978 P. 171-195.
18. Kraessig, H. A. Svenski Papperstidning. 1971. 74, 417.
19. Nakamura, Y.; Negishi, M.; Kakinuma. T. J. Polym. Sci., 1968, C. 23 629.
20. Nakamura, Y.; Arthur, J. C.; Negishi, M.; Doi, K.; Kakeyama, E.; Kudo, K. J. Appl. Polym. Sci., 1970, 14 929.
21. Sakurada, I.; Okada, T.; Kaji, K. J. Polym. Sci., 1972. C.37,1
22. Williams, J. L.; Stannett, V.; Roldan, L. G.; Sello, S. B.; Stevens, C. V. Int. J. Appl. Rad. and Isotopes. 1975. 26 169.
23. Morris, N. J.; Blouin, F. A.; Arthur, J. C. J. Appl. Polym. Sci. 1968, 12, 373
24. Stannett, V.; Wellons, J. D.; Yasuda, H., J. Polym. Sci., 1964, C.4 551.
25. Huang, R. Y-M. J. Appl. Polym. Sci. 1966, 10, 325.
26. Rogovin, Z. A., J. Polym. Sci. 1972, C.37. 221
27. Chapiro, A.; Stannett, V. Int. J. Appl. Rad. and Isotopes. 1960, 8, 164.
28. Goldstein, I. S.; Dreyer, W. A.; Jeroski, E. B.; Nielson, K. F.; Oberly, W. J.; Weaver, J. W. Ind. Eng. Chem. 1959, 51 1313
29. Blouin, F. A.; Morris, N. J.; Arthur, J. C. 1968 Text. Res.J. 38, 710.

30. Negishi, M.; Nakamura, Y.; Hosoi, M. J. Appl. Polym. Sci. 1968, 12 1209.
31. Williams, J. L.; Woods, D. K.; Stannett, V.; Sello, S. B.; Stevens. Int. J. Appl. Rad. and Isotopes, 1975, 26, 159.
32. Williams, J. L.; Stannett, V. J. Polym. Sci., 1970, B8, 711.
33. Williams, J. L.; Stannett, V. Polymer Letters, 1972, 10, 665.
34. Arthur, J. C. Adv. in Chem. Series, 1969, 91, 574.
35. Williams, J. L.; Stannett, V. J. Appl. Polym. Sci. 1979, 23, 1265.
36. Zahran, A. H.; Williams, J. L.; Stannett, V. J. Appl. Polym. Sci., 1980, 25 535.
37. Czikovsky, T. Atomic Energy Review, 1968, 6, 3.
38. Mott, W. E.; Stannett, V. Isotopes and Rad. Tech., 1969, 6 329.
39. Phillips, R. B.; Quere, J., Guiroy, G.; Stannett, V. Tappi, 1972, 55, 858.
40. Jayme, G.; Hebbel, G. W. Das Papier, 1971, 25, 113.
41. Schwab, E.; Stannett, V.; Rakowitz, D. H.; Magrane, J. K. Tappi, 1962, 45, 390.
42. Vitta, S. B.; Stahel, E. P.; Stannett, V. Unpublished results.
43. Su, C. J.; Goldstein, I. S. Tappi, 1972, 55, 1318, 1322.
44. Richards, G. N.; White, E. F. T. J. Polym. Sci. 1964, C.4, 1251.
45. Riande, E.; Mateos, A. M.; Guzman, G. M. Europ. Polym. J. 6, 437.
46. Riande, E.; Mateos, A. M.; Guzman, G. M. Europ. Polym. J. 1970, 6, 1247.
47. e.g. Chemical Week, 1974, July 24, P. 21-22.
48. e.g. Chem. Eng. News 1979, Nov. 5, P. 23-24.
49. Lepoutre, P.; Hui, S. H.; Robertson, A. A. J. Appl. Polym. Sci.1973, 17, 3143.
50. Lepoutre, P.; Hui, S. H. J. Appl. Polym. Sci., 1975, 19, 1257.
51. Lepoutre, P.; Hui, S. H.; Robertson, A. A. J. Macromol. Sci. Chem. 1976, A10
52. Lepoutre, P.; Robertson, A. A., Tappi, 1974, 57, 87.
53. Kobayashi, A.; Phillips, R.B.; Brown, W.; Stannett, V., Tappi, 1971, 54, 215.
54. Gangneux, A.; Wattiez, D.; Marechal, E., Europ. Polym. J., 1976, 12, 535, 543.
55. Gangneux, A.; Wattiez, D.; Marechal, E. Europ. Polym. J. 1976, 12, 551

RECEIVED January 6, 1982.

2

Free-Radical Initiated Graft Polymerization of Vinyl Monomers onto Cellulose

JETT C. ARTHUR, JR.

Technical Consultant, 3013 Ridgeway Drive, Metairie, LA 70002

An interpretative review of the reactions initiated by macrocellulosic free radicals with vinyl monomers to yield block and graft copolymers of fibrous cellulose was made. Macrocellulosic radicals are usually formed by interactions with radiation or chemical redox systems. Important factors in these heterogeneous reactions are lifetimes and accessibilities of the radicals and interactions of solutions of monomer with fibrous cellulose. Changes in organochemical, macromolecular, and morphological structures in cellulosic fibers through formation of copolymers are made.

Celluloses, with high molecular weights are naturally occurring polymers which are useful as clothing, housing, and industrial products. Natural cellulose occurs morphologically as fibers, such as cotton or wood pulp fibers. Fibers are defined here as relatively flexible, macroscopically homogeneous bodies having a high ratio of length to thickness and a small cross section. Cellulose is polymorphic and, in nature, crystallizes into different forms. Cellulose lattice type I is found in cotten fibers and consists of about 70-80 percent highly ordered or crystalline regions, as recorded by x-ray methods. Cellulose lattice type II may be found in wood pulp fibers and consists of about 60 percent highly ordered or crystalline regions. The chemical structure of cellulose consists of repeating units of cellobiose; however, the degree of polymerization of cellulose is usually expressed as the number of repeating anhydroglucose units and may range up to several thousand. The morphology, crystallinity, and chemical properties of cellulose have been reviewed (1, 2). For purposes of this discussion, it is emphasized that properties and reactions of cellulosic fibers are related to their organochemical structure, macromolecular structure, molecular orientation, and morphological structure.

0097-6156/82/0187-0021$6.00/0
© 1982 American Chemical Society

Both chemical and macromolecular modifications of celluloses to increase their usefulness in applications in films, fibers, and plastics have been reported (1-7). In the context of this discussion, the macromolecular modification of cellulose by graft or block polymerization with vinyl monomers will be considered. A graft or block copolymer of cellulose is defined here as a combination of cellulose and polymer that is difficult to separate by solvent extraction without first degrading the cellulose. Furthermore, only proposed mechanisms and reaction methods of free-radical initiated graft and block polymerizations of vinyl monomers with cellulose are discussed.

Free-Radical Initiation

The formation of graft or block copolymers of fibrous cellulose involves contacting vinyl monomer in vapor phase or in solution with cellulose. Then, initiation of copolymerization of monomer with cellulose by ionic (8, 9), charge-transfer (10), or free-radical processes (3) has been reported. Uncatalyzed graft polymerization has also been reported (10). It is generally agreed that the intitiating free radical is a macrocellulosic radical. The sites for the attachment of grafted polymers are usually considered to be the site of the macrocellulosic radical. It has also been proposed that the initiation of polymerization of vinyl monomer involves the formation of a cellulosic radical; however, it was proposed that the sites for attachment of grafted polymers are not necessarily the sites of the cellulosic radicals. It was suggested that grafting on cellulose is a termination reaction of a growing polymer chain on cellulose (10).

Macrocellulosic free radicals may be formed through dehydrogenation, depolymerization, or oxidation of cellulose. Commonly used methods of free-radical initiation include: ionizing radiations (both high energy and ultraviolet radiations), chemical redox systems, decomposiition of peroxy compounds, and chemical modification. Corona or arc discharge, electrochemical, mechanical, ultrasonic, and thermal methods have also been used (3). Chemically, formation of macrocellulosic radical involves the oxidative depolymerization of cellulose with an increase in acidic and reducing groups. The increase in concentration of reducing groups is much greater than the increase in acidic groups. When ionizing radiations are used as initiators, evolution of carbon dioxide, carbon monoxide, and hydrogen also occur. Because of the molecular orientation in cellulosic fibers, oxidative depolymerization of cellulose in fibers does not immediately lead to large losses in fibrous tensile and mechanical properties (11).

Macrocellulosic radicals initiated by chemical redox systems are temperature sensitive and short-lived (12, 13). Vinyl monomers (M) in solution in contact with cellulose (cell-H), when the radicals are formed, increases the possibility of grafting reactions. Homopolymerization is usually a problem, particularly when an intermediate radical is formed. The intermediate radical reacts with either cellulose or monomer initially. Reaction with monomer initiates with cellulose can also lead to graft polymerization (10). In the case where an intermediate radicals is formed, the following reactions could occur.

$$H_2O_2 + Fe^{2+} \longrightarrow OH^- + {}^{\cdot}OH + Fe^{3+} \quad (1)$$

$$\text{cell-H} + {}^{\cdot}OH \longrightarrow \text{cell}\cdot + HOH \quad (2)$$

$$\text{cell}\cdot + nM \longrightarrow \text{cell-}(M)_{n-1}M\cdot \quad (3)$$

or

$$M\text{-}H + {}^{\cdot}OH \longrightarrow M^{\cdot} + HOH \quad (4)$$

$$M\cdot + nM \longrightarrow M_nM^{\cdot} \quad (5)$$

$$M_nM^{\cdot} + M\cdot \longrightarrow M_{n+2} \quad (6)$$

$$M_nM\cdot + \text{cell-H} \longrightarrow \text{cell-}M_{n+1} + H\cdot \quad (7)$$

Evidence for the formation of free radicals in these types of reactions has been obtained by electron spin resonance (e.s.r.) spectroscopy (13).

The reaction of ceric ion in aqueous solution with cellulose cleaves the anhydroglucose ring between carbon C_2 and C_3 with formation of a short-lived radical on carbon C_2 and oxidation of carbon C_3 to a reducing group. Grafting occurs by reaction of the radical at carbon $C_2^{\cdot}$ with monomer. Radical termination can occur by reaction of carbon $C_2^{\cdot}$ with Ce^{4+} to yield Ce^{3+} and oxidation of carbon C_2 to a reducing group. Of course, it is equally likely that the roles of carbon C_2 and C_3 in the reactions could be reversed. Oxidative depolymerization of cellulose also occurs and could yield short-lived intermediate homopolymerization. In the case where an intermediate radical is not formed, the following reactions could occur to form the initiating macrocellulosic radical:

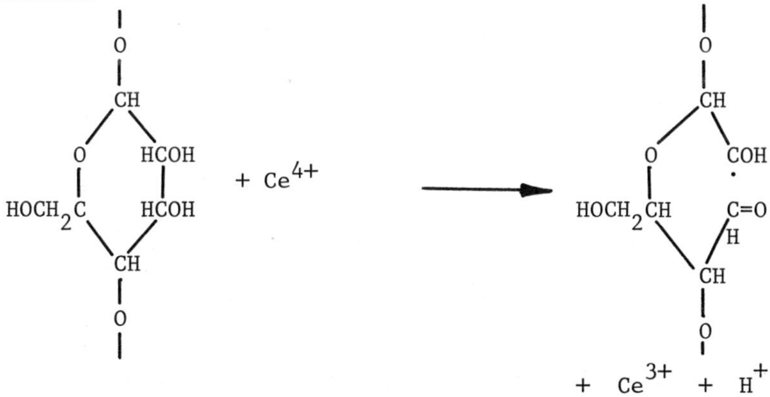

(8)

These free-radical reactions have been followed by e.s.r. spectroscopy (12).

Many other chemical redox systems have been reported as initiators of macrocellulosic radicals and as catalysts for graft polymerization. One variation has been to modify cellulose chemically to increase its reactivity with selected oxidizing and reducing agents which on reaction yield macrocellulosic radicals (14, 15).

Macrocellulosic radicals initiated by ionizing radiation can be divided into two classes: one class in which the radiant energy is selectively absorbed by the cellulose molecule and another class in which the radiant energy is randomly absorbed by the cellulose molecule. In both cases, after localization of the absorbed radiant energy, oxidative depolymerization of cellulose is initiated, and macrocellulosic radicals are formed (3).

Ultraviolet radiation is selectively and weakly absorbed by purified cellulose to initiate a macrocellulosic radical that generates a single-line e.s.r. spectrum. The mechanism of the interaction of ultraviolet radiation with cellulose has been discussed (16). If monomer is present when cellulose is photolyzed, an e.s.r. spectrum characteristic of chain propagating radical for the graft polymerizing monomer is generated. Photolysis of monomer neat or in solution in the absence of cellulose does not usually generate polymerization. The propagating radical for photoinitiated grafting poly(methyl methacrylate) onto cellulose should be:

$$-\underset{\substack{|\\ H}}{\overset{\substack{H\\ |}}{C}} - \underset{\cdot}{\overset{\substack{CH_3\\ |}}{C}} - COOCH_3 \qquad (9)$$

A five-line e.s.r. spectrum is recorded. Evidently, the grafted polymer on the photoirradiated cellulose had a conformation that restricted rotation about the $C_\alpha - C_\beta$ bond, so that only one of the methylene hydrogens and the freely rotating methyl group interact with the unpaired electron (17).

In model compound studies, when methyl methacrylate in rigid glasses of methanol that contained ferric chloride were photolyzed at 77-100 K, similar five-line e.s.r. spectra of propagating chain radicals were generated. A second propagating chain in which neither of the methylene hydrogens interacted with the unpaired electron evidently occurred. A four-line e.s.r. spectrum from the interaction of the freely rotating methyl group with the unpaired electron was generated. With two propagating chains in two different conformations, one generating a five-line e.s.r. spectrum and the other generating a four-line e.s.r. spectrum, a composite nine-line e.s.r. spectrum was recorded. It can be concluded that photoinitiated grafting at a methanolic radical site in rigid glasses at less than 100 K also restricted the conformation of the propagating chain (18).

In photolysis, the yield of macrocellulosic radicals increased by addition of photosensitizers either covalently bonded to cellulose or added to the monomer solution (19). Photoinitiated grafting of monomers onto cellulose has been recently reviewed (20).

Randomly absorbed radiant energy from high-energy ionizing radiation localizes in cellulose to intiate oxidative depolymerization of cellulose and a relatively large number of macrocellulosic radicals, when compared with the number of radicals generated by ultraviolet radiation. The mechanisms of the interaction of high-energy radiation with cellulose has been discussed (11). High-energy radiation may originate from either machine or nuclear sources. The yield of macrocellulosic radicals is dependent on the absorbed dosage of radiation and, to a lesser extent, on the type and source of high-energy radiation. Both short-lived macrocellulosic radicals and trapped, long-lived radicals are formed (21). Although the interaction of high-energy radiation with cellulose is random and chemically nonspecific, the final localization of energy and initiation of the radical site on cellulose is dependent on the relative amount of delocalizing energy available at each carbon atom in the anhydroglucose unit.

As noted above, when high-energy radiation interacts with cellulose, trapped, long-lived radicals are formed in the highly molecularly oriented areas of cellulosic fibers. This offers the possibility of separating the steps of formation of macrocellulosic radicals and of initiation of graft polymerization with vinyl monomers. The morphological structure and molecular orientation of cellulosic fibers can be selectively altered by immersing fibers in solutions (22, 23). When irradiated cellulose is immersed in solutions, the short-lived radicals, which are evidently located in the less molecularly oriented or amorphous areas of the fibers, are rapidly scavenged. If irradiated cellulosic fibers are immersed in monomer solutions that swell their structure and/or alter their molecular orientation, then long-lived trapped radicals are made accessible to vinyl monomers. The extent of changes in morphological structure and molecular orientation of fibers is dependent on the composition of the solutions (24, 25). Graft polymerization is then initiated, with a minimum of homopolymerization, at the site of the cellulosic radical.

As recorded in the case of photoirradiated cellulose, the conformation of grafted polymer on high-energy irradiated cellulose is evidently restricted. The e.s.r. spectra of propagating radicals of methacrylates grafted on high-energy irradiated cellulose also had only five lines. This indicated, as shown in radical 9, that rotation about the $C_\alpha - C_\beta$ bond was restricted, so that only one of the methylene hydrogens and the freely rotating methyl group interact with the unpaired electron. However, in high-energy irradiated cellulose, the concentration of unreacted cellulosic radicals was much greater than that of the unreacted radicals in photoirradiated cellulose. The time-averaging computer attachment of the e.s.r. spectrometer was used to subtract the spectra generated by unreacted cellulosic radicals from the composite spectra. Then the e.s.r. spectrum generated for grafting methacrylate onto cellulose was recorded as a five-line spectrum(26).

Living Polymer Radicals

The trapped, long-lived macrocellulosic radicals could be considered as living polymer radicals. These radicals are initiated on the interaction of high-energy radiation with cellulosic fibers probably through a dehydrogeneration reaction. They are evidently located in the highly molecularly oriented areas of the fibers. Immersion of irradiated fibers in monomer solution increases the accessibility of these radicals to scavenging or terminating reactions and to initiating of grafting reactions. As previously reported, this is not an all-or-none phenomenon. The composition of the solution and its interaction with irradiated cellulosic fibers determine the increase in accessibility of these radicals. The life-times of trapped radicals in irradiated, dried (less than about 2 percent moisture) cellulose appear to be indefinite. Immersion in solutions that strongly interact with cellulosic fibers in which both their morphology and molecular orientation are changed does not necessarily scavenge or terminate all of the trapped radicals. However, immersion of irradiated cellulose in liquid ammonia reduced molecular orientation and terminated all of the trapped radicals (21, 24, 25).

E.s.r. evidence for a living (propagating) polymer radical in grafting reactions of vinyl monomers with irradiated cellulose has also been reported (26).

Reaction Methods

The grafting reaction of vinyl monomer onto cellulosic fiber is a heterogeneous reaction system. Cellulosic fibers are in a solid phase, and vinyl monomers are in a vapor phase or, as a solute, in a solution phase. Two general methods for initiating grafting reactions are, as follows: (1) one-step method: monomer is in contact with cellulose when the initiating macrocellulosic radicals are formed; or (2) two-step method: after formation of the radicals, monomer is contacted with the activated cellulose. The one-step method can be used when either short-lived or long-lived, trapped radicals are formed. The two-step method can be used when long-lived trapped radicals are formed (3).

When vinyl monomers are in a vapor phase, the extent and rate of the grafting reaction are increased, if the cellulosic fibers are wetted with solvents for the monomers. There is a very low extent of reaction between activated, dried cellulose and vinyl monomer neat (3).

When vinyl monomers are solutes in solution phase, the morphology and molecular orientation of cellulose can be selectively changed dependent on the composition of the solution. Also, the extent and rate of grafting are influenced by the composition of the solution which is referred to as the Trommsdorff effect (3, 27).

When the reaction adds polymer as a side chain to the cellulosic molecule, grafted polymers are formed. When the reaction adds polymer to the end of the cellulosic molecule, particularly when oxidative depolymerization occurs, block polymers are formed. However, it is not possible to distinguish between the two reaction products. The molecular weight of grafted and/or block polymer is large and often exceeds the molecular weight of cellulose, particularly the molecular weight of the partially depolymerized cellulose to which the polymer is bonded. In either grafted or block polymer addition to cellulose initiated by macrocellulosic radicals, the ratio of the number of molecules of polymer to the number of molecules of cellulose is much less than one. Because of the small number of covalent chemical bonds between polymer and cellulose, direct evidence has not been reported for such bonding (28). Differential solubility changes in cellulose, polymer, and cellulose interacted with polymer appear to be evidence for bonding. E.s.r. evidence for model reactions of vinyl monomers with activated substrates (18) and for reactions of monomers with activated cellulose (20, 26) indicate that the initiating radical is decreased in intensity and the propagating polymer radical is increased in intensity during grafting reactions. This could indicate bonding at the radical site.

Grafted block copolymers of activated cellulose (cell·) are formed by immersing the activated cellulose in a solution of two or more monomers (M and m). Typical reactions include:

$$\text{cell}\cdot + \text{M} \longrightarrow \text{cell-M}\cdot \qquad (10)$$
$$\text{cell}\cdot + \text{m} \longrightarrow \text{cell-m}\cdot \qquad (11)$$
$$\text{cell-M}\cdot + \text{m} \longrightarrow \text{cell-M-m}\cdot \qquad (12)$$
$$\text{cell-m}\cdot + \text{M} \longrightarrow \text{cell-m-M}\cdot \qquad (13)$$
$$\text{cell-M}\cdot + \text{M} \longrightarrow \text{cell-M-M}\cdot \qquad (14)$$
$$\text{cell-m}\cdot + \text{m} \longrightarrow \text{cell-m-m}\cdot \qquad (15)$$

In the predominating reactions, the number of different types of monomer units and their sequences are determined by their relative molecular reactivities for the macrocellulosic radicals and the monomer reactivity ratios. These types of reactions are useful in that less reactive monomers can be included in copolymers to add selected organochemical and macromolecular properties to the modified cellulosic products. In cases where vinyl monomers have been reacted to form oligomers, these reactions are useful in increasing the reactivity of oligomers with macrocellulosic radicals (29, 30, 31).

In summary, some of the important factors in polymerization reactions of vinyl monomers initiated by macrocellulosic radicals are, as follows: (1) the lifetimes of the macrocellulosic radicals; (2) the accessibility of the radicals to the monomers; (3) accelerating effects of the solutions of the monomers on rate and extent of polymerization reactions; and (4) the effects of the solution of the monomers on molecular orientation and morphology of the cellulosic fibers. Technical reports on grafting reactions on cellulose have been world-wide and number in the hundreds of contributions since 1960; space does not permit extensive citations. These investigations are continuing to be an active research and development area. This area of work was extensively reviewed in 1970 (3) and 1972 (4).

Characterization of Products

The characterization of graft or block copolymers of cellulosic fiber products depends on determination of changes in organochemical, macromolecular, and morphological structures and molecular orientations of the products from those of unmodified cellulosic fibers. Formation of macrocellulosic radicals involves the oxidative depolymerization of cellulose. Increases in carbonyl groups and in carboxyl groups on cellulose are measured. The formation of carbonyl groups is several times greater than the information of carboxyl groups. Grafting of selected vinyl monomers onto cellulose changes surface properties, resistances to microbial degradation, chemical reactivities, thermal stabilities, and related properties of cellulosic fibers. These changes in organochemical properties of cellulose are measured by the usual techniques and have been reported (3).

Macromolecular properties of grafted cellulosic fibers usually measured are: differential solubility in either polymeric or cellulosic solvents, mechanical or physical properties, and abrasion resistances. The molecular weights of the grafted or block polymers and of cellulose, both before and after formation of macrocellulosic radicals, have been determined. The number of grafted or block polymer molecules per cellulose molecule calculated has usually been much less than one. Grafted cellulosic fibers exhibit second order transition temperatures, dependent on the composition of the grafted polymer (3, 4).

Molecular orientations of grafted celluloses can be changed to a small extent. Interactions of monomer solutions with cellulosic fibers to decrease crystallinities and to change lattice type of the products from those of unmodified fibers have been reported. X-ray diffraction methods are used to determine the changes (24).

Morphological or supermolecular structure is the most easily changed property of cellulosic fibers. Interactions of selected monomer solutions with fibers can yield grafted products without change in their original cross sections to grafted products with highly swollen and rounded cross sections. Grafted polymer distribution in the fibrous cross sections range from concentrations of polymer in the surfaces of the fibers to concentrations of polymer uniformly distributed in fibrous cross sections. Concentrations of grafted polymer in the fibrous surface can be attached as coatings or encapsulations of the fibers. In cotton fibers which have concentric layered structures of cellulose, concentrations of polymer in the lumen area or center if the fibrous cross section can be obtained. Light, transmission electron, and scanning electron microscopy and energy dispersive (EDX) x-ray analysis have been used to determine the distribution of grafted polymer in fibrous products. For polymers which contain elements such as phosphorus, EDX analysis clearly records the distribution of grafted polymer in cellulosic fiber cross sections (3, 22, 30, 31).

Useful Processes and Products

Modifications of the macromolecular properties of cotton and regenerated cellulosic fibers by graft polymerization with selected vinyl monomers impart new and improved textile properties. For example, at low degrees of grafting of poly(acrylonitrile), elastic recovery properties of cotton fibers are increased. At high degrees of grafting of poly(ethyl acrylate) cellulosic fibers have elastomeric properties with high degrees of recovery from deformations (6, 7). Cellulosic textiles usually require some chemical finishing treatments to impart properties to make them competitive with durable-press properties of man-made textiles. These treatments usually decrease the abrasion resistances of cellulosic textiles. However, grafting of vinyl monomers onto cellulosic textiles changes their morphology and increases their abrasion resistances while simultaneously improving their durable-press properties. Grafting of vinyl monomers onto cellulosic fibers yields products with second order transition temperatures lower than the decomposition temperature of cellulose. These modified cellulosics are thermoplastic and heat moldable (32). Fibers are also encapsulated by grafting to yield durable coatings with reactive groups (33, 34).

Modifications of the organochemical properties of cellulosic fibers by graft polymerization with selected monomers impart new chemical properties. The microbiological and light resistances of cellulosic fibers to degradation are increased by grafting (3). Surface properties of modified fibers are changed to impart soil-release (35), dyeability (36), and flame-resistance (37, 38) properties.

Paper products have been grafted to increase bursting strength and to impart hydrophobic, hydrophilic, thermoplastic, and ion exchange properties. The dielectric properties of grafted paper give products with increased electrical insulation properties. Grafted wood products have increased dimensional stability and compressive, impact, and bending strengths. Grafted cellulosic films are useful as osmotic membranes and battery separators (3).

Literature Cited

1. Arthur, J. C., Jr. "Cotton" in Encyclopedia of Polymer Science and Technology, H. F. Mark, N. G. Gaylord, and N. M. Bikales, Eds., John Wiley and Sons, Inc., New York, 1966, Vol. 4, pp. 244-272.
2. Bikales, N. M., and Segal, L., Eds. "Cellulose and Cellulose Derivatives", High Polymers Vol. V, Parts IV and V, Wiley-Interscience, New York, 1971.
3. Arthur, J. C., Jr. "Graft Polymerization onto Polysaccharides" in Advances in Macromolecular Chemistry, W. M. Pasika, Ed., Academic Press, London, 1970, Vol. 2, pp. 1-87.
4. Arthur, J. C. Jr., Ed. "Proceedings of the Symposium on Graft Polymerization onto Cellulose", Interscience, New York, 1972, Polymer Symposia No. 37.
5. Arthur, J. C., Jr., Advan. Chem. Ser. 1969, No. 91, 574-591.
6. Arthur, J. C., Jr., Advan. Chem. Ser. 1971, No. 99, 321-339.
7. Arthur, J. C., Jr., "Special Properties of Block and Graft Copolymers and Applications in Fiber Form" in Block and Graft Copolymers, J. J. Burke and V. Weiss, Eds., Syracuse University Press, Syracuse, New York, 1973, pp. 295-323.
8. Avny, Y., Yom-Tov, B., Zilkha, A. J. Appl. Polym. Sci. 1965, 9, 817.
9. Simionescu, C. I., Rusan, V. J. Polym. Sci., Part C, 1972, No. 37, 173-186.
10. Gaylord, N. J. Polym. Sci., Part C, 1972, No. 37, 153-172.
11. Arthur, J. C., Jr. "Radiation Effects on Cellulose" in Energetics and Mechanisms in Radiation Biology, G. O. Phillips, Ed., Academic Press, London, 1968, pp. 153-181.
12. Arthur, J. C., Jr., Baugh, P. J., Hinojosa, O. J. Appl. Polym. Sci. 1966, 10, 1591-1606.
13. Arthur, J. C., Jr., Hinojosa, O., Bains, M. S. J. Appl. Polym. Sci. 1968, 12, 1411-1421.
14. Faessinger, R. W., Conte, J. S. Belgian Patent 1964, 646, 284.
15. Faessinger, R. W., Conte, J. S. U.S. Patent 1967, 3, 330, 787; 3, 359, 224.
16. Baugh, P. J. "Photodegradation and Photooxidation of Cellulose" in Developments in Polymer Photochemistry, N. S. Allen, Ed., Applied Science, Barking, 1981, Vol. 2, pp. 165-214.
17. Reine, A. H., Hinojosa, O., Arthur, J. C., Jr. J. Appl. Polym. Sci. 1973, 13, 3337-3343.
18. Harris, J. A., Hinojosa, O., Arthur, J. C., Jr. J. Polym. Sci., Polym. Chem. Ed. 1973, 11, 3215-3226.

19. Reinhardt, R. M., Arthur, J. C., Jr., Hinojosa, O. J. Appl. Polym. Sci. 1979, 24, 1691-1699.
20. Arthur, J. C., Jr. "Photoinitiated Grafting of Monomers Onto Cellulose Substrates" in Developments in Polymer Photochemistry, N. S. Allen, Ed., Applied Science, Barking, 1980, Vol. 1, pp. 69-92.
21. Baugh, P. J., Hinojosa, O., Arthur, J. C., Jr. J. Appl. Polym. Sci. 1967, 11, 1139-1153.
22. Rollins, M. L., Cannizzaro, A. M., Blouin, F. A., Arthur, J. C., Jr. J. Appl. Polym. Sci. 1968, 12, 71-105.
23. Reinhardt, R. M., Arthur, J. C., Jr., Muller, L. L. J. Appl. Polym. Sci. 1979, 24, 1739-1745.
24. Hinojosa, O., Nakamura, Y., Arthur, J. C., Jr. J. Polym. Sci., Part C, 1972, No. 37, 27-46.
25. Reine, A. H., Hinojosa, O., Arthur, J. C., Jr. J. Polym. Sci., Part B, 1971, 9, 503-507.
26. Hinojosa, O., Arthur, J. C., Jr. J. Polym. Sci., Part B, 1972, 10, 161-165.
27. Goynes, W. R., Jr., Carra, J. H. in Characterization of Metal and Polymer Surfaces, L.-H. Lee, Ed., Vol. 2, Academic Press, New York, 1977, p. 251.
28. Morris, N. J., Blouin, F. A., Arthur, J. C., Jr. J. Appl. Polym Sci. 1968, 12, 373-380.
29. Harris, J. A., Arthur, J. C., Jr. J. Appl. Polym. Sci. 1970, 14, 3113-3128.
30. Harris, J. A., Arthur, J. C., Jr., Goynes, W. R., Jr. J. Appl. Polym. Sci. 1979, 23, 2555-2565.
31. Harris, J. A., Arthur, J. C., Jr., Goynes, W. R., Jr. J. Appl. Polym. Sci. 1979, 24, 201-210.
32. Arthur, J. C., Jr., J. Appl. Polym. Sci., Appl. Polym. Sym. 1981, No. 36, 201-208.
33. Harris, J. A., Arthur, J. C., Jr. J. Appl. Polym. Sci. 1979, 24, 1767-1770.
34. Harris, J. A., Arthur, J. C., Jr. U.S. Patent 1981, 4, 253, 841.
35. Mares, T., Arthur, J. C., Jr., Harris, J. A. U.S. Patent 1977, 4, 063, 885.
36. Harris, J. A., Arthur, J. C., Jr. Textile Res. J. 1976, 46, 219-223.
37. Harris, J. A., Arthur, J. C., Jr., Goynes, W. R., Jr. J. Appl. Polym. Sci. 1979, 23, 2555-2565.
38. Harris, J. A., Arthur, J. C., Jr., Goynes, W. R., Jr. J. Appl. Polym. Sci. 1979, 24, 201-210.

RECEIVED December 17, 1981.

3

Graft Copolymerization of Vinyl Monomers onto Cellulosic Fibers

B. RÅNBY and L. GÄDDA

The Royal Institute of Technology, Department of Polymer Technology,
S-100 44 Stockholm, Sweden

Graft copolymerization of vinyl monomers - in this case acrylonitrile - onto cellulosic fibers is initiated by Mn^{3+} ions as pyrophosphate complex in aqueous solution at pH 1.5 to 2 and about 30°C under a nitrogen blanket in a process similar to what we previously reported for grafting onto starch. The following substrates are used: dissolving pulp (MDC) and grease-proof pulp (GP), both bleached sulfite pulps from spruce, ethyl cellulose (EC) and hydroxyethyl cellulose (HEC) of low degree of substitution (0.25, 0.40 and 0.60). At low Mn^{3+}-complex concentration (1-2 mmole/l) and substrates in never-dried form the reaction is completed in 2 to 3 h giving high conversion to polymer (70 to 85%) and high grafting efficiency (80 to 98% of the polymer is grafted). The grafted substrates retain their fiber morphology and their liquid water retention (5 to 8 g/g). After hydrolysis of the graft copolymers in aqueous NaOH solution (1% NaOH at 100°C), which saponifies the nitrile groups to amide and carboxyl groups, the retention of liquid water increases 5 to 10 times. The grafted and hydrolyzed MDC and GP samples, containing native cellulose lattice (cellulose I) reach saturation in water retention in a few minutes, while the corresponding EC and HEC samples, containing disordered mercerized cellulose lattice (cellulose II), show gradually increased water retention for a long time (> 60 min.).

Manganic ions (Mn^{3+}) in aqueous solution have been developed in our laboratories as an initiator for grafting vinyl monomers onto various polysaccharides, e.g. starch (1), cellulose (2) and guaran (3). This research is part of a general program on graft copolymerization onto polysaccharide substrates as described in a review paper (4). The Mn^{3+} grafting method has been applied as

0097-6156/82/0187-0033$6.00/0
© 1982 American Chemical Society

an efficient process for preparation of superabsorbent starch products with unusual thermodynamic properties (5). Using acrylonitrile as monomer, the grafted starch can be hydrolyzed in aqueous alkali to give a product which from humid air absorbs moderate amounts of water (about 1 part by weight) at relative humidities of 95 to 98%. The same product in contact with liquid water retains very large amounts of water (up to 800 parts per part of grafted hydrolyzed starch). The product is indeed a very efficient "super-slurper" (6). The water retention is reduced considerably at low and high pH (< 4 and > 11), and when dissociated salts, e.g. NaCl and $CaCl_2$, are added (5). The water molecules absorbed from humid air are strongly hydrogen bonded to the substrate while the large amounts of retained liquid water are loosely bonded, probably as clusters of water molecules (5).

The Mn^{3+} initiator, applied grafting of acrylonitrile and other acrylic and methacrylic monomers onto starch, gives a most efficient grafting process. At Mn^{3+} concentrations of 1 to 3 mmole/liter, and the Mn^{3+} ions complexed with 1 to 3 moles of $K_4P_2O_7$ per mole of Mn^{3+} to prevent precipitation, about 90% conversion of these monomers to polymer is obtained in 1 to 3 hrs at 30°C in an acid aqueous medium (pH 1.5 to 2). The grafting efficiency is as high as 98%, i.e. only 2% homopolymer is formed. The only precaution to be taken in the process is removal of dissolved oxygen by bubbling pure nitrogen in a slow stream through the reacting solution.

Cellulose is a partly crystalline and highly hydrogen-bonded substrate. Consequently it is much less accessible to grafting than starch (2). The present paper describes grafting experiments with the Mn^{3+}-initiator applied to never-dried pulp fibers from wood and to fibers from cellulose derivatives of low degrees of substitution, i.e. cellulosic substrates known to be more accessible to chemical reactions than other cellulose fibers after drying.

Experimental

Materials. The following cellulose substrates were used in never-dried state:

Dissolving pulp, a bleached sulfite pulp from spruce, MoDoCell H3430 from Mo och Domsjö AB, Örnsköldsvik, Sweden.

Grease-proof pulp, a bleached sulfite pulp from spruce, with high hemi-cellulose content, from Billerud-Uddeholm AB, Säffle, Sweden.

Ethyl cellulose, three samples prepared from alkali cellulose and ethyl chloride to low degrees of substitution (DS): about 0.25, 0.4 and 0.6 ethyl ether groups per glucose unit.

Hydroxyethyl cellulose, three samples prepared from alkali cellulose and ethylene oxide (oxirane) to low degrees of molar substitution (MS): about 0.25, 0.4 and 0.6 moles of oxirane substituted per mole of glucose units (some oxirane units form substituted oligo-ethylene glycol chains).

The samples of ethyl and hydroxyethyl cellulose were specially prepared and supplied by Berol Kemi AB, Stenungsund, Sweden. The following chemicals were used:
Manganese (II) sulfate, $MnSO_4 \cdot H_2O$, p.a. (Merck).
Potassium permanganate, $KMnO_4$, p.a. (Merck).
Sodium pyrophosphate, $Na_4P_2O_7 \cdot 10\ H_2O$, p.a. (Merck).
Sulfuric acid, conc., p.a. (Merck).
Acrylonitrile (AN), synthesis grade (Merck).
The AN monomer was purified by washing with water and distillation at reduced pressure under nitrogen atmosphere. Other chemicals were used as received from the supplier.

Grafting Method. The initiator solutions of Mn^{3+} ions were prepared from aqueous base solutions of $MnSO_4$, $KMnO_4$, $Na_4P_2O_7$ and H_2SO_4. The concentrations of chemicals in the reaction vessel (total liquid volume 500 ml) were 2.0 mmole/l Mn^{3+} (in most cases), 2.0, 4.0 or 6.0 mmole/l $Na_4P_2O_7$, and 80 mmole/l H_2SO_4 (which gives pH 1.7). The Mn^{3+} ions are stable in aqueous solution as pyrophosphate ion complex.

The cellulose substrates added in each experiment were 5, 10 or 15 g (given on dry basis). The amounts of acrylonitrile added were varied from 6 to 25 ml, as listed in the tables for the various experiments.

The grafting experiments were started by dispersion of the cellulose substrates in acidified water in the reaction vessel, the AN was added to the suspension and the initiator solution poured into the funnel, and N_2 was bubbled through funnel and vessel for 30 minutes. The temperature was adjusted to 30°C and the reaction started by emptying the monomer from the funnel into the reaction vessel. The reaction solution was stirred by a propeller through a mercury seal continuously under N_2 blanket. The reaction was stopped after 3 hrs by letting oxygen (air) into the reaction vessel and adding on equal volume of methanol or acetone to the solution. The grafted samples were washed with water to remove monomer and electrolytes, extracted with dimethylformamide (DMF) to remove AN homopolymer and for analysis of grafting efficiency, and then dried in air and vacuum at 40°C in a Soxhlet for 24 hrs.

In graft copolymerization the following definitions (based on weight) are used:

$$\text{Conversion (\%) (of monomer to polymer)} = \frac{\text{Polymer formed}}{\text{Monomer charged}} \times 100$$

$$\text{Add on (\%)} = \frac{\text{Synthetic polymer}}{\text{Graft copolymer}} \times 100$$

$$\text{Grafting efficiency (\%)} = \frac{\text{Grafted polymer}}{\text{Polymer formed}} \times 100$$

Molecular Weight of Grafted PAN Chains. A homopolymer-free graft copolymer sample was hydrolyzed in 2 M HCl for 48 h at room temperature to depolymerize the cellulose chains. The residue, mainly polyacrylonitrile (PAN) chains, was washed with water and ethanol, dried in air and extracted for 24 h at 20°C in DMF to dissolve the PAN chains. After filtration, the PAN solutions were used for molecular weight determinations in an Ubbelhode viscometer at a series of concentrations. The $\overline{M}_n$-values were calculated from the relation (7):

$$[\eta] = 3.92 \cdot 10^{-4} \cdot \overline{M}_n^{\,0.75}$$

The hydrolyzed samples were not completely soluble. The $\overline{M}_n$-values represent the soluble part of the previously grafted PAN chains. From the $\overline{M}_n$-values obtained for PAN, the number of anhydroglucose units per grafted PAN chain was calculated. Corrections were introduced for the weight of the anhydroglucose units of EC and HEC according to the degree of substitution.

Hydrolysis in Aqueous Alkali. Samples of the grafted celluloses were hydrolyzed in aqueous alkali to obtain superabsorbing materials. The conditions were 2 g of grafted sample (on dry basis) dispersed in 200 ml 1% NaOH at 100°C for 1.5 h. After hydrolysis the samples were precipitated in 600 ml methanol, washed with methanol and ethanol, dried and weighed. The aqueous alkali hydrolysis of PAN is known to give a gradual conversion of all the nitrile groups to amide and carboxyl groups. Intermediate ringclosed compounds form as sidegroups and appear as red-brown colour. At the completion of the hydrolysis, the sidechains are a copolymer of acrylamide and acrylic acid in certain proportions related to the conditions of the hydrolysis (8).

Water Retention of Hydrolyzed Samples. The water retention of the AN-grafted samples after hydrolysis in aqueous alkali were determined using three different methods:

A1. 0.5 g sample (on dry basis) was swollen in 100 ml distilled water for 10 minutes. The fiber suspension was poured into a sintered glass filter (porosity 1) and sucked at 700 mm Hg pressure. The volume of filtrate was measured and the water retention calculated as g of water per g of dry material. The measurement was repeated after drying the hydrolyzed sample in an oven at 60°C.

A2. The water retention for 0.1 g sample (on dry basis) was determined in a glass equipment, in which the sample on a sintered glass filter is sucking up the water from a U-shaped burette (7). The retained water (g/g) is measured as a function of time. Recorded A2-values are usually given for 20 minutes exposure time.

A3. The water retention for 0.1 g sample (on dry basis) was determined by swelling in 10 ml distilled water for 10 minutes in a graded centrifuge tube. The suspensions were then centrifuged at 900 G (2500 rev./min) for 20 minutes. A sharp boundary between gel phase and supernatant developed. The water retention in g/g was calculated from the volume of the gel (= the bottom layer).

Results and Discussion

The initiation reaction in grafting of vinyl monomers with Mn^{3+} ions to polysaccharides is a radical formation by oxidation, presumably of aldehyde and/or 1,2-diol groups (1) as further indicated by studies of model compounds of low molecular weight (9). Grafting of acrylonitrile to regenerated cellulose using Mn^{3+} has previously been reported (10).

The grafting process gives reproducible conversion and grafting efficiency (Table I) both for ethyl cellulose (EC, DS=0.60) and grease-proof pulp (GP). The measured grafting efficiency for the EC samples is lowered by a partial water solubility of the substrate.

The monomer conversion is increased to a level of 85% at high concentrations of substrate (EC with DS 0.40) and monomer (AN) in the reacting solution (Table II). The grafting efficiency remains high (> 95%) at all levels of conversion.

The degree of substitution (DS) of the EC substrate has a pronounced effect (Table III). At high DS values, conversion of monomer increases from about 70 to the level 80%, but the grafting efficiency decreases at DS=0.60. This may be a blocking effect of the ethyl ether groups and to some extent an effect of increased water solubility of the substrate after grafting. The intermediate DS-level (0.40) is most favorable for grafting.

Substrates of hydroxyethyl cellulose (HEC) samples of different molar substitution (MS) of oxirane give similar results in grafting with AN as EC (Table IV). At MS-values of 0.40 and 0.60, the water solubility influences the results, especially the grafting efficiency. HEC-samples of low MS (0.25) give the best results in grafting.

The dissolving pulp (MoDo Cell) and grease-proof pulp (GP) grafted with Mn^{3+} initiation give 70 to 71% conversion of AN to polymer of which up to 95% is grafted to the substrate (Table V). This is a good result compared with other grafting methods. The two never-dried native wood pulps are almost as accessible to grafting as the low-substituted cellulose derivatives studied (EC and HEC) and the starch substrates in previous work (1). It is assumed that only the amorphous and otherwise accessible part of the cellulosic fibers react in the initiation of grafting.

The water retention of the EC and HEC samples before grafting is of the order 6 to 7 g/g. After grafting with AN to various levels of add-on (30 to 60%), the water retention remains at about the same level (5-8 g/g) measured as A1 and A2 values. Alkaline hydrolysis gives a sharp increase in water retention to a level 6 to 8 times higher. This is illustrated by data for three grafted EC samples hydrolyzed for 1.5 hrs at $100^{\circ}C$ in 1% aqueous NaOH solution (Table VI). It is obvious that an increased degree of grafting (measured as add-on) gives a pronounced increase in water retention of the hydrolyzed samples. Retention values of 50 to 70 g/g are of considerable interest for commercial development of water-retaining materials.

Table I. Grafting of acrylonitrile (AN) onto ethyl cellulose (EC, DS=0.60) and grease-proof pulp (GP) using Mn^{3+} initiation. 12.0 ml (9.70 g) AN and 10.0 g substrate in all experiments. Initiator: Mn^{3+} 2.0 mmole/l, $Na_4P_2O_7$ 2.0 mmole/l, pH ~ 1.7, time 3 hrs, temperature 30°C.

Substrate	Conversion of monomer (%)	Add-on (%)	Grafting efficiency(%)
EC	82.2	39.8	83.1
EC	80.0	39.4	83.8
GP	70.2	30.4	94.1
GP	70.1	39.2	94.6

Table II. Grafting of acrylonitrile (AN) onto ethyl cellulose (EC) of DS=0.40 at increasing substrate and monomer concentrations. pH ~ 1.7, time 3 hrs, temperature 30°C.

EC (g)	AN (g)	Initiator mmole/l Mn^{3+}	Initiator mmole/l $Na_4P_2O_7$	Conversion of AN (%)	Add-on (%)	Grafting efficiency (%)
5	4.9	2.0	6.0	62.8	37.3	97.6
10	9.7	2.0	6.0	70.7	39.9	96.6
15	15.0	2.0	6.0	79.4	43.6	97.5
10	9.7	2.0	4.0	80.5	42.9	96.4
15	15.0	2.0	4.0	83.6	44.6	96.5
10	20.3	2.0	4.0	84.8	62.2	95.6

Table III. Grafting of acrylonitrile (AN) onto ethyl cellulose (EC) of different degrees of substitution (DS = 0.25, 0.40 and 0.60). In all experiments, 10.0 g EC; 12.0 ml (9.70 g) AN; pH ~ 1.7; temperature 30 °C; time 3 hrs.

DS of EC	Initiator mmole/l Mn^{3+}	$Na_4P_2O_7$	Conversion of AN (%)	Add-on (%)	Grafting efficiency (%)	
0.25	4.0	4.0	70.2	38.6	92.1	
0.40	4.0	4.0	75.9	41.3	95.3	
0.60	4.0	4.0	76.8	40.1	89.5	
0.25	2.0	6.0	69.5	39.8	98.1	
0.40	2.0	6.0	70.7	39.9	96.6	
0.60	2.0	6.0	75.8	37.6	81.8	
0.25	2.0	4.0	76.5	41.5	95.8	(1)
0.40	2.0	4.0	80.5	42.9	96.4	(2)
0.60	2.0	4.0	82.2	39.8	83.1	(3)

For the last three samples (1,2,3), the $\overline{M}_n$-values for grafted PAN-chains were measured to be 50 200, 81 600 and 77 600, respectively which gives grafting frequencies of 306, 486 and 447 anhydroglucose units per grafted PAN-chain respectively.

Table IV. Grafting of acrylonitrile (AN) onto hydroxyethyl cellulose (HEC) of different molar substitution (MS): 0.25, 0.40 and 0.60. In all experiments, 10.0 g HEC; 12.0 ml (9.70 g) AN; pH ~ 1,7; temperature 30 °C; time 3 hrs.

MS of HEC	Initiator mmole/l Mn^{3+}	$Na_4P_2O_7$	Conversion of AN (%)	Add-on (%)	Grafting efficiency (%)	
0.25	4.0	4.0	76.5	41.2	94.1	
0.25	2.0	6.0	72.2	40.2	97.1	
0.25	2.0	4.0	73.9	41.2	97.6	
0.25	2.0	4.0	73.9	41.2	97.6	(1)
0.40	2.0	4.0	71.9	37.8	87.1	(2)
0.60	2.0	4.0	70.8	35.3	79.4	(3)

For the last three samples (1,2,3), the $\overline{M}_n$-values for grafted PAN-chains were measured to be 52 300, 87 800 and 60 600 respectively, which gives grafting frequencies of 312, 504 and 365 anhydroglucose units per grafted PAN-chain respectively.

Table V. Grafting of acrylonitrile (AN) onto MoDoCell dissolving pulp (MDC) and grease-proof pulp (GP) using Mn^{3+} initiation. 10.0 g of substrate, pH~1.7, temperature 30°C and 3 hrs in all experiments.

Substrate	Initiator mmole/1 Mn^{3+}	$Na_4P_2O_7$	AN g	Conversion of AN (%)	Add-on (%)	Grafting efficiency (%)	
MDC	2.0	4.0	9.70	71.5	38.1	88.5	(1)
GP	2.0	4.0	9.70	70.2	39.4	95.1	(2)
GP	2.0	4.0	14.60	71.6	49.9	95.3	(3)
GP	4.0	4.0	19.40	71.4	56.6	91.7	(4)

For the four grafted samples (1,2,3 and 4) the $\overline{M}_n$-values for the PAN-chains were measured to be 75 900, 71 800, 121 000 and 126 000 respectively, which gives grafting frequencies of 483, 457, 511 and 401 anhydroglucose units per grafted PAN-chain respectively.

Table VI. Water retention values for three AN-grafted EC-samples (DS = 0.40), hydrolyzed with aqueous NaOH-solution, measured using the three methods A1, A2 and A3. When two retention values (g/g) are given, the first (1) refers to a never-dried sample and the second (2) to a once-dried sample. The yield of hydrolysis (YH) is given as hydrolyzed product in % of grafted sample.

Add-on (%)	YH (%)	Water retention in g/g A1	A2	A3
37.3	122	25 (1)	28 (1)	43
		17 (2)	24 (2)	
43.6	124	36 (1)	32 (1)	51
		24 (2)	30 (2)	
62.2	137	72 (1)	37 (1)	68
		36 (2)	39 (2)	

The water retention of the original grafted samples (before hydrolysis) were 7.5 (A1) and 7.4 (A2) g/g.

For grafted hydroxyethyl cellulose (HEC), dissolving pulp (MDC) and grease-proof pulp (GP), the water retention after alkaline hydrolysis reached A1, A2 and A3 values of the same order of magnitude as the corresponding EC samples (of Table VI). Three examples are given as illustration (Table VII). The grafted MDC pulp shows great losses in alkaline hydrolysis due to partial solubility of degraded material.

The water retention is time-dependent to various degrees for the samples studied. This is shown for A2-values vs. time in Figure 1. The EC and HEC samples are of DS=0.60 and MS=0.60, respectively, grafted with equal weight of AN and hydrolyzed in 1% aqueous alkali for 1.5 hrs at 100°C. The GP and MDC samples, containing native fiber structure, reach equilibrium much faster than the EC and HEC samples. All these measurements refer to once-dried samples. At 15-20 minutes exposure, the water retention values are of the same magnitude for the four samples.

Conclusions

Grafting of dissolving pulp (MDC) and grease-proof pulp (GP) (both are bleached sulfite pulps from spruce) and low-substituted ethyl cellulose (EC) and hydroxyethyl cellulose (HEC), all samples in never-dried state, with acrylonitrile (AN) using Mn^{3+} ions, complexed with pyrophosphate ions ($Na_4P_2O_7$), as initiator gives high conversion of monomer (70 to 85%) and high grafting efficiency (80 to 98%). The grafted products retain both fiber morphology and their original water retention values (5 to 8 g/g).

After hydrolysis of the grafted samples in 1% aqueous NaOH at 100°C, the water retention increases to a 5 to 10 times higher level, i.e. 25 to 80 g/g, for never-dried samples and somewhat lower values for once-dried samples.

The grafted and hydrolyzed MDC and GP samples reach saturation in water retention after a short time (a few minutes) while the corresponding samples of cellulose derivatives (EC and HEC) show increased retention for an extended time of exposure (more than 60 minutes).

Grafting and hydrolysis proceed as efficiently for cellulosic substrates as previously found for starch and water soluble polysaccharides. The grafted and hydrolyzed cellulose-based products prepared by the methods described, are of potential technical interest for water retention in various applications.

Table VII. Water retention of grafted HEC, dissolving pulp (MDC) and grease-proof pulp (GP) after alkaline hydrolysis, measured by three methods (A1, A2 and A3). Add-on (%) in grafting and yield in hydrolysis (YH in %) characterize the samples.

Sample	Add-on (%)	YH (%)	Water retention g/g A1	A2	A3
HEC (MS=0.40)	37.8	124	78 (1) 36 (2)	36 (1) 31 (2)	51
MDC	38.1	86	58 (1) 38 (2)	27 (1) 16 (2)	65
GP	39.4	113	61 (1) 44 (2)	31 (1) 25 (2)	75

The water retention values (1) refer to never-dried samples after hydrolysis and (2) to once-dried samples.

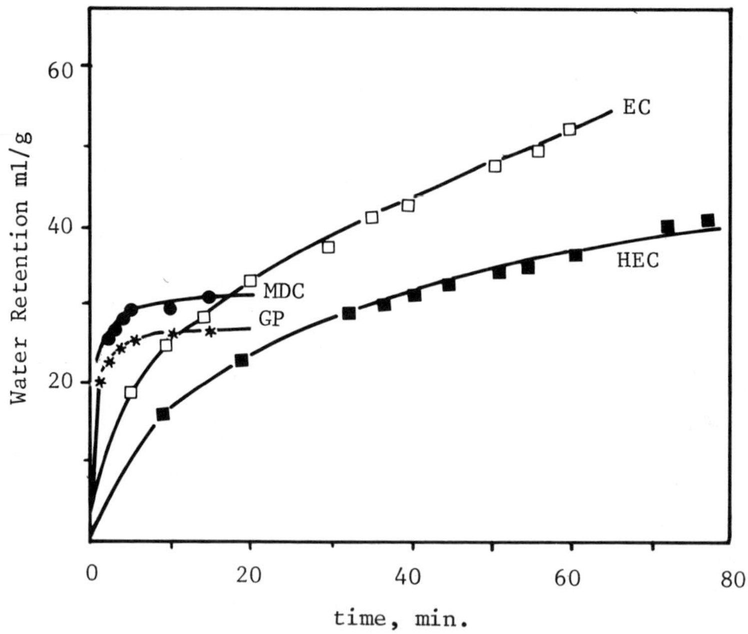

Figure 1. Water retention of grafted and hydrolyzed cellulosic samples measured by method A2. Key: MDC, dissolving pulp; GP, greaseproof pulp; EC, ethylcellulose; and HEC, hydroxyethylcellulose.

Acknowledgement

These investigations have been made possible by a grant from "Carl Tryggers Stiftelse för Vetenskaplig Forskning", which is gratefully acknowledged.

Literature Cited

1. R. Mehrotra and B. Rånby, J. Appl. Pol. Sci. 1977, 21, 1647, 3407; ibid. 1978, 22, 2991, 3003.

2. B. Rånby, "New Methods for Graft Copolymerization onto Cellulose and Starch", in "Modified Cellulosics" (Eds. R.M. Rowell and R.A. Young), Acad. Press, New York 1978; p. 171-195.

3. B. Rånby and D. Zuchowska, unpublished work.

4. B. Rånby and C. Rodehed, "New Graft Copolymerization onto Cellulosic Fibers", in "Review and Preview in Polymer Science. A Tribute to Herman Mark" (Ed. G.A. Stahl), ACS Mon. Ser. 1981.

5. B. Rånby and C. Rodehed, Polymer Bulletin, 1981, 5, 87-94.

6. M.O. Weaver, L.A. Guglimelli, W.M. Doane and C.R. Russel, Appl. Polymer Symp. 1977, 25, 97.

7. P.F. Onyon, J. Polymer Sci. 1959, 37, 315.

8. Method developed by Hercules Inc., USA, for Aqualon^{R} absorbent fiber: Capillary Absorbancy Test. Hercules Bulletin VC-494.

9. B. Rånby, S. Stone-Elander and T. Doba, unpublished work.

10. R.M. Livshitz, T.S. Sydykov and Z.A. Rogovin, Cellulose Chem. Technol. 1968, 2, 3 (Russ.). C.A. 1968, 69, 44513 y.

11. N. Grassie and I.C. McNeill, J. Polymer Sci., 1959, 39, 211.

RECEIVED January 19, 1982.

4

The Ceric Ion Method of Grafting Acrylic Acid to Cellulose

D. J. McDOWALL, B. S. GUPTA, and V. STANNETT

North Carolina State University, Fiber and Polymer Science, School of Textiles, Raleigh, NC 27650

The ceric ion method of grafting acrylic acid to cellulose was extended to rayon filaments. Experimental procedures were designed to optimize grafting yields and minimize homopolymer formation. The fiber sample was pretreated with an aqueous solution of ceric ammonium sulfate. This was followed by a washing step to remove excess ceric solution at the fiber surfaces. The grafting reaction was carried out in toluene to reduce homopolymer formation. Grafting yields of approximately 40% were obtained with grafting efficiencies in excess of 92%. Among the variables examined were the ceric ammonium sulfate concentration, acrylic acid concentration, reaction time and reaction temperature.

Grafting reactions provide the potential for significantly altering the physical and mechanical properties of a substrate material. Grafting possesses great potential for tailoring material properties to specific end uses. The natural abundance and a number of attractive properties of cellulose appear to make it an ideal grafting substrate. In grafting studies, the focus generally is to improve selected properties without significantly altering others.

Numerous methods have been developed for the grafting of vinyl monomers to cellulose. These have been reviewed in a number of papers and a recent monograph (1, 2, 3). Among the most promising and practical of these is the ceric ion technique. This method of initiation was first proposed by Mino and Kaizerman in 1958 (4) for alcohols, thioles, glycols, aldehydes and amines. Schwab et al (5) were among the first to extend this method to cellulosic substrates.

A great deal of interest exists in grafting acrylic acid and methacrylic acid onto cellulose due to the possibility of

0097-6156/82/0187-0045$6.00/0
© 1982 American Chemical Society

developing enhanced ion exchange and water sorbency properties. Unfortunately, such acid monomers have proven to be very difficult to graft due to their sensitivity, which leads to homopolymerization and irreproducibility. Among others, Richards and White (6), Cumberbirch and Holker (7), and Mansour and Schurz (8, 9, 10) have grafted acrylic acid onto cellulose using a ceric ion technique with rather low grafting yields.

More recently, Gangneux et al (11) developed a method for grafting acrylic acid onto cellulose powder, "Solka Floc," for use in textile waste treatment. The cellulose was treated with ceric ion in aqueous solution prior to its reaction with acrylic acid. A benzene-acrylic acid solution was used for grafting to reduce homopolymerization. Presumably, the hydrated ceric ion would not diffuse into the monomer solution to initiate homopolymerization although the reverse could still take place. They obtained grafting yields up to approximately 70% accompanied by 45% homopolymer. In the present investigation, their method is extended to fibers and additional emphasis is placed on the reduction of homopolymerization.

Experimental

A 50 denier rayon monofilament (~90μ diameter), courtesy of American Enka, was used as the grafting substrate. The samples were Soxhlet extracted in methanol prior to use.

A reagent grade acrylic acid, obtained from Fisher Scientific Co. and inhibited with p-methoxyphenol, was employed in this study. The acrylic acid was purified prior to each use with a Falling Film Still.

The ceric ammonium sulfate, toluene, methanol, sodium hydroxide, sulfuric acid and hydrochloric acid used were all of reagent grade (Fisher Scientific Co.). A low oxygen grade of nitrogen (less than 0.5 ppm oxygen), obtained from Airco Industrial Gases, was employed.

The grafting experiments were carried out on rayon samples weighing about 0.3 grams. The fiber sample was placed in a 50 milliliter flask, stoppered, evacuated with an aspirator for 1 minute and then flushed with nitrogen. This procedure was repeated three times. Approximately 25 milliliters of a 0.025 M ceric ammonium sulfate (CAS) solution in 1 N H_2SO_4 was injected into the reaction vessel. This was followed by three more evacuations and nitrogen flushings. The system was left under a positive nitrogen atmosphere and kept in the dark at room temperature (21°C) for the desired soaking period. This allowed the ceric solution to diffuse into the rayon fibers prior to the grafting reaction.

After the incubation period, excess CAS solution was rinsed from the fiber surface. This step lessened the amount of surface graft and homopolymer, which may inhibit the grafting reaction. The sample was dried on a coarse fritted glass crucible by passing air over it for 2 minutes and then rinsed in a solvent for 30 seconds. Excess solvent was removed by again air drying the sample for one minute.

The sample was then placed in a clean reaction vessel and evacuated with an aspirator and flushed with nitrogen three times. Approximately twenty milliliters of an acrylic acid in toluene solution was injected into the system. The system was evacuated and purged with nitrogen. The grafting reaction proceeded under a positive nitrogen atmosphere for the desired time at temperature under mild agitation in a constant temperature water bath.

A nonaqueous medium was chosen in this procedure, as in the work of Gangneux et al (11), to inhibit diffusion of CAS out of the fiber, thus lessening homopolymer formation in the solution, and to increase the availability of the acrylic acid. Selective experiments were performed to examine the potential of homopolymer formation. A mixture of CAS solution, water and acrylic acid was found to form homopolymer throughout whereas a mixture of CAS solution, toluene and acrylic acid formed homopolymer only in the aqueous CAS phase. This offers evidence of the advantages of using a two phase system. Toluene was used instead of benzene, as employed in the French work (11), for safety considerations.

After grafting, the sample was washed successively in methanol, 1 N NaOH, 1 N H_2SO_4, 1 N NaOH and several times in water followed by an overnight water soak with some agitation to remove homopolymer. The samples were converted to the acid form by soaking in 0.1 N HCl for 2 hours. This was followed by rinsing with water and another overnight soak. The samples were rinsed in methanol, then dried under vacuum at 40°C. The majority of samples were extracted in methanol to remove any remaining homopolymer. Selected samples were also extracted in water. No significant weight change was observed after these extractions. Percent "graft" was calculated from the dried weights as the increase in weight divided by the original weight multiplied by 100. Graft is given in quotation marks since in spite of the rigorous extraction methods employed it may still include some occluded homopolymer. The typical grafting procedure used in this investigation is outlined in Table I. The actual percent homopolymer was determined in a number of cases by precipitating it from the washing liquor in acetone according to the method of Chapiro and Sommerlatte (12). With the higher percent "grafts," 25%, 35% and 40%, the homopolymer represented only 4%, 2% and 8%, i.e., grafting efficiencies were 96%, 98% and 92% respectively.

Results and Discussion

Several different solvents were examined as rinsing agents for the fiber sample after soaking in the CAS solution. Their

TABLE I

Typical Experimental Procedure

1. Soaked in CAS solution for 2 hrs. under N_2 at 21°C.
2. Rinsed in 10/90 MeOH/Toluene.
3. 1 M Acrylic Acid in Toluene added to substrate under N_2 and reacted for 6 hrs at 35°C with agitation.
4. Product washed with MeOH, 1 N NaOH, 1 N H_2SO_4, 1 N NaOH, and finally water.
5. Soaked overnight in water with agitation.
6. Converted to acid form with 0.1 N HCl, washed with water and soaked overnight.
7. Rinsed in MeOH and dried under vacuum at 40°C.

purpose was to reduce the concentration of CAS and the aqueous medium at the fiber surface. As seen in Table II, methanol, acetone and butanol resulted in slightly negative values. This may be due to very effective removal of CAS including sorbed salt or inhibition of the grafting reaction by the solvents. Although toluene yielded the highest graft value, the fibers appeared visually to have polymer on the surfaces. Since toluene and the CAS solution are not very soluble, these results are not surprising. A mixture of 10/90 methanol/toluene by volume yielded high grafting results with low surface deposition. This mixture appeared to be sufficiently polar to remove surface CAS solution yet hydrophobic enough so as not to remove that sorbed into the fibers. All remaining samples were rinsed in this solvent mixture.

TABLE II
Percent "Graft" for Solvent Washed Samples

Solvent	Percent "Graft"
Methanol	-0.2
Acetone	-2.6
Butanol	-0.6
10/90 MeOH/Toluene	39.2
Toluene	50.2

Conditions: 0.025 M CAS, 2 hrs. soaking time, 1 M AA, Reaction time of 6 hrs. at 35°C.

The effect of oxygen on the grafting procedure was also examined. Oxygen in the CAS soak had no effect on the percent "graft" whereas the presence of oxygen during the grafting reaction reduced grafting from 39.2% to 1.3%. As is true with all radical reactions, the effect of oxygen is important due to its pronounced inhibiting effect. In an attempt to remove remaining traces of oxygen, a mechanical vacuum pump was used in place of simple aspiration in the evacuation procedure prior to the grafting reaction. No increase in grafting yields were obtained using the mechanical vacuum pump, therefore, aspiration was used for all ensuing experiments.

The effect of soaking time in the CAS solution on percent "graft" is shown in Figure 1. Initially, the grafting exhibits an increase and then levels off around 2 hours. Even the extremely long soaking periods of 1 week (165 hrs.) and 2 weeks (330 hrs.) showed no improvement in grafting. The two hour period was thus adopted as the standard experimental soaking time.

Figure 2 shows the effect of reaction time on percent "graft." Grafting increased initially and then leveled off before six hours. Since there is a large excess of acrylic acid available even after the longer reaction times, this leveling off is presumably due to catalyst exhaustion. A six hour reaction time was used for the remaining experiments.

Monomer concentration versus percent "graft" is examined in Figure 3. Grafting increases with the concentration of acrylic acid and reaches a maximum around 1 molar. Further increases in concentration lead to decreases and finally a leveling off of grafting values. This trend is in agreement with the work of Gangneux et al (11) and the earlier work reported by McDowall et al (13) in a benzene medium. Although the reasons for this trend are not clear, it may be due to the limited solubility of acrylic acid in toluene. There may be an association into micelles of the acrylic acid in the toluene at higher concentrations possibily reducing the effective monomer concentration accessible for grafting. The acrylic acid in toluene provides a reservoir for the grafting reaction and the rate of diffusion of the acrylic acid will depend upon its concentration. This concentration would initially increase and then diminish as the acrylic acid associates. In the French work (11), the maximum monomer concentration was found to be 2 M acrylic acid in benzene. This larger value may be due to the enhanced solubility of acrylic acid in benzene compared with toluene.

Ceric Ammonium Sulfate concentration and percent "graft" are plotted in Figure 4. The percent "graft" increased with increasing ceric concentration and reached a maximum value at about 0.040 molar. After this concentration, the percent "graft" decreased and began to level off. This effect was previously observed for other cellulosic systems by Schwab et al (5) and Cumberbirch and Holker (7). The decrease in grafting with increased ceric ion concentration is believed to be due to an increase in the radical termination step involving the Ce^{4+} ion.

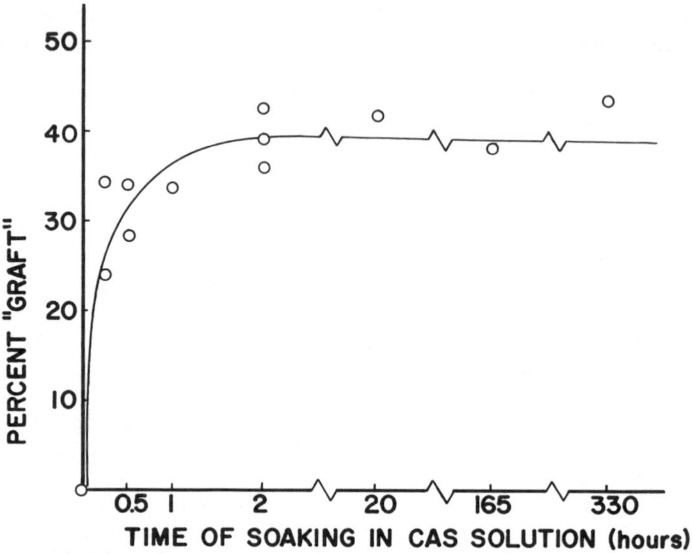

Figure 1. The effect of the soaking time in CAS solution on percent "graft." Conditions: 0.025 M *CAS, 1* M *AA, reaction time of 10 h at 35°C.*

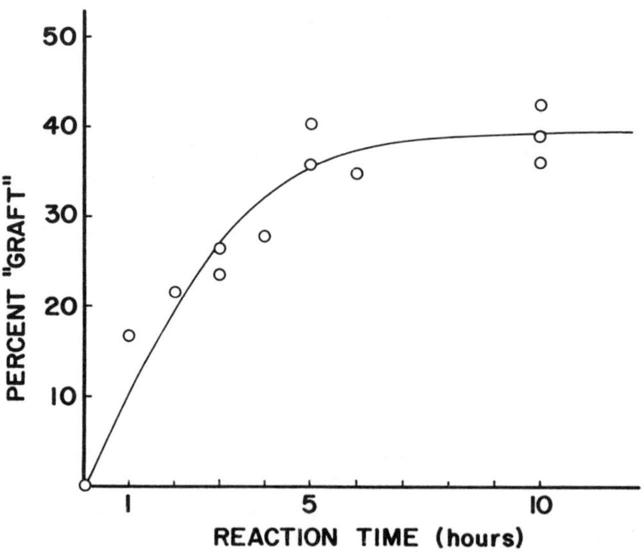

Figure 2. The effect of reaction time on percent "graft." Conditions: 0.025 M *CAS, 2 h soaking time, 1* M *AA, reaction temperature of 35°C.*

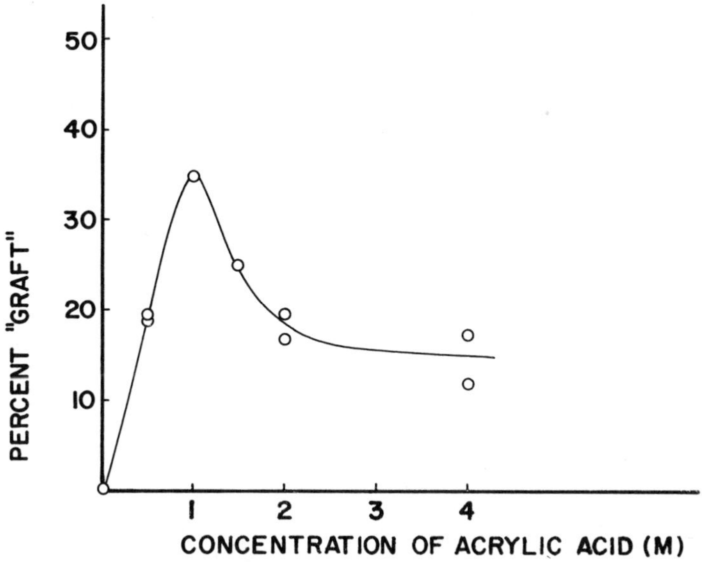

Figure 3. The effect of the concentration of AA on percent "graft." Conditions: 0.025 M *CAS, 2 h soaking time, reaction time of 6 h at 35°C.*

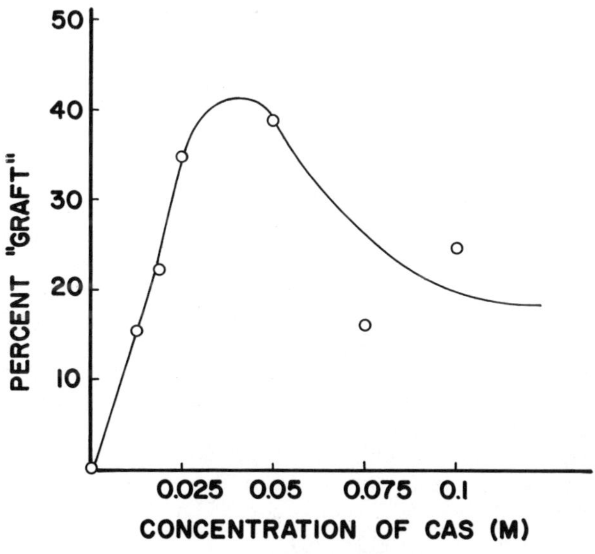

Figure 4. The effect of the concentration of CAS on percent "graft." Conditions: 2 h soaking time, 1 M *AA, reaction time of 6 h at 35°C.*

Figure 5 shows the effect of reaction temperature on percent "graft." Grafting yields are very low at 0°C and increase with temperature up to about 35°C. At 45°C, grafting decreases and at 55°C it is further reduced with obvious degradation occurring. The degradative process, although occurring at lower temperatures, becomes more prominent at elevated temperatures. Radical termination may also be increasing with temperature. This phenomena has also been reported by Hebeish and Mehta (14) with Ceric Ammonium Nitrate (CAN) and acrylonitrile, and by Schwab et al (5) for both CAS and CAN. Huque et al (15) found a decrease in percent graft with CAS and methyl methacrylate onto jute as the temperature was increased from 30 to 50°C.

The natural log of the percent "graft," after 6 hours, versus 1/T may be plotted for the initial portion of the curve (0°C to 35°C) in Figure 5. These values fall on a straight line, as seen in Figure 6. The calculated activation energy was 3.8×10^4 J/mol (9.1 kcal/mol). This overall activation energy reflects initiation, termination and propagation together with degradation.

The acrylic acid used in this study was normally distilled once prior to each use. However, in an attempt to increase grafting yields, the effect of redistilling was examined. Somewhat higher yields were obtained with the redistilled acrylic acid. For example, 47.5% grafting was obtained by redistilling as opposed to 39.2% for single distillation. Further distillations were not performed due to the susceptibility of uninhibited acrylic acid to spontaneous polymerization. The effect of monomer purity will be examined more extensively in future work.

Conclusion

The procedure, as developed from that of Gangneux et al (11), gave grafting yields up to 40 percent with very little homopolymer. The optimum experimental conditions found in the present work are summarized in Table III. Somewhat higher grafting yields, up to 47.5%, were obtained using redistilled acrylic acid. The important features of the procedure include pretreating the fiber with CAS, removing excess CAS solution from the fiber surface with a rinsing procedure and grafting in a hydrocarbon with limited water solubility. The effects of varying monomer concentration, ceric ion concentration and reaction temperature on this system are in agreement with what has been reported in the literature with other vinyl monomers and cellulose (5, 7, 11, 14, 15).

It should also be noted that the rayon used in this study was a large denier fiber and that higher grafting yields would be expected with smaller diameter fibers, and other cellulosic

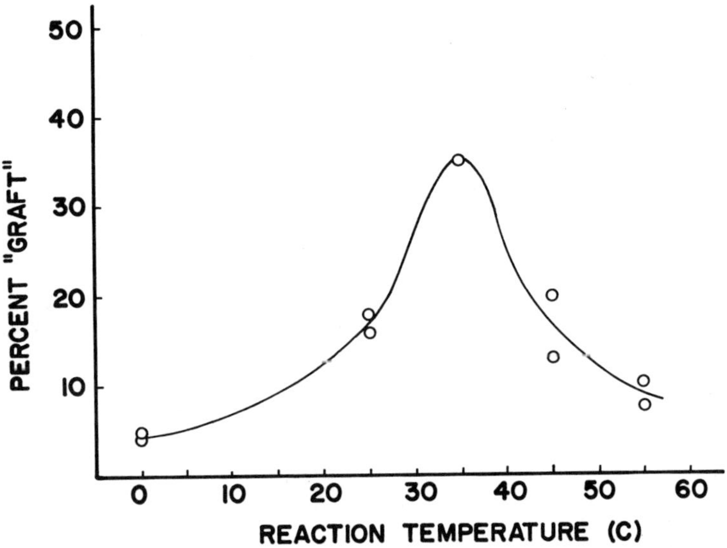

Figure 5. The effect of reaction temperature on percent "graft." Conditions: 0.025 M *CAS, 2 h soaking time, 1* M *AA, reaction time of 6 h.*

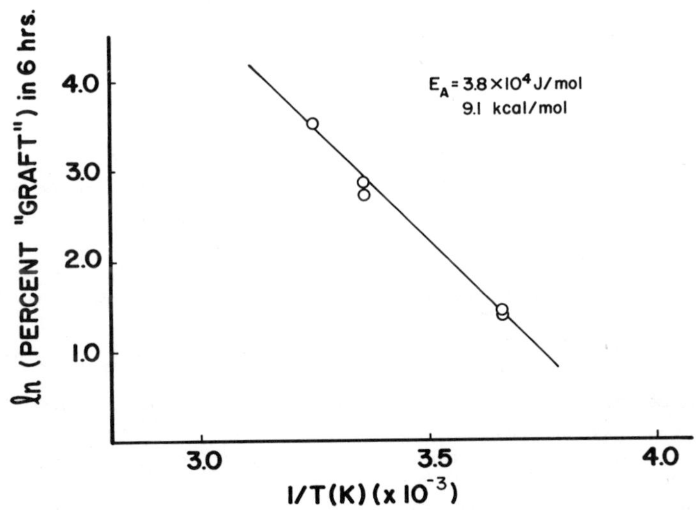

Figure 6. The natural log of percent "graft" in 6 h versus 1/T. Conditions: 0.025 M *CAS, 2 h soaking time, 1* M *AA, reaction time of 6 h.*

Table III

Summary of Reaction Variables Studied

CONDITIONS VARIED	RANGE USED	OPTIMUM
CAS Conc (M)	0.013-0.100	~0.040
Soaking Time in CAS (hrs)	0.25-330	2
AA Conc (M)	0.5-4.0	1.0
Reaction Temp (°C)	0-55	35
Reaction Time (hrs)	1-10	~5

materials with greater surface area. A large denier was selected in order to facilitate the examination of the location of the graft and morphological changes. These together with other properties, such as the water sorbency and related properties of the sodium salt, will be studied and compared with similar samples made by a preirradiation grafting technique (13, 16).

Acknowledgements

This work was supported by organized research funds of the School of Textiles, North Carolina State University. The rayon fiber sample used for grafting was specially manufactured and supplied by Dr. Tom Allen of the American Enka Company. The authors gratefully acknowledge their assistances.

Literature Cited

1. Krässig, H. A.; Stannett, V. Adv. Polym. Sci. 1965, 4, 111-56.
2. Arthur, J. C., Jr. Adv. Macromol. Chem. 1970, 2, 1-87.
3. Hebeish, A.; Guthrie, J. T. "The Chemistry and Technology of Cellulosic Copolymers"; Springer-Verlag: New York, 1981.
4. Mino, G.; Kaizerman, S. J. Polym. Sci. 1958, 31, 242-43.
5. Schwab, E.; Stannett, V; Rakowitz, D. H.; Magrane, J. K. Tappi 1962, 45, 390-400.
6. Richards, G. N.; White, E. F. T. J. Polym. Sci., Part C 1963, 4, 1251-60.
7. Cumberbirch, R. J. E.; Holker, J. R. J. Soc. Dyers Colour. 1966, 82, 59-63.
8. Mansour, O. Y.; Schurz, J. Sven. Papperstidn. 1973, 76, 258-62.
9. Mansour, O. Y.; Schurz, J. Sven. Papperstidn. 1973, 76, 288-91.
10. Mansour, O. Y.; Schurz, J. Sven. Papperstidn. 1973, 76, 415-18.

11. Gangneux, A.; Wattiez, D.; Marechal, E. Eur. Polym. J. 1976, 12, 535-41.
12. Chapiro, A.; Sommerlatte, T. Eur. Polym. J. 1969, 5, 707-24.
13. McDowall, D. J.; Gupta, B. S.; Stannett, V. T.; Williams, J. L.; Zahran, A. H. "Grafting of Acrylic and Methacrylic Acid to Cellulose Materials." National Joint Pacific Conference of ACS, Honolulu, Hawaii, April 1979.
14. Hebeish, A.; Mehta, P. C. J. Appl. Polym. Sci. 1968, 12, 1625-47.
15. Huque, M. M.; Habibuddowla, M. D.; Mahmood, A. J.; Jabbar, M. A. J. Polym. Sci., Polym. Chem. Ed. 1980, 18, 1447-58.
16. Zahran, A. H.; Williams, J. L.; Stannett, V. T. J. Appl. Polym. Sci. 1980, 25, 535-42.

RECEIVED December 17, 1981.

5

Grafting of Rayon Fabrics in Cold Plasma Conditions

C. I. SIMIONESCU, M. M. MACOVEANU, S. PERCEC, G. CAZACU, and A. IOANID

"Petru Poni" Institute of Macromolecular Chemistry, 6600 Jassy, Romania

The rayon fabrics were grafted with aromatic or aliphatic phosphorus and nitrogen containing polymers. Thus, the fabrics impregnated with solutions of these polymers at different concentration (10-25%) have been treated 15 minutes under RF plasma conditions at 770 W radiofrequency power. The resulting materials were extracted with solvents and hydrolised in acid and basic medium in order to evidence the polymer-polymer grafting. The separated fractions have been studied by IR spectroscopy, TGA, electron microscopy, elemental analysis, etc. The flammability tests carried out on the grafted rayon fabrics indicated the improvement of their flame retardant properties. The best results are obtained when an aliphatic nitrogen and phosphorus containing polymer were used.

The grafting process in electrical discharge conditions is more often utilised in order to obtain grafted copolymers of cellulosic fibers and fabrics. This is due to the following reasons: - the possibility of generating active species (macroradicals and/or polyradicals) in plasma conditions on the prefabricated article and fabrics (1-3) - the large range of grafting agents (4,5) - the rigorous control of grafting parameters (the power applied on the electrodes, the plasma time action, the thicknesses of the solid grafting agent layers deposited on the polymer support; (6) - the thickness of fiber layer affected by plasma action is lower than in the case of other penetrating energies (7) - the process is economical due to the absence of the solvents, catalysts and the possibility of a continuous grafting procedure.

In order to limit extensive side reactions (advanced destruction of the support-polymer, modification of grafting monomers structure) which usually accompany grafting reactions in electrical discharge conditions, we have elaborated the solid-solid grafting procedure in cold plasma. Thus, the rayon fabric was grafted with halogen and phosphorus containing polymers in

0097-6156/82/0187-0057$6.00/0
© 1982 American Chemical Society

cold plasma conditions in order to obtain flame-retardant compounds (8).

There are, of course, other methods for treating fibers and rayon fabrics with organo-phosphorus compounds (9-13) in order to improve their flame-retardant properties. However, the modified fibers present the changed physical and mechanical properties due to large additives utilised and often these additives are eliminated after repeated washings.

In this paper, the surface grafting of rayon fabrics with nitrogen and phosphorus containing polymers in cold plasma is studied. The analytical data (IR spectroscopy, TGA, electron microscopy, elemental analysis, etc.) indicate the formation of grafted copolymers. The grafted rayon fabrics present improved flame-retardant properties, the best behavior was proved by those grafted with polyurea of phosphinic acid.

Experimental

Phosphorus and nitrogen containing polymers were synthesized according to literature data (14-16). The chemical structure and some physical and chemical properties of these polymers together with the modifications occurring in cold plasma are presented in Table I.

The rayon fabrics (3 x 21 cm) extracted with solvents in order to remove the textile ingredients were impregnated for an hour with grafting solutions of different concentrations (10-25%), then were dried at room temperature.

All of the grafting reactions were carried out in a static plasma grafting device presented in Figure 1. In a typical experiment, the ingredient free impregnated textile material as we have already described was introduced in the plasma reactor (1). After several washing cycles of reactor with nitrogen, the preselected pressure was established (0.7 mmHg). The radiofrequency power (770 W) was transferred to the discharge tube through external semicylindrical electrodes (2).

The compounds obtained were extracted 48 hours at room temperature with various solvents (acetone, methanol, water) in order to remove the unreacted polymers and the side-products resulting from grafting agents in cold plasma conditions.

The extracted rayon fabrics were analysed by spectral, physical and chemical methods in order to prove that the grafting reactions took place and what kind of new properties the fabrics have obtained.

Results and Discussion

The experimental conditions and the phosphorus and nitrogen contents (elemental analysis; 17, 18) of grafted fabrics in cold plasma are presented in Table II. It can be noticed that increasing of the grafting agent solution concentration up to 20 percent, the phosphorus and nitrogen content grows but does not exceed 1.5 percent. These small values of the phosphorus content indicate low grafting degrees of the fabrics but these are

Table I
Physical and chemical properties of the phosphorus and nitrogen containing polymers used in the grafting plasma processes.

Polymers	Structure	Content of P (%)	Content of N(%)	Solubility Aspect
Polyphosphonate P-1	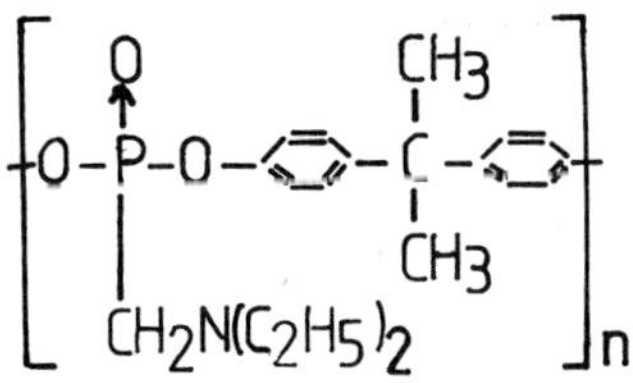	11.80	3.90	s (acetone) white powder
Plasma treated polymer		10.00	3.00	s; ins yellow
Polyamidophosphate P-2	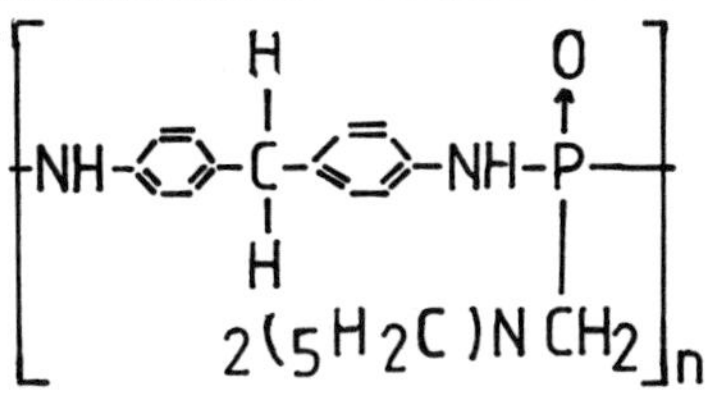	8.60	8.80	s(methanol) orange powder
Plasma treated polymer		8.24	8.12	s; ins brown
Polyamidophosphate P-3	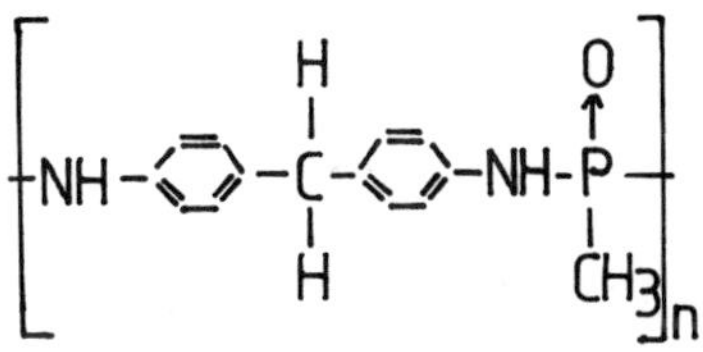	13.20	5.80	s(methanol) brown powder
Plasma treated polymer		12.20	4.77	s; ins brown
Polyurea of phosphinic acid P-4	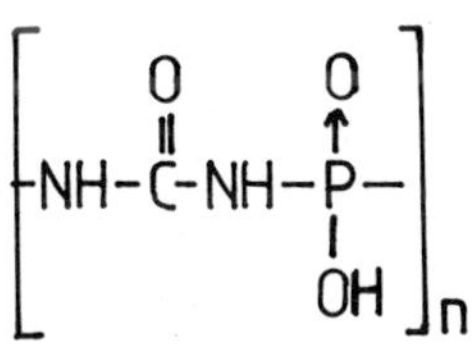	20.30	24.50	s (water) white powder
Plasma treated polymer		20.00	24.00	s; ins brown

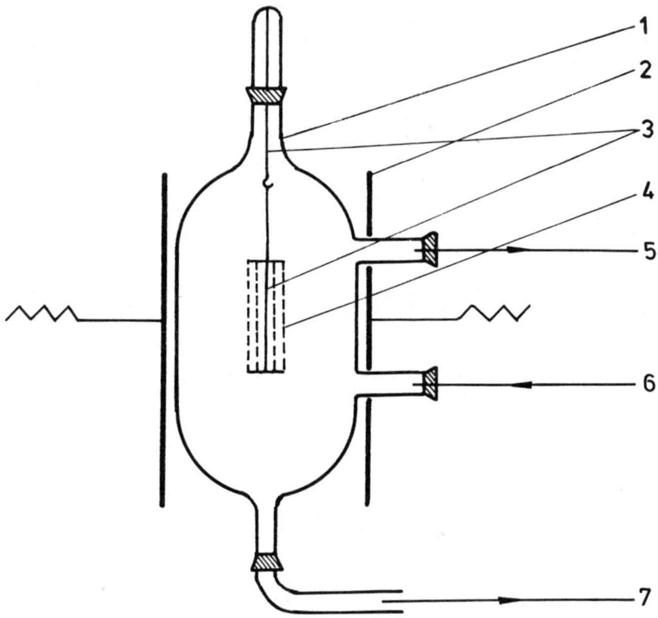

Figure 1. Schematic diagram of the static plasma grafting device. Key: 1, glass reactor; 2, copper electrodes connected to HF generator; 3, device for polymer sustaining; 4, fabrics to be treated; 5, outlet to vacuum pump; 6, gas supply; 7, outlet to vacuum gauge.

Table II

Grafting experimental conditions and phosphorus and nitrogen contents of the plasma grafted rayon fabrics.

N_o	Experimental conditions			Elemental analysis	
	Polymer support	Grafting agents	Solution concentration (%)	P (%)	N (%)
1.	Rayon fabric	P-1	10	0.270	0.100
2.	Rayon fabric	P-1	15	0.432	0.140
3.	Rayon fabric	P-1	20	0.705	0.250
4.	Rayon fabric	P-1	25	0.870	0.350
5.	Rayon fabric	P-2	10	0.291	0.540
6.	Rayon fabric	P-2	15	0.360	0.640
7.	Rayon fabric	P-2	20	0.580	0.500
8.	Rayon fabric	P-2	25	0.380	0.830
9.	Rayon fabric	P-3	10	0.170	0.330
10.	Rayon fabric	P-3	15	0.440	0.400
11.	Rayon fabric	P-3	20	0.330	0.500
12.	Rayon fabric	P-3	25	0.270	0.780
13.	Rayon fabric	P-4	10	0.380	0.300
14.	Rayon fabric	P-4	15	0.410	0.320
15.	Rayon fabric	P-4	20	1.230	1.700
16.	Rayon fabric	P-4	25	1.000	1.140

sufficient to modify the flammability resistance. At high ratios of grafting agent (up to 25 percent) one can ascertain that the values of phosphorus content are decreasing probably due to limited penetration of the electrons into the deposited polymer layer, according to results obtained in our previous works (6). It is clearly observed that the rayon fabrics grafted with P-4 presented the highest nitrogen and phosphorus content. This can be explained, firstly by the highest phosphorus and nitrogen content of the grafting agent P-4 in comparison with the other grafting agents P-1, P-2, and P-3 and secondly by a good stability of this P-4 in cold plasma conditions (see Table I).

The formation of the graft copolymers was proved by the analysis of the fractions I-IV obtained through extraction and H_2SO_4 and NaOH treatment, according to the schematic diagram presented in Figure 2. The IR spectra of the insoluble fractions II and soluble fraction III present absorption bands (19,20) characteristic of those of the control rayon and of the polymers P-1 - P-4, as it can be seen in Table III and Figure 3, respectively. In the same time, we have found that the phosphorus and nitrogen contents of the grafts (soluble fraction III) are similar to those of the grafting agents, a fact which proves the occurrance of polymer to polymer grafting reactions in cold plasma conditions.

The IR spectra of insoluble fraction IV resulting after basic hydrolysis, indicate the cellulosic structure presence. The elemental analysis also shows low nitrogen and phosphorus contents of these fractions, due probably to bound or enclosed polymer fragments into the structure of the rayon fabrics and to the fact that the grafting agents are sufficiently resistant at the hydrolysis.

Electron microscopy data of the standard rayon and grafted rayon fabrics bring additional proofs for the polymer-polymer grafting. The existence of uniform coatings having various morphological aspects is clearly observed from the surfaces (carbon-paladium technique) of the grafted fabrics with phosphorus and nitrogen containing polymers. Thus, in the case of P-1 grafted rayon fabric continuous lamellar formations which cover all the fiber surface have been remarked (Figure 4b). In the case of rayon fabric grafted with P-2, a pellicular cover with small cavities of uniform distribution on the fiber surface has been noticed (Figure 4c). The rayon fabrics grafted with P-3 present an uniform coating with globose aspect disposed as longitudinal bands on the fiber surface (Figure 4d). The P-4 grafted rayon fabrics have a major quantitative cover in comparison with the other grafted rayon fabrics, the grafts being arranged as longitudinal bands oriented along the fiber (Figure 4e). The coatings are rough, substantial and uniform. On the other hand, the images of grafted fiber cross sections obtained at the electron microscope (techniques of including in the blocks with epoxides sectioned at the ultramicrotom) in comparison with control

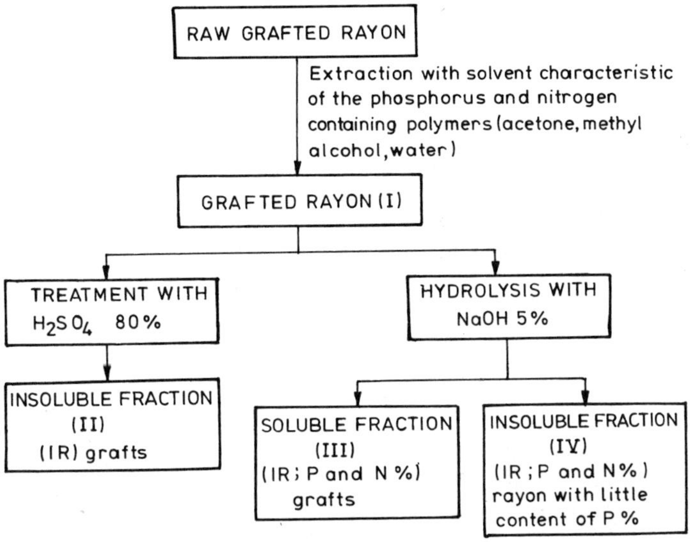

Figure 2. Schematic diagram of the procedure for the characterization of the raw material and its derived fractions.

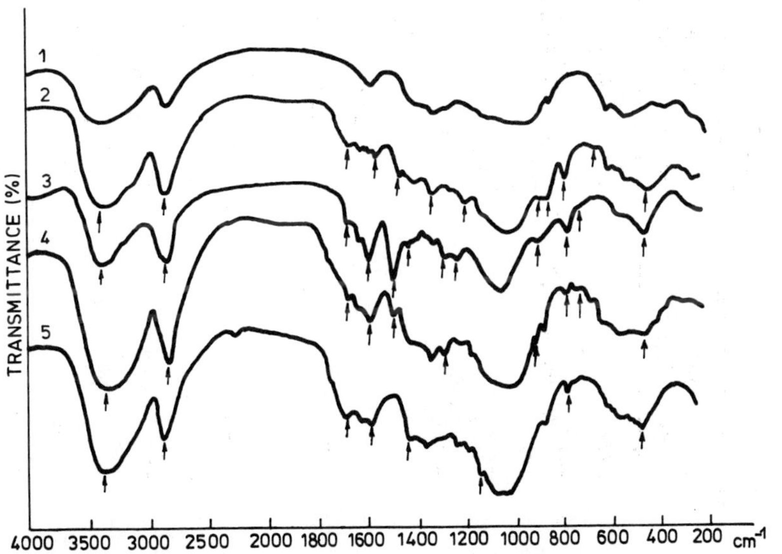

Figure 3. IR spectra of the soluble fraction III. Key: 1, standard rayon; 2, P-1 grafted rayon fabric; 3, P-2 grafted rayon fabric; 4, P-3 grafted rayon fabric; 5, P-4 grafted rayon fabric.

Table III

Infrared spectra of the insoluble fraction II.

Wavenumber (cm^{-1})	P-1 grafted rayon fabric	P-2 grafted rayon fabric	P-3 grafted rayon fabric	P-4 grafted rayon fabric
3400 - 3200	-OH	-OH; $-NH_2$	-OH	-OH; -NH; $-NH_2$
2980 - 2920	$-CH_{arom,aliph}$	$-CH_{arom,aliph}$	$-CH_{arom,aliph}$	$-CH_{aliph}$
1720 - 1700		-C=O	-C=O	-C=O
1650 - 1600	-NH; -N=	-NH	-NH	-NH-CO
1500	-P-O-Ph	-P=N	-P=N	
1450				-NH
1480 - 1360		-P-N-Ph	-P-N-Ph; -P=N	
1300 - 1200	-N=; -P-O	-P-O;-P-OH;-P=N	-P=N	
1150				-P-O
930	-P-O-C			
910 - 900		-P=N	-P=N	
835 - 800	$-P-c_{alkyl}$	$-P-C_{alkyl}$	$-P-C_{alkyl}$	-P-NH
760 - 730	$-CH_{arom,aliph}$	$-CH_{arom,aliph}$	$-CH_{arom,aliph}$	
500 - 460	-P-O-C	-P-O-C	-P-O-C	-P-O-P

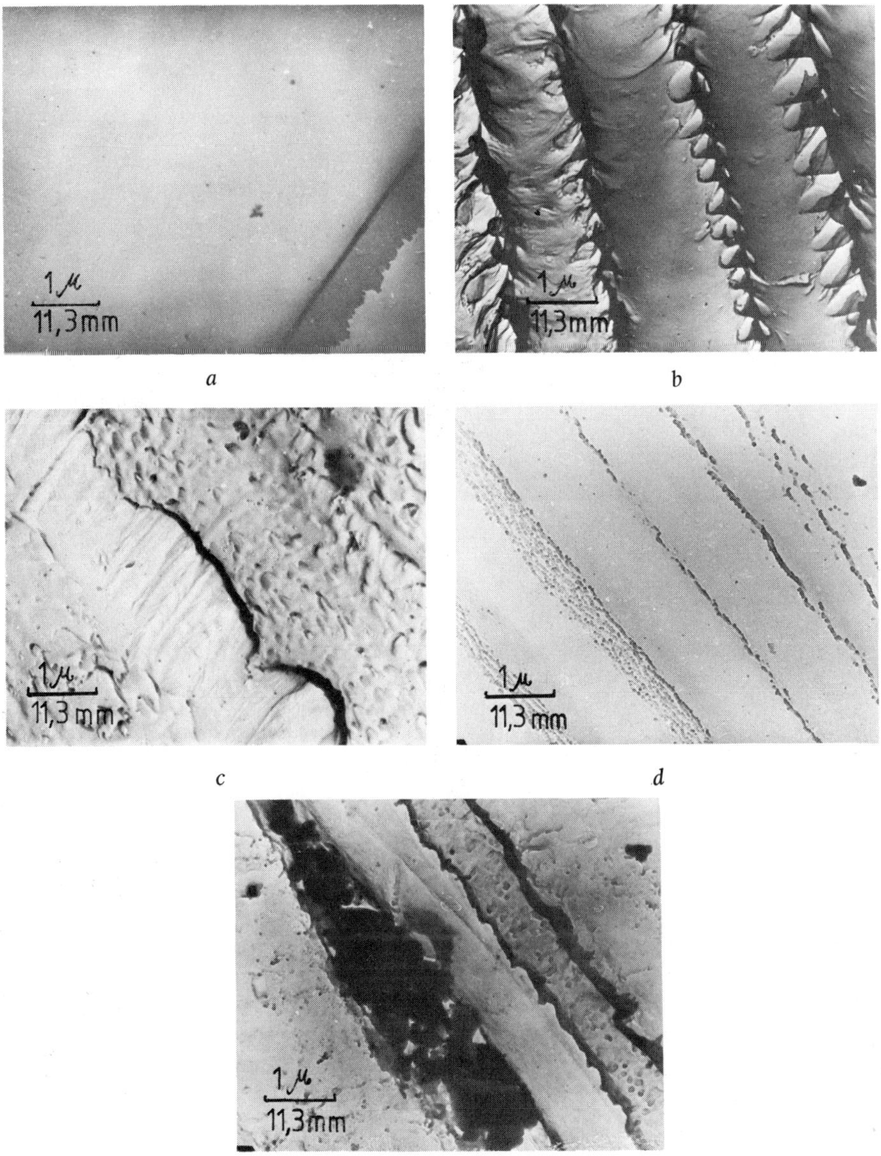

Figure 4. Electron microscope images for a, surface of the control rayon; b, P-1 grafted rayon fabric; c, P-2 grafted rayon fabric; d, P-3 grafted rayon fabric; e, P-4 grafted rayon fabric. Continued.

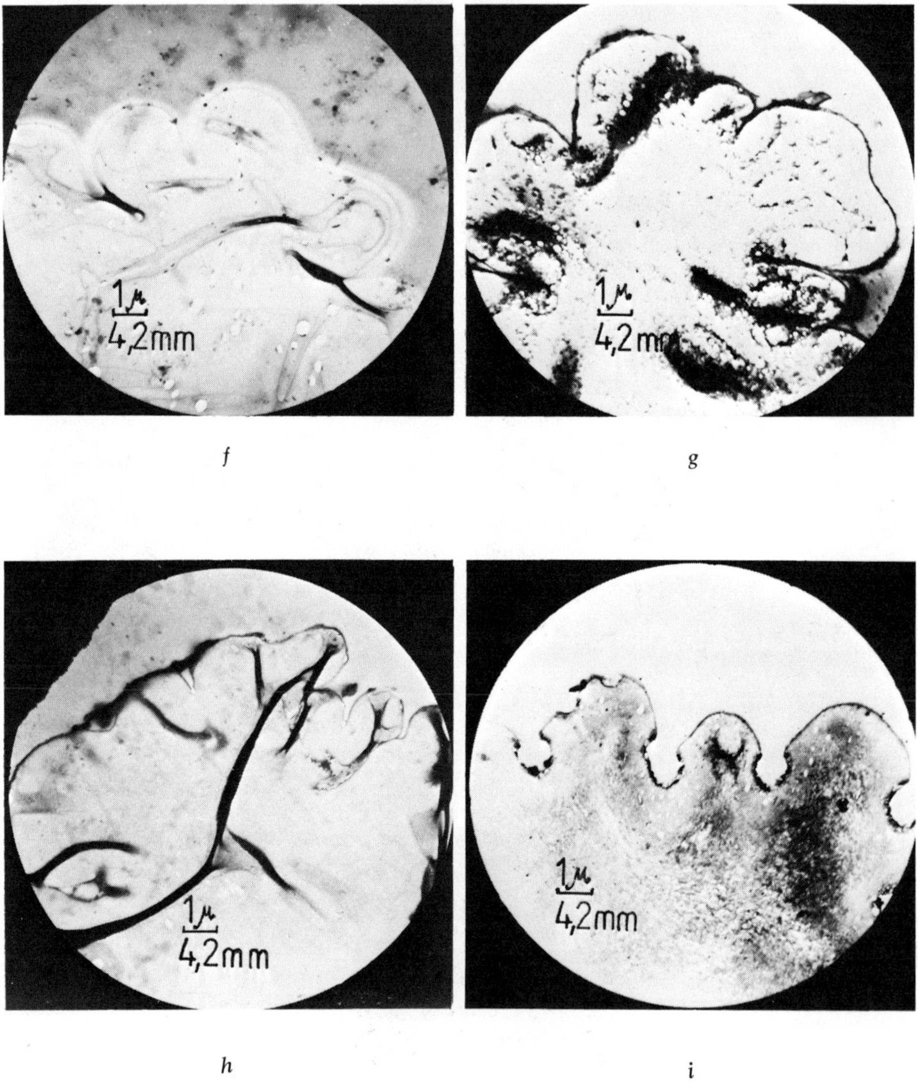

Figure 4. Continued. Electron microscope images for f, cross-section of the control rayon fiber; g, P-1 grafted fiber; h, P-2 grafted fiber; i, P-4 grafted fiber.

rayon, present a continuous thin graft layer on the lobate edge of the rayon (Figure 4f-i).

For the burning test, pieces of the grafted rayon fabrics were burned on a semicircle tester according to DIN 54 331. The results are given in Table IV and Figure 5.

It can be remarked that the rayon fabrics grafted with P-1, P-2 and P-3 have more reduced extinguishing angles and burning times than those of standard rayon. The macroscopical aspects of the grafted rayon fabrics are not changed.

The rayon fabrics grafted with P-4 ignited only on the end of the sixth attempt and immediately self-extinguished. This good burning behavior is due to the major and uniform cover of grafts on the surface fibers and the higher nitrogen and phosphorus contents of the polymer P-4.

These results are in good agreement with the thermo-destructive behavior of the P-4 grafted rayon fabrics in comparison with control rayon and other grafted rayon fabrics (Table V, Figure 6). The thermal analyses were registered on a Paulik-Paulik Erdely MOM Budapest (in air, 12°/min heating rate, 25-600 °C temperature range). The activation energy and the order of degradation reaction have been calculated according to Coats method (21).

It is well known that flame retardant polymers decrease the beginning temperature of thermal degradation. These phenomena associated with their endothermic decomposition have as the consequence the decrease of the pyrolysis rate. In addition, the gaseous pyrolysis products can influence the chain depropagation reactions which usually take place in the flame. The data from Table V shows that the temperature of the beginning of thermal decomposition is lower for grafted rayon fabrics (165 °C and 160 °C for rayon grafted with P-1 and P-4 respectively) in comparison with the standard rayon (190 °C) and also, the decrease of the volatile content (weight losses, ΔW, %) from 63 percent for control rayon to 47 percent for rayon grafted with P-4 can be observed.

The TG curves in Figure 6 indicate that the phosphorus presence in the grafted fabrics increase the thermal decomposition rate (curves 2,3,4 and 5 have different slopes in comparison with curve 1) facilitating the char formation (curve 4) and decreases the flammable volatile content.

Our investigations suggest that grafted copolymers are obtained probably by the recombination reactions of the macroradicals generated both on the cellulosic support and on the nitrogen and phosphorus containing polymers.

In our previous works (1-3) the possibility of the generation of macro- and polyradicals on the cellulosic support, under the cold plasma conditions, by the dehydrogenation of OH groups or the cleavage of 1,5 glucoside linkages was evidenced.

It has been also shown that under cold plasma action some of the organo-phosphorus compounds (triphenyl phosphate or tricresyl

Table IV
Flammability tests by standard laboratory procedure carried out on the grafted rayon fabrics.

N_o	Sample	Burning angle (degree)	Burning time (seconds)	Remarks
1.	Rayon fabric[x)] 35 x 180 x 0.8 mm	90	220	It continuously burns with high flame; the macroscopical aspect is changed.
2.	Rayon fabric[xx)] 30 x 180 x 2.0 mm	80	225	Ibidem.
3.	P-1 grafted rayon[x)] fabric	70	182	It burns uniform with small flame and smoke; the macroscopical aspect of fabric is not changed.
4.	P-2 grafted rayon[x)] fabric	80	196	Ibidem.
5.	P-3 grafted rayon[x)] fabric	80	183	Ibidem.
6.	P-4 grafted rayon[x)] fabric	40	16	It is self-extinguishing after 16 seconds.
7.	P-1 grafted rayon[xx)] fabric	80	170	It burns uniform with small flame and smoke; does not present after glow. The macroscopical aspect of fabric is not changed.
8.	P-2 grafted rayon[xx)] fabric	55	116	Ibidem.
9.	P-3 grafted rayon[xx)] fabric	55	143	It burns after the second test with small flame. The macroscopical aspect is not changed.
10.	P-4 grafted rayon[xx)] fabric	0	0	It does not burn after six tests. It is immediately self-extinguishing.

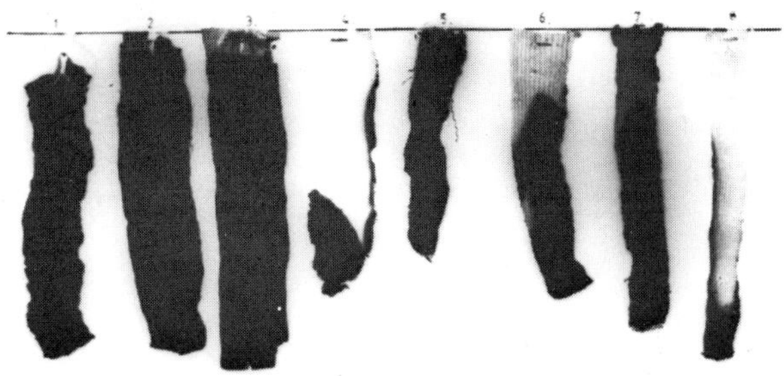

Figure 5. The aspect of the grafted rayon fabrics after the flammability test. Key: left to right, P-1 grafted compact rayon fabric; P-2 grafted compact rayon fabric; P-3 grafted rayon fabric; P-4 grafted rayon fabric; P-1 grafted lower compact rayon fabric; P-2 grafted lower compact rayon fabric; P-3 grafted lower compact rayon fabric; P-4 lower compact grafted fabric.

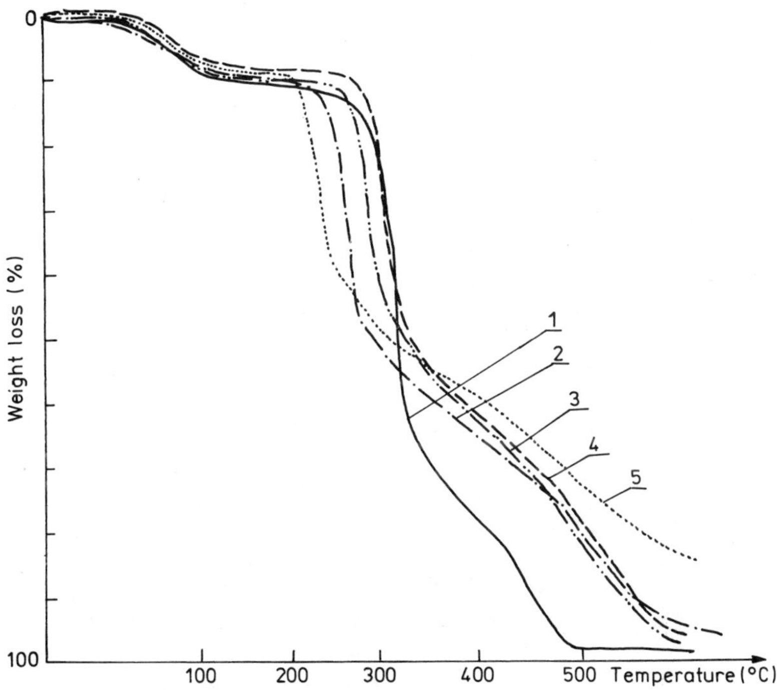

Figure 6. TG curves for 1, control rayon fabric; 2, P-1 grafted rayon fabric; 3, P-2 grafted rayon fabric; 4, P-3 grafted rayon fabric; 5, P-4 grafted rayon fabric.

Table V
Thermo-oxidative destruction behaviour of the grafted rayon fabrics.

N_o	Sample	TGA Peak I T_b	Peak I T_m	Peak I T_e	Peak I ΔW%	Peak II T_b	Peak II T_m	Peak II T_e	Peak II ΔW%	Peak III T_b	Peak III T_m	Peak III T_e	Order of reaction n	Activation energy kcal/mol
1.	Rayon control	30	75	155	9.0	190	330	395	63	410	470	510	0	36.86
2.	P-1 grafted rayon fabric	25	75	165	9.4	165	270	365	50	300	520	583	0	43.01
3.	P-2 grafted rayon fabric	25	75	155	9.0	190	285	370	48	305	525	590	0	41.39
4.	P-3 grafted rayon fabric	25	75	175	9.0	195	305	375	53	415	510	565	0	49.82
5.	P-4 grafted	25	70	160	9.0	160	235	335	47	410	510	565	0	54.78

phosphate) lead to the development of polymers by the dehydrogenation and/or through aromatic cycle opening mechanisms (5).

In addition, the active species of plasma give intensive dehydrogenation and dehydrochlorination reactions of the phosphorus and halogen containing polymers with the generation of macroradicals (8).

It can be presumed that under cold plasma conditions, the nitrogen and phosphorus containing polymers can lead also to the macroradicals through dehydrogenation of CH aromatic and aliphatic or NH groups.

Because the polymer-polymer grafting reactions in cold plasma conditions are superficial, the physico-mechanical properties of the grafted rayon fibers are similar to those of standard rayon.

Conclusions

Rayon fabric has been grafted in cold plasma conditions with phosphorus and nitrogen containing polymers (P-1, P-2, P-3 and P-4) in order to achieve improved flame-retardant properties. Spectral, physical and chemical analysis proves the graft-copolymers formation. Polymer (rayon fabrics) to polymer (phosphorus and nitrogen containing macromolecular compounds) grafting processes can be carried out under plasma conditions. The graft-copolymers are obtained probably by the recombination reactions of the macroradicals generated on the cellulosic support and on the nitrogen and phosphorus containing polymers.

Literature Cited

1. Simionescu, C. I.; Macoveanu, M. M.; Olaru, N. Cellulose Chem. Technol. 1976, 10, 197.
2. Simionescu, C. I.; Macoveanu, M. M. Cellulose Chem. Technol. 1977, 11, 87.
3. Simionescu, C. I.; Macoveanu, M. M.; Olaru, N. Acta Polymerica, 1979, 30, 241.
4. Simionescu, C. I.; Dénes, F.; Macoveanu, M. M.; Totolin, M. "5th Bratislava IUPAC sponsored International Conference on Modified Polymers Preprints of Short Contributions", Bratislava, 1979, 138.
5. Simionescu, C. I.; Dénes, F.; Macoveanu, M. M.; Cazacu, G.; Totolin, M.; Popescu, Fl.; Percec, S. Cellulose Chem. Technol. 1979, 13, 463.
6. Simionescu, C. I.; Dénes, F.; Macoveanu, M. M.; Cazacu, G.; Totolin, M.; Segal, A. Cellulose Chem. Technol. 1978, 12, 603.
7. Simionescu, C. I.; Dénes, F. Cellulose Chem. Technol. 1980, 14, 285.
8. Simionescu, C. I.; Dénes, F.; Macoveanu, M. M.; Cazacu, G.; Totolin, M.; Percec, S.; Balaur, D. Cellulose Chem. Technol. 1980, 14, 869.

9. Pantil, K. C.; Vittal, J. P.; Patel, C. C. J. Fire Retard. Chem. 1980, 7, 3.
10. Beninate, J. V.; Moreau, J. P.; Dvake, G. L., Jr. J. Fire Retard. Chem. 1980, 7, 36.
11. Pandya, H. B.; Chanban, R. S.; Bhagwat, M. M. J. Fire Retard. Chem. 1980, 7, 27.
12. Marom, M. A.; Behar, D. J. Appl. Polym. Sci. 1980, 25, 691.
13. Harris, J. A.; Arthur, J. C., Jr.; Goynes, W. R., Jr. J. Appl. Polym. Sci. 1979, 23, 2555.
14. Percec, S.; Popescu, Fl. Rom. Pat. 64155.
15. Percec, S.; Popescu, Fl. Rom. Pat. 70877.
16. Asakura, Isao; Ueno, Keisuke, Japan 7 242 080.
17. Roth, H. Mikrochem. 1951, 36, 1002.
18. Bradstreet, R. B. "The Kjeldahl Method for Organic Nitrogen," Academic Press, New York, 1965; p. 200.
19. Avram, M. "Spectroscopia in infraroşu, aplicaţii în chimia organica," Editura Tehnica, Bucureşti, 1966; p. 500.
20. Thomas, L. C. "The Identification of Functional Groups in Organophosphorus Compounds," Academic Press, Inc. London, LTD, 1974; p. 52, 62, 65.
21. Coats, A. W.; Redfern, J. T. Nature, (London) 1964, 201, 68.

RECEIVED December 17, 1981.

6

Grafting of Methyl Methacrylate onto KPM Rayon and Jute Fiber

Effect on the Properties of Grafted Fibers

MD. HABIBUDDOWLA

BCSIR Laboratories, Research Division, Dacca-5 Bangladesh

Graft copolymerisation reactions were carried out with KPM rayon and jute fiber in presence of air using ceric ammonium sulphate as the initiator. The tenacity and stiffness of methyl methacrylate grafted KPM rayon, defatted and bleached jute fiber were studied and compared with those of ungrafted fibers. Grafting resulted in the decrease in tenacity substantially. The effect of grafting on the tensile properties of fiber was found to be rather obscure to propose any definite mechanism. Of the grafted fibers, KPM rayon grafted at 40°C showed a maximum of about 19% decrease in tenacity. In general, lower temperature seemed to favor lesser decrease in tenacity for MMA grafted fibers. Grafting appreciably increased the stiffness for both KPM rayon and defatted jute fiber, but for bleached jute, the stiffness was found to decrease. A maximum of about 17% increase in stiffness was observed for grafted KPM rayon and a maximum of about 5% decrease in stiffness was observed for grafted bleached jute.

Recently, efforts are being made in Bangladesh for the commercial utilization of jute fiber (1-2) and KPM (Karnafuli Paper Mill) rayon in the textile field to partially meet the total fabric requirements of the country as well as to minimise import of synthetic fiber. Due to low tensile strength, high water absorbancy and difficulty in dyeing, the use of KPM rayon in fact has remained restricted in this field and the consequential effect is that few textile mills of Bangladesh are in a position to manufacture blended yarn for fabric production. On the other hand, although jute fiber possesses high dimensional stability (3) certain unfavorable textile properties, e.g., high stiffness, very low elasticity, susceptibility towards sunlight etc., have tremendously limited its use (4). With a view to minimising their

0097-6156/82/0187-0073$6.00/0
© 1982 American Chemical Society

undesirableness and enhancing their effectiveness for intensified textile uses, graft copolymerisation reactions were carried out onto KPM rayon and jute fiber in presence of air using ceric ammonium sulphate (CAS) as the initiator. Although, laboratory research on chemical modification of rayon and jute by grafting has received considerable attention in various countries (5-12) little is known about the properties of grafted KPM rayon and jute for their use as textile fibers. Since encouraging results have been reported by only a few authors (13-14), regarding the extensibility, dyeability and tenacity of the grafted jute, the object of this work has therefore been towards investigating into some of the tensile properties of the grafted jute and KPM rayon.

Experimental

Chemicals. Reagent grade chemicals were used in the study of grafting. Acetone, ethyl alcohol and benzene were redistilled before use. Monomer, methyl methacrylate (MMA), was used fresh after purification and distillation. It was purified following the method adopted by Gupta and Nandy (15). CAS was used after standardization by an usual method (16). All CAS solutions used for the study was acidified with 1% H_2SO_4.

Substrate and its Preparation. The substrate used for grafting purpose were jute fiber (defatted and bleached) and KPM rayon yarn. Jute fiber under investigation was of Corchorus Capsularis variety. It was arbitrarily divided into three portions. The middle portion was chosen for the purpose of study.

Defatted jute was prepared by treating it with alcohol-benzene (1: 2v/v) mixture in a Soxhlet apparatus, washed well with alcohol and air dried. Bleaching of jute fiber was carried out with 0.7% sodium chlorite following a special method (17). KPM rayon sample was prepared by treating it with dilute soap solution, washed well and dried.

Grafting Procedure and Calculation of Percent Grafting. Graft copolymerisation treatment of KPM rayon yarn and jute fiber were carried out in a well stoppered, Erlenmeyar flask. Fifty millimeter CAS solution of known concentration was kept at a desired temperature. One millimeter MMA was added following immediately by 1.0 gm of KPM rayon yarn, after which the contents were stirred intermittently. After the end of the reaction, the flask were removed from the thermostat and methanol was immediately added. The mixture was then poured into a Buchner funnel and suction applied. The homopolymer in the filtrate was separated from the solution by filtering through previously weighed sintered crucible (G-3). It was washed repeatedly with hot water and dried at 60°C. The grafted fiber on the funnel was extracted with acetone for 10-12 hours in a soxhlet apparatus. Similar procedure was also adopted for jute fiber.

Percent grafting has been calculated using the standard formula (18). Gain in weight of the grafted fiber/weight of the monomer X 100.

Determination of Tenacity and Stiffness. Tenacity has been expressed as grams/denier. The denier of both grafted and ungrafted KPM rayon was derived from the standard formula at standard conditions (14). After conditioning the jute sample, it was combed and the fiber aggregates of uniform length was taken, weight and length determined. The tenacity of KPM rayon yarn was determined by means of a Frank Tensile Strength Tester, while for jute yarn a Zweigles Strength Tester was used.

Stiffness or tensile modulus at break was determined from the value of tenacity and elongation at break using the following relation:

$$\text{Stiffness at break} = \frac{\text{Tenacity at break}}{\text{Elongation at break}}$$

Results and Discussion

In order to study the effects of grafting on the tenacity and stiffness of the fiber, various samples of KPM rayon, defatted jute and bleached jute were grafted in presence of 1% H_2SO_4 following the procedure as described earlier. The results have been presented in Tables 1 and 2 and also graphically shown in Figures 1, 2 and 3 which show that graft copolymerisation reaction considerably influences the tenacity and stiffness of KPM rayon and jute fiber. From Figure 1 it is quite clear that no regularity in the relationship exist between % grafting and % decrease of tenacity for KPM rayon and defatted jute. However for bleached jute linear pattern do exist when the graft load has been augmented from 19% to 40.5%. So far as the effect of grafting on the tenacity of the fiber is concerned, three different relation pattern exist for three different types of fiber grafted between the temperature range of 30°C and 50°C (Figure 2).

For grafted KPM rayon and grafted bleached jute, % decrease of tenacity follows linear pattern. For defatted jute % decrease of tenacity is much less pronounced at 30°C, after which increases sharply to about 15% as the temperature has been increased to 50°C. In the case of KPM rayon, rise of temperature from 30°C to 40°C enhance % decrease of tenacity linearly from about 13% to 19% and falls sharply to about 12% as the temperature has been increased to 50°C. In general lower temperature seems to favor lesser decrease in tenacity for MMA grafted fiber. A maximum of about 19% decrease of tenacity has been observed for grafted KPM rayon at 40°C and a minimum of about 1.5% decrease of tenacity has been found for MMA grafted defatted jute at 30°C. In general, this decrease of tenacity is quite understandable from the mechanistic viewpoint of graft copolymerisation reactions of

Table - I. Effect of grafting on the tensile properties of MMA grafted KPM rayon and jute fiber. Continued on next page.

Dessicator atmosphere conditioned at R.H. of 75% and a temperature of 26°C. Twist/cm (jute fiber) = 1. Lignin content of bleached jute = 10%.

Sample	Grafting %	Elongation at break (B.L.) (%)	Tenacity (gms/denier)	Tensile modulas at break (Stiffness)
1% H_2SO_4 treated rayon (30°C)	-	13.80	0.86	6.23
1% H_2SO_4 treated defatted jute (30°C)	-	1.52	2.01	132.20
1% H_2SO_4 treated bleached jute (30°C)	-	2.15	1.85	86.04
MMA grafted rayon (30°C)	50.70	10.30	0.75	7.28
MMA grafted defatted jute (30°C)	12.85	1.35	1.98	146.60
MMA grafted bleached jute (30°C)	30.50	2.12	1.80	84.90
1% H_2SO_4 treated rayon (40°C)	-	13.50	0.83	6.15
1% H_2SO_4 treated defatted jute (40°C)	-	1.49	1.95	130.10

Table - I. Continued. Effect of grafting on the tensile properties of MMA grafted KPM rayon and jute fiber.

Sample	Grafting %	Elongation at break (B.L.) (%)	Tenacity (gms/denier)	Tensile modulas at break (Stiffness)
1% H_2SO_4 treated bleached jute (40^oC)	-	2.13	1.82	85.44
MMA grafted rayon (40^oC)	39.80	10.80	0.67	6.20
MMA grafted defatted jute (40^oC)	13.66	1.39	1.91	137.41
MMA grafted bleached jute (40^oC)	21.19	2.19	1.78	81.27
1% H_2SO_4 treated rayon (50^oC)	-	13.40	0.81	6.04
1% H_2SO_4 treated defatted jute (50^oC)	-	1.47	1.92	130.61
1% H_2SO_4 treated bleached jute (50^oC)	-	2.05	1.75	85.36
MMA grafted rayon (50^oC)	37.0	10.80	0.69	6.38
MMA grafted defatted jute (50^oC)	5.66	1.22	1.63	133.60
MMA grafted bleached jute (50^oC)	19.08	2.08	1.72	82.69

Table - II

Tensile properties of MMA grafted KPM rayon and jute fiber.

Sample	Grafting %	%decrease breaking elongation	%increase of breaking elongation	%decrease of tenacity	%increase of stiff-ness	%decrease of stiff-ness
Rayon (30°C)	50.7	25.36	-	12.79	16.85	-
Rayon (40°C)	39.8	20.00	-	19.27	0.81	-
Rayon (50°C)	37.0	19.40	-	14.81	5.63	-
Defatted jute (30°C)	12.85	11.18	-	1.49	10.8	-
Defatted jute (40°C)	13.66	6.71	-	2.05	5.62	-
Defatted jute (50°C)	5.66	17.0	-	15.10	2.29	-
*Bleached jute (30°C)	30.50	1.39	-	2.70	-	1.32
*Bleached jute (40°C)	21.19	-	2.81	2.19	-	4.88
*Bleached jute (50°C)	19.08	-	1.46	1.71	-	3.13

*Chlorite bleached jute prepared under minimum cellulosic degradation (Lignin content = 10%).

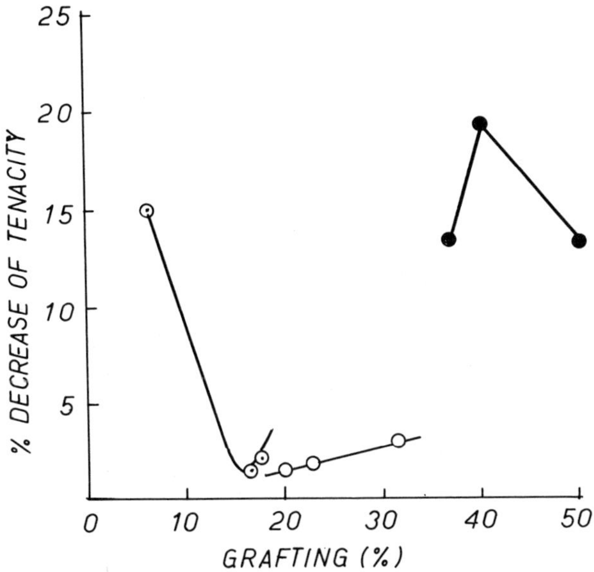

Figure 1. Plot of percent decrease of tenacity vs. percent grafting. Key: ●, rayon yarn; ⊙, defatted jute; ○, bleached jute.

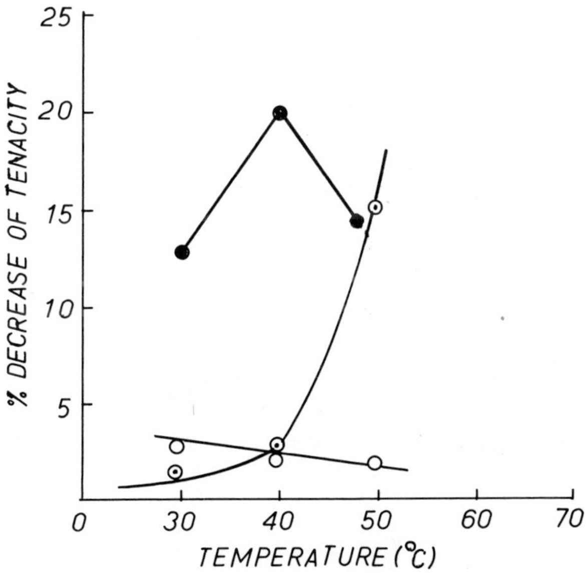

Figure 2. Changes of tenacity of MMA grafted fiber. Key: ●, rayon yarn; ⊙, defatted jute; ○, bleached jute.

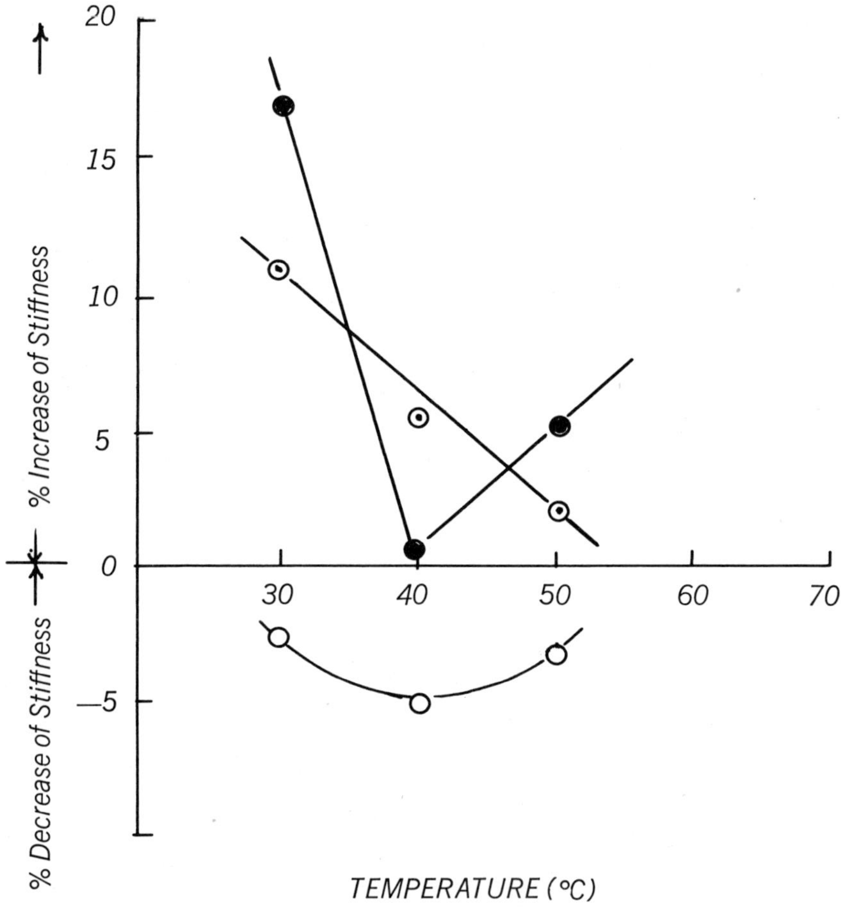

Figure 3. Changes of stiffness of MMA grafted fiber. Key: ●, rayon yarn; ⊙, defatted jute; ○, bleached jute.

cellulosic fiber with vinyl monomer (14). Certainly, the presence of air has a direct effect on the cellulosic chain of the fiber favoring oxidation reactions and resulting in profound decrease of tenacity.

As far as the stiffness is concerned, grafting has an appreciable influence on the stiffness of both KPM rayon and defatted jute fiber. However for bleached jute, the stiffness has been found to decrease. The results have been shown in Tables 1 and 2 and graphically represented in Figure 3 which represents three different relation patterns for three different kinds of samples. In the case of defatted jute, rise of temperature from 30°C to 50°C reduces % increase of stiffness linearly from about 11% to 2%. In other words, higher temperature has been found to favor lesser increases in stiffness. For KPM rayon increase of temperature from 30°C to 40°C decreases % increase of stiffness substantially from about 17% to 1% but as the temperature is raised from 40°C to 50°C increase of stiffness is not so prominent. Opposite results have been obtained for bleached jute where a decrease of stiffness was observed when grafting was carried out at a temperature of 30, 40 & 50°C. A maximum of about 5% decrease in stiffness has been observed for grafted bleached jute. The overall effect of grafting on the tensile properties of fiber was found to be rather obscure to propose any definite mechanism.

Acknowledgements

The author wishes to express sincere thanks to the authorities of the B.C.S.I.R. Laboratories, Dacca for the arrangements made for determining the properties of the fiber. Mr. Monsur Ahmed, T. O., Arts & Crafts section of B.C.S.I.R Labs., also deserve thanks for preparing the graphs and figures.

Literature Cited

1. Jute and Fabrics, Bangladesh; 1980, July, 6, No. 7.
2. Jute and Fabrics, Bangladesh; 1980, June.
3. Science News, series 54, 1-55.
4. Little, Arthur D. "The Engineering Properties of Fibers" Incorporated Cambridge, Massachusetts, 1966.
5. Richards, G. N.; J. Soc. Dyers, Colourists, 1964, 80 (12), 640-2 (1964)
6. Yoshitake, I; Obuchi, S; Tsuneda, T; Maeda, H; (Nippan Rayon Co., Ltd, Kyoto, Japan) Sen-i-Gakkashi, 1965, 19 (7), 540-6.
7. Morin, B. P.; Stanchenko, G-I; Rovenkova, T. A; Gracheva, Y. L.; Bershova, N. W (Mosk, Tekst, Inst, Moscow, USSR) Khim Volknan, 1975, 2, 22-4
8. Nakamura, Y; Negeshi, M.; Kakimuma, T.; (Gumma Univ, Kiryn, Japan) J. Poly. Sci. Part C, 1966, Pub 69, No 23 (Pt 2), 623-45.

9. Livshits, R. M. ; Stanchenko, G-I; USSR, Naush Issled, Tr. Mosk, Tekst, Inst, 1969, 22, 317-23.
10. Rogovin, Z. A.; (Text Inst. Moscow, USSR), Tappi, 1974, 57 (7), 65-8.
11. Siddique F.R.Al; Zaman, M. A.; Nuclear Science and Application, Series B, 1976, 9, 80-82.
12. Huque, M. M.; Habibuddowla, Md.; Mahmood, A. J.; Mian, A. J.; J. Polym. Sci., Polymer Chemistry Ed., 1980, 18, 1447-1458.
13. Trivedi, I. M.; Mehta, P. C.; Cellulose Chem. Technol, 1973, 7, 401-416.
14. Haque, M. M. and Habibuddowla, Md; Bangladesh J. Sci. Ind. Res, 1980, 15, No. 1-4, 64-70.
15. Gupta, S. N.; Nandi, U. S.; J. Polym. Sci., A-1; 1970, 8, 3019-3031.
16. Vogel, A-I; "Quantitative Inorganic Analysis," ELBS & Lognman, 3rd Ed, 1961, p. 319.
17. Rahman, M. H. and Haque, M. M.; Pak, J. Sci. Ind. Res 1970, No. 13, 303.
18. Kojima K; Iwabuchi, S; Kojima, J. Polym. Sci.; Polym. Letter Ed, 1971, 9, 25-29.

RECEIVED December 23, 1981.

7

Photochemical Grafting on Wood Cellulose

J. P. FOUASSIER

Laboratoire de Photochimie Générale, Equipe de Recherche Associée au CNRS, Ecole Nationale Supérieure de Chimie, 68093 Mulhouse Cedex, France

UV light induced grafting onto wood cellulose is reported for several vinyl monomers. The reaction is initiated by free radical initiators such as phenylacetophenone and benzophenone derivatives. Percent grafting-time conversion curves are determined as a function of the initiators, monomers, pulps and additives. Additional typical results obtained in IR spectroscopy, GPC and thermal analysis are reported. A discussion about the use of a photochemical procedure in obtaining cellulose graft copolymers is presented.

1. INTRODUCTION

Photochemical and photophysical processes in cellulose and related compounds have received considerable attention during the last decades, resulting in research work concerned with the improvement of cellulosic materials via physical and chemical modifications. One method was to apply a copolymer between the cellulose and a synthetic polymer which are generally grafted by free radical reactions.

Two routes can be explored with a view to modify the surface properties of cellulosics. The first one involves chemical reactions involving the functional groups of cellulose. The second route can be achieved through irradiation or thermal grafting of monomers onto cellulose.

Several radical reactions have been extensively studied as a function of the excitation e.g. γ irradiation, U.V. irradiation, redox chemical initiators, heating, dye-sensitized reaction. Some of these investigations are reported in this volume.

Until recently, few studies of UV light induced grafting of cellulosics have been reported (e.g.1-7 and references therein).

It further appears that sensitized photografting reactions are more attractive, since longer wavelengths of excitation may be used, thus increasing the possibility of a potentially valuable

0097-6156/82/0187-0083$6.00/0
© 1982 American Chemical Society

commercial exploitation. Moreover, less degradation of cellulose is obtained. However, this kind of reaction requires a knowledge of the photoreactivity of the initiator with respect to the chemical nature of the monomer and the macromolecule.

Our previous papers in this field dealt with photosensitized free radical generation on cellulose (8) and photochemical grafting of various monomers in presence of a large variety of initiators onto cotton (9) and starch (10). Several parameters of the grafting reaction have been investigated : incident light intensity, irradiation wavelength, temperature, oxygen effect, irradiation time, photo-initiator concentration, monomer concentration and presence of additives. The results obtained make it possible to determine the most suitable conditions for grafting and to discuss the mechanism of reaction.

The present report extends our work to the photografting of various monomers, using several photo-initiators, onto wood cellulosic materials. Moreover, it gives the possibility to discuss some problems related to the photochemical grafting.

2. EXPERIMENTAL

a) Materials

Several samples of commercial dissolving pulp from softwood were used : bleached pulp (whiteness 92 %), almost entirely free of lignin, containing about 94 % α cellulose (I) ; paper pulp (II) (whiteness : 91,5 %) ; mechanical yellow pulp (III) (containing lignin), Kraft Laponia pulp (IV), whiteness : 92 %).

Benzoin derivatives and benzophenone derivatives were commercial products of the highest purity grade. They are 2,2 - dimethoxy-2-phenyl acetophenone (DMPA), benzophenone (BP), 4-benzoylbenzoic acid (BBA), 3,3',4,4'-benzophenone and tetracarboxylic dianhydride (BTCD).

Methyl methacrylate (MMA) was washed with a solution of sodium carbonate, dried over calcium chloride and distilled under reduced pressure and nitrogen. Acrylonitrile was washed with sulfuric acid (5 %), sodium carbonate (5 %) and water ; after drying, the monomer was distilled under reduced pressure and nitrogen. Styrene was washed with sodium hydroxyde (10 %) and water, dried and distilled. THF was purified by washing and distillation.

- *Graft copolymerization*

The cellulose sample was generally treated according to the procedure described previously (9). In some experiments, the cellulose sample was treated with an aqueous solution of sodium hydroxide (17,6 %) for 2 h at room temperature and washed with water to neutrality for 24 h. The swollen sample was obtained after filtration.

In the photosensitized experiments, the cellulose sample was treated with a solution of the initiator in acetone for 1 h at room temperature and the solvent evaporated in air. When this kind of adsorption was not possible, the initiator was dissolved in THF and added to the reactant medium.

Graft copolymerization was carried out at a chosen temperature, for a given time in nitrogen, by irradiating reaction cells containing 80 ml of distilled water, 0.5g of treated or untreated cellulose sample and a variable amount of monomer. The U.V. light source was a high pressure mercury lamp (HPK 125 W). For the sensitized experiments, suitable wavelengths were chosen through a pass band filter centered around 366 nm. The incident light intensity was determined through actinometry measurements. Typical values were in the range of 2×10^{16} photons $cm^{-2}s^{-1}$ on the sample.

The irradiated samples were extracted by refluxing for 24 hours (we verified that subsequent extraction did not modify the results) with the following solvents, according to the monomer : acetone (methyl methacrylate) ; benzene (styrene) ; N,N-dimethylformamide (acrylonitrile).

The percent grafting is considered (see discussion) as the weight increase of the dried cellulose sample after extraction. The percent homopolymer is calculated as the ratio weight of homopolymer over the weight of original monomer.

3. RESULTS AND DISCUSSION

3.1. Grafting experiments

a) Investigation of the general parameters

The results of the graft copolymerization on cellulose in various conditions are shown as a function of the irradiation time, initiator concentration, monomer concentration, presence of additive.

Table I shows the evolution of the percent grafting and percent homopolymer as a function of the irradiation time. A fast increase in grafting was observed for short irradiation times whereas longer ones resulted in a decreased percent grafting.

Table II shows an optimum value of the initiator concentration. The percent grafting may be expected to increase as a function of the amount of initiator up to a critical concentration. At higher concentrations, the percent grafting decreases presumably in connection with secondary reactions induced by the initiator radicals.

The percent grafting and the percent homopolymer are a function of the monomer concentration (Table III). An optimum value is obtained. The maximum of the percent grafting is strongly dependent on the comparative values of the MMA and initiator concentrations.

Table I : Percent grafting and percent homopolymer as a function of time.
MMA : 5g ; THF : 5 cm^3 ; sample (I) : 0,5g ; water : 80g ; λ = 365 nm ; T = 20°C.

Initiator	Initiator amount (mg)	Irradiation time (h)	% grafting	% homopolymer
BP	50	1	13,4	-
	50	2	28,2	-
	50	3	12	1,5
DMPA	50	1	14,4	-
	50	2	30,2	10,2
	50	3	10,4	13,6
BBA	50	1	7,4	-
	50	2	23	-
	50	3	16	-
BBA	25	1	21,6	-
	25	1h30	29,2	2,2
	25	2	30,8	8,9
	25	3	36,4	9,7
	25	4	52,2	20,8
	25	5h15	30	54
	25	6	38	61,5
BTCD	50	1	47,4	-
	50	2	57,2	30,6
	50	3	26	57,5

Table II : Percent grafting versus initiator amount ; sample (I) ; monomer : MMA ; initiator : BTCD ; water (80 cm^3) + THF (5 cm^3)

Initiator amount (mg)	% grafting	% homopolymer	Irradiation time (h)
5	12	-	2
10	40	-	2
25	81	25,3	2
50	57	30,6	2

Table III : Percent grafting versus time ; monomer : MMA ; initiator : BTCD

Initiator amount (mg)	Time (h)	Monomer amount (g)	% grafting
25	1	1	0
25	1	2,5	0
25	1	5	4
25	2	1	2
25	2	2,5	19,2
25	2	5	40
25	3	1	17,4
25	3	2,5	30
25	3	5	55,4
25	4	5	34
50	1	1	2
50	1	2,5	36,4
50	1	5	47,4
50	2	5	57,2
50	3	5	26

When using photoinitiators such as phenyl acetophenone derivatives, addition of an hydrogen donating solvent (Table IV) to the system improved ultimately the percent grafting and, in addition, the concentrations of initiator required were lower. Moreover, the presence of THF is necessary with benzophenone derivatives. These results are in agreement with similar observations with respect to MMA polymerization in solution ; in fact, the rate of polymerization was enhanced in the presence of THF.

However, addition of amine molecules (triethylamine TEA) did not increase the percent grafting.

b) Grafting in the presence of various initiators

The results of the graft copolymerization of MMA onto cellulose in the presence of various water insoluble initiators are shown in Figure 1 which display the time dependence on the percent grafting. In a general way, one observes a rapid increase of the percent grafting within two hours of irradiation with wavelength of 365 nm. For longer irradiation times, the percent grafting curves level off ; in some cases the percent grafting decreases. This was also reported in the previous paper (9).

At first sight, benzophenone compounds appear to be the most efficient photo-initiator for the copolymerization reaction, whereas phenylacetophenone derivatives do not rank among the best. In comparison, 2,2-dimethoxy-2-phenylacetophenone seems even poorly efficient, although this compound is generally considered as very reactive in UV curing (11) or in solution photopolymerization of vinyl monomers (12).

Table IV: Typical effects ot the amount of THF on the percent grafting. Sample I ; monomer : MMA ; water : 80 cm^3.

Initiator	Initiator amount (mg)	THF amount (cm^3)	max % grafting
DMPA	50	0	25
	50	5	30
	25	0	18
	25	5	25
BBA	25	0	0
	25	5	52

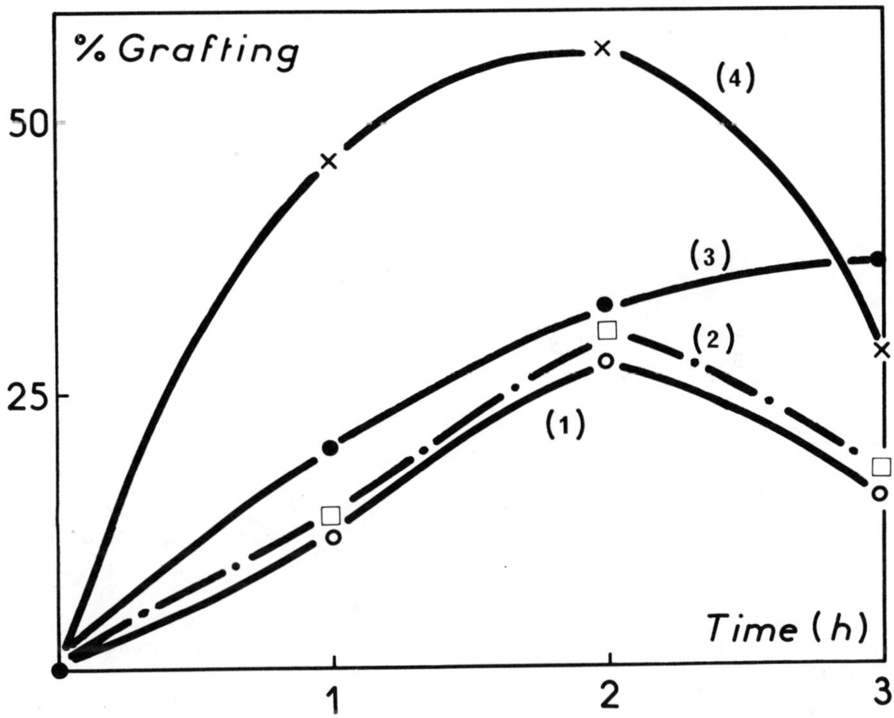

Figure 1. Percent grafting vs. time. The monomer was MMA; the initiators were BP (1), DMPA (2), BBA (3), and BTCD (4).

Addition of THF is necessary with benzophenone derivatives and enhances the efficiency of phenyl acetophenone derivatives as shown by the following reactions (the photochemistry of these compounds are very well known in homogeneous solutions) :

Benzophenone derivatives :

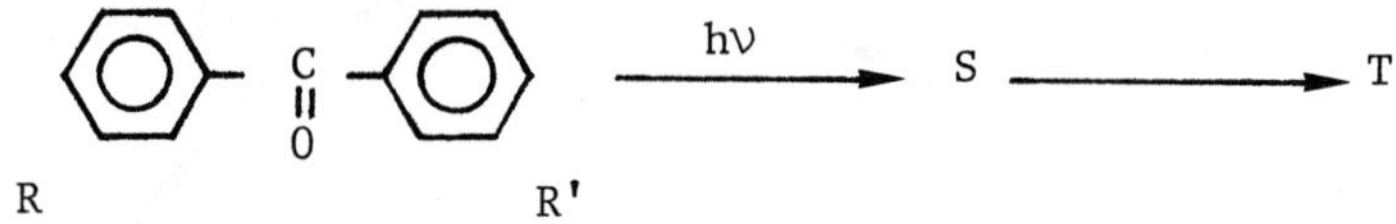

(1)

T $\xrightarrow{\text{Cell-H}}$ ketyl radical $\left(-\overset{\bullet}{\underset{\text{OH}}{\text{C}}}-\right)$ + Cell$^\bullet$ (2)

T $\xrightarrow{\text{THF}}$ ketyl radical + THF$^\bullet$ (3)

DMPA

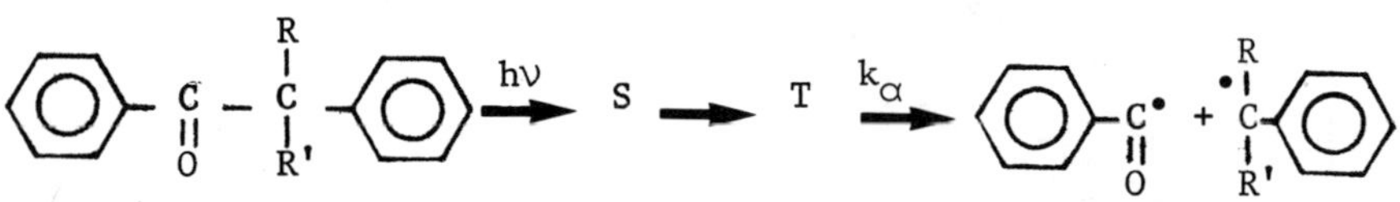

R = R' = OCH_3 α-cleavage (4)

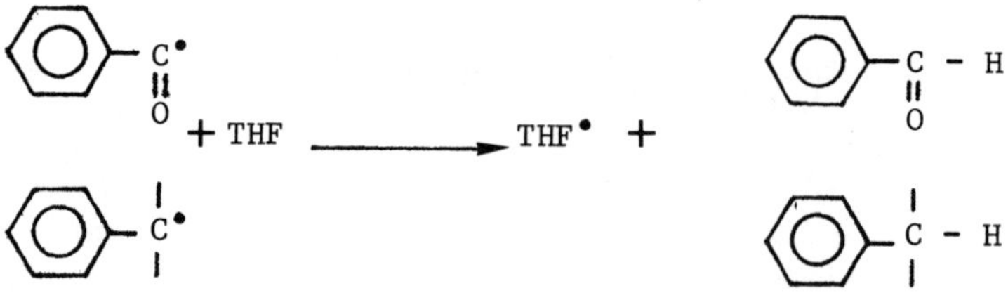

(5)

A better solubilization of the initiator in the mixture water-THF may also play a part in the improved efficiency.

c) Grafting of various monomers onto cellulose

The experiments were carried out with three monomers. Some typical results are reported in Table V. In a general way, various percent graftings were obtained depending on the monomer-initiator couple. The percent grafting may be adjusted to a given value by modification of the experimental conditions.

Table V : Typical maximum percent grafting as a function of the nature of the initiator, monomer and pulp.

Sample	Initiator	Monomer	% grafting
(I)	BTCD (25 mg)	MMA (5g)	40 %
(I)	BTCD (50 mg)	MMA (5g)	57
(I)	BTCD (25 mg)	AN (3g)	13
(II)	BTCD (25 mg)	MMA (5g)	60
(III)	BTCD (25 mg)	MMA (3g)	20
(IV)	DMPA (25 mg)	Styrene(2g)	46

The differences in reactivity are due presumably to specific quenching of the reactive species by the monomer and to the specific reactivity between the monomer and the radicals created upon the cellulosic backbone.

As reported in literature (13) the irradiation of a cellulose sample at 254 nm results in glucosidic bond scission and C - H bond scission. Recent results in photosensitized experiments (15) show that the same behaviour can be obtained :

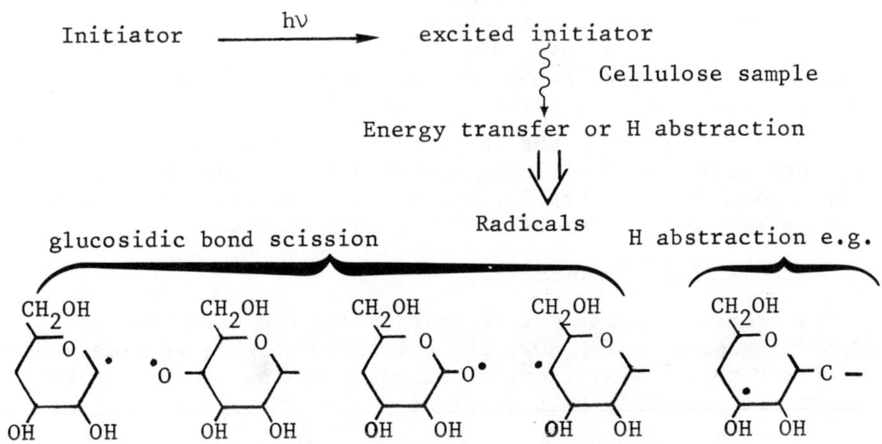

thus, graft copolymer should occur through :

- *recombination reaction*

THF ~~~~~~~~ •M + •Cell ⟶ THF ~~~~~~ M - Cell

- *glucosidic scission or C-H bond scission, initiation, propagation, eg :*

Cell• + M ⟶ Cell - M• ⟶ Cell ~~~~~~~~~~~~~•

- *occlusion of homopolymer in the cellulosic chain*

d) Grafting onto various pulps

Recent typical results are represented in Figure 2 and Table V. They show a strong influence of the nature of the pulp ; specially in the rate of the reaction (Figure 2). More results are needed for a detailed discussion of these effects.

3.2. Analysis of the grafted samples

a) Thermal analysis

The results are reported in Figure 3 and show the typical behaviour of the grafted sample compared to that of PMMA and cellulose. From the experimental curves, it is concluded that the polymer chain is chemically linked onto the cellulosic backbone. In curve B, the first and second peaks correspond to the unsaturated chain end and to the saturated chain end of the polymer respectively. In curve C, one observes that the peak corresponding to polymer chains terminated through disproportion have disappeared ; this points out the fact that the cellulosic macromolecules could be crosslinked by PMMA chains.

b) Infra red spectroscopy

A typical IR spectrum is shown by Figure 4. One can see the differencial spectrum between cellulose and a blend homopolymer cellulose. The IR spectrum of the grafted sample exhibits the characteristic bands of cellulose and PMMA.

Moreover, a plot of the IR absorbance shows that the extinction coefficient of the blends are higher than that of the grafted copolymer (Figure 5). These two arguments are in favour of the presence of chemically linked PMMA chains.

c) Analysis of the grafts

The grafted polymer was separated from the cellulose by hydrolyzing with 72 % H_2SO_4. The molecular weights were determined by GPC after neutralization, washing and drying. The molecular weights are generally high (Typically 900 000) and the polydis-

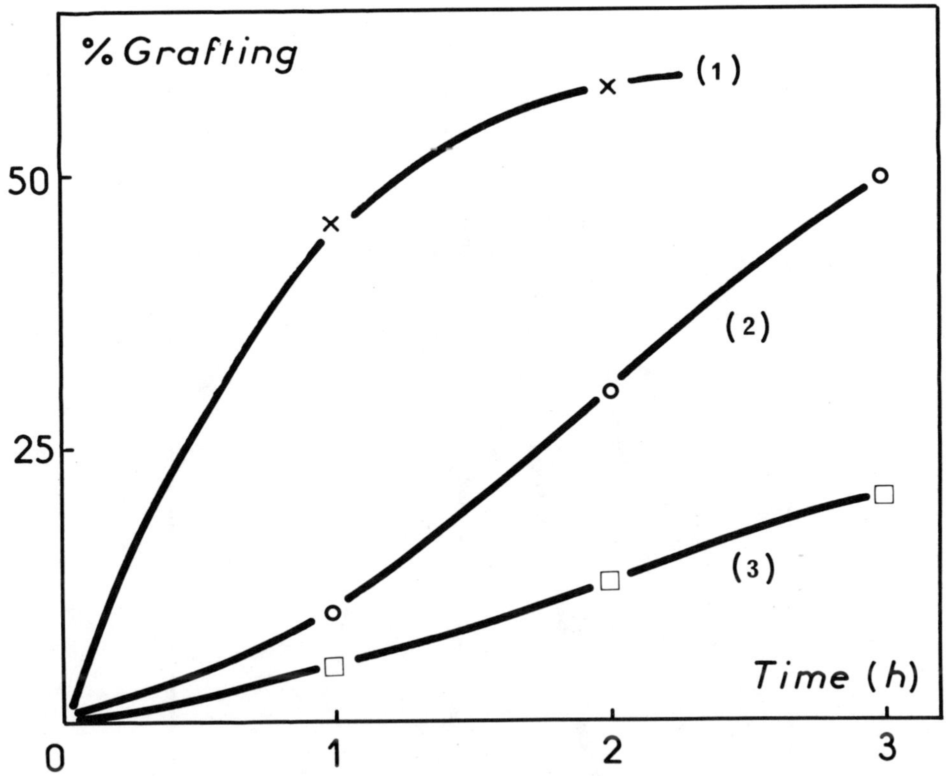

Figure 2. Percent grafting vs. time. The initiator was BTCD (25 mg); the monomer was MMA (3 g); and the three pulps I, II, III (1–3).

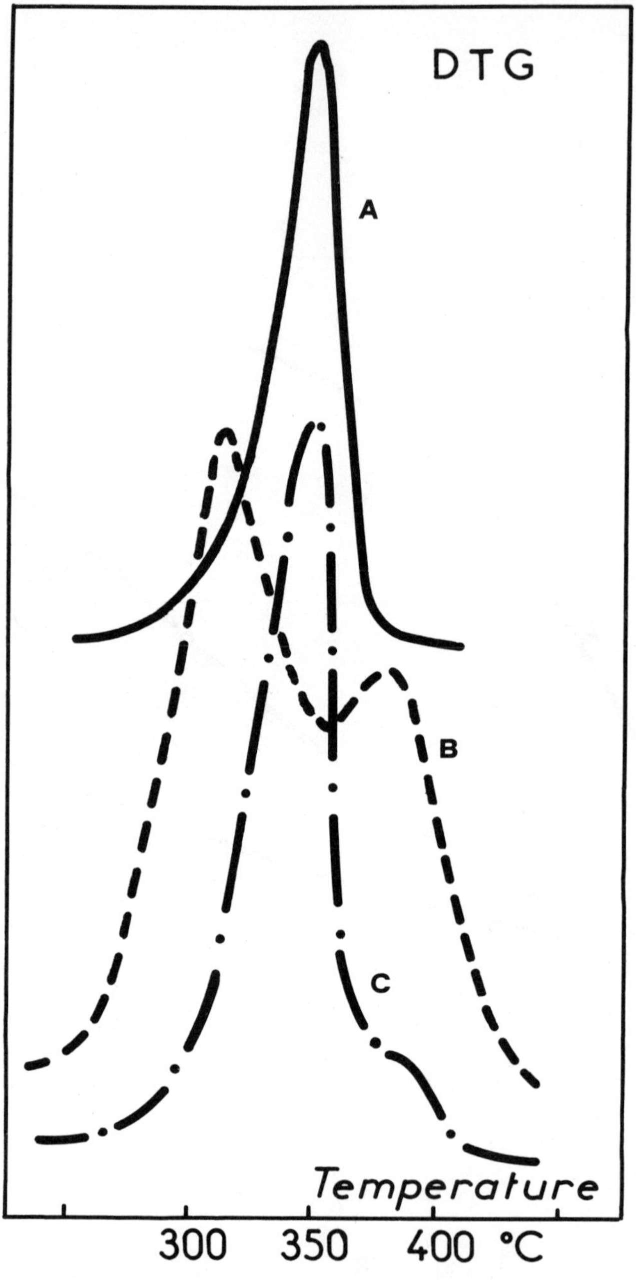

Figure 3. Typical DTG thermogram obtained in thermal analysis. Key: A, cellulose; B, PMMA; C, graft copolymer.

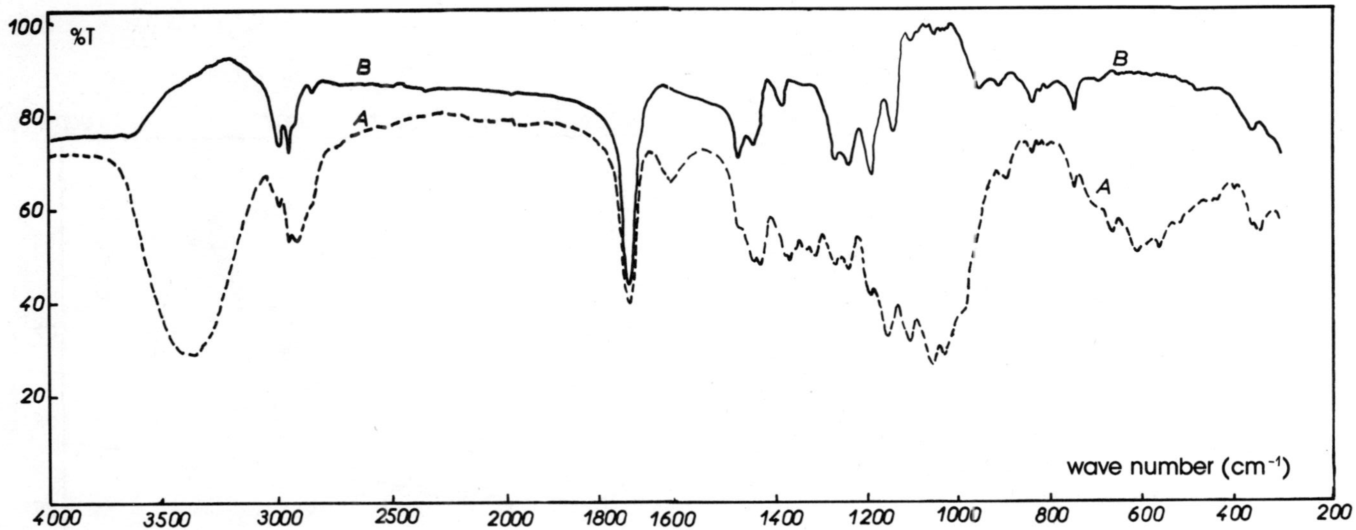

Figure 4. A: IR spectrum of the graft copolymer; B: differential IR spectrum of PMMA. Continued.

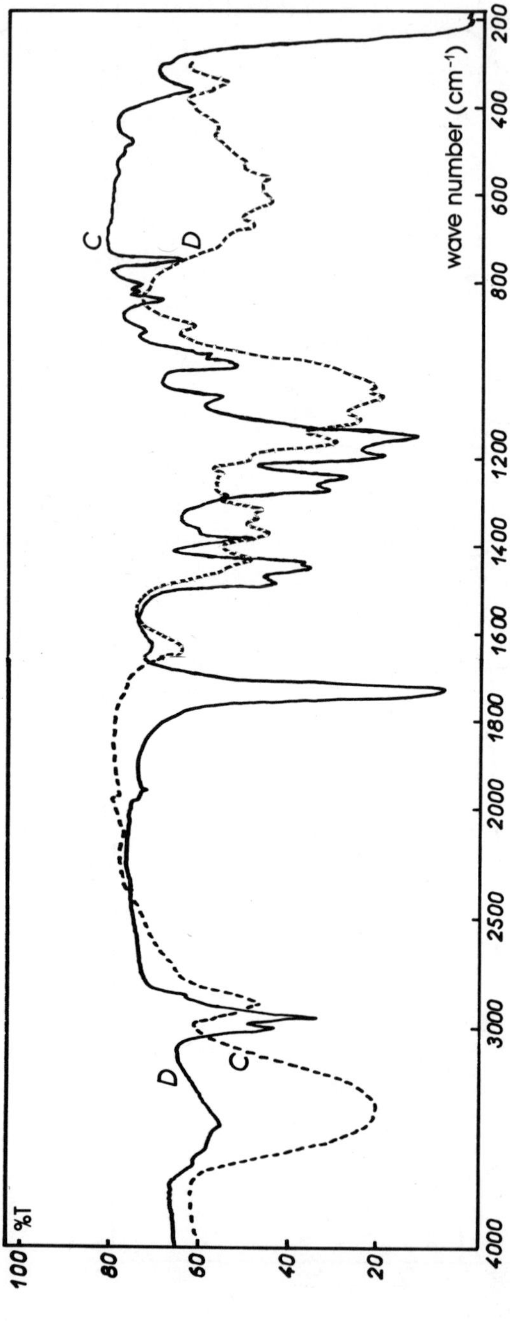

Figure 4. Continued. C: IR spectrum of cellulose. D: IR spectrum of PMMA.

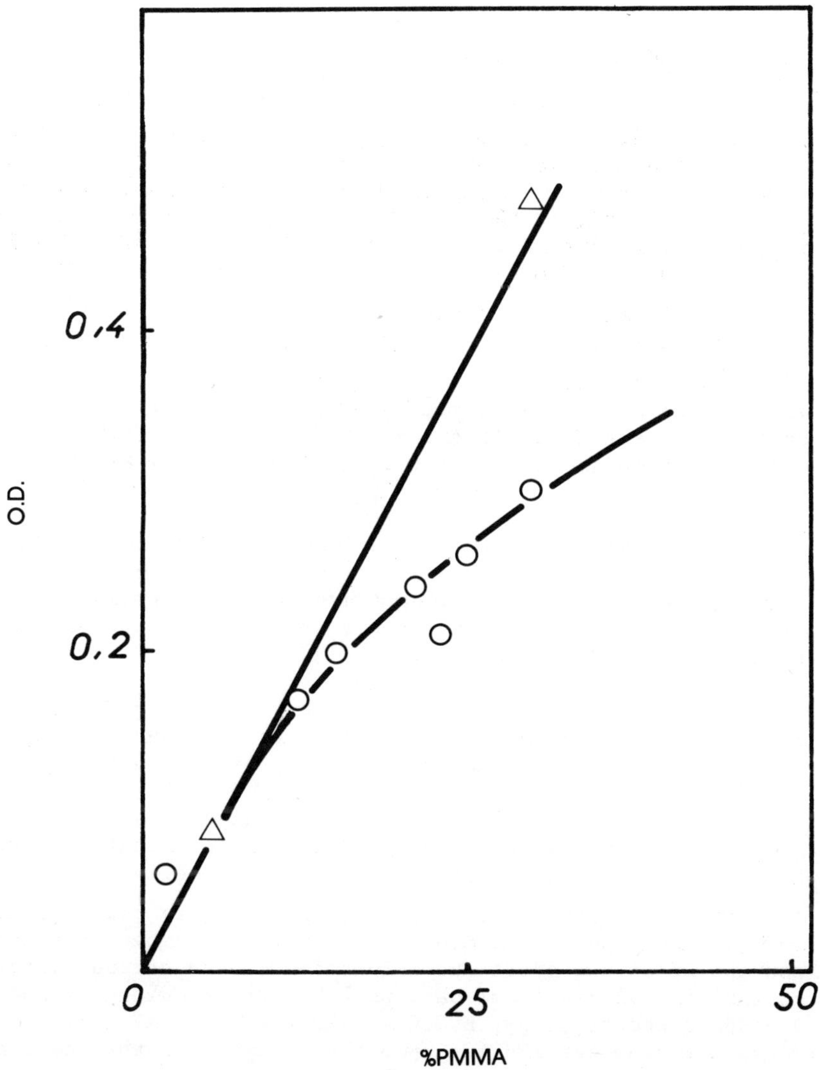

Figure 5. IR absorbances ($\bar{\nu}$ = 1730 cm^{-1}) as a function of PMMA content in a graft sample (○) and in a blend copolymer (△).

persity has the usual value obtained in radical homopolymerization. It seems that there are few linkages between the polymer and the cellulosic chain. Complete results on the analysis of the grafts will be published later (14).

4. DISCUSSION AND CONCLUSION

The aim of this work was to obtain graft copolymers of cellulose with monomers and to study the different possibilities of a photo-induced radical reaction. The use of UV light seems very promising, due to the large possible variety of experimental grafting conditions. The results presented in this paper appeared in the perspective of a series dealing with an investigation of the grafting reactions onto natural substrates.

The following points have been demonstrated (a more detailed investigation will be published later (14)):

a) The photochemical grafting of monomers onto commercial paper pulps can be developed as a laboratory process. In this case, the treated pulp corresponds to 0.5g dry weight. The experimental procedure involves the irradiation of a heterogeneous suspension of pulp and monomer in water.

b) As shown in previous experiments (9) conducted with other substrates, the grafting reaction is dependent on a lot of factors : experimental conditions (wavelength of irradiation, incident light intensity, temperature, oxygen, pH ...), monomer, initiator, pulp, additive ...

c) In such experiments, the rates of grafting (% grafting mn^{-1}) can reach values about 1 % mn^{-1} for a power density of illumination in the range of 15 mW cm^{-2} at λ = 366 nm.

d) Preliminary results in thermal analysis and infra red spectroscopy suggest that the grafted sample contains polymer chains which are chemically linked onto the cellulosic backbone.

Extensive works (14) are now in progress to obtain complete informations upon the number of grafts, the number of scissions and the DP of the grafts as a function of the various conditions of grafting. Additionnal data will be also provided by infra red and Raman spectroscopy, E.S.C.A. and thermal analysis. Other works are also carried out upon the analysis of the physical properties of the grafted samples.

Last but not the least, we would like to discuss some of important points which can be pointed out in the use of a photochemical procedure in obtaining graft copolymers of ligno cellulosic materials.

a) The first problem is related to the discussion concerning the problem between a graft copolymer sample and a sample containing occlusion of homopolymer in the substrate without chemical bonding. It seems that homopolymerization predominates in a presence of a photo-initiator in contrast of a direct UV excitation of the substrate. (However, the photosensitized grafting is more attractive since it gives the possibility to select the characteristics of the samples (9). In our case, we show that the polymer, remaining even after a drastic solvent extraction, is strongly fixed on the cellulose.

In many experiments, it appears that such drastic solvent extraction can remove the major part of the synthetic polymer, showing that the grafted polymer presents, in point of fact, a good "adhesive" bondability. However, in the case of surface modification, a heavy grafting is not necessary and the degree of permanence is a function of the insolubility of the homopolymer in the solvents used in the common course of subsequent treatments. So it can be inferred that this notion of grafting may be dependent on the efficiency of the solvent extraction.

b) The second problem is related to the improved physical properties of the grafted pulp as a function of the % grafting and the characteristics of the grafts (DP, molecular weight, number of grafts, number of bonds between the cellulose and the grafted polymer. In a general way, it is known that a radical type polymerization yields very long growing chains and consequently, the DP are high and the number of grafts (G) is low. The question is to have a substantial amount of grafted samples in order to (i) obtain the DP and G dependence on the experimental conditions, to (ii) determine the physical properties of the samples and to (iii) try to establish a structure property relation ship. This type of work is important since it could define additional experiments in which the DP of the grafts would be adjusted. However, many experimental results (e.g.2) support the fact that an improvment of the physical properties can be achieved through graft cellulose even with long chain polymers.

c) The third problem concerns the transposition of these experiments to a pilot scale or an industrial process. We are trying to develop a photo chemical reactor in which pulp sample of about 50g can be treated. Our first attempts do not always yield to successfull results and up to now, the reasons of that are not very clear (even if we take into account the problem of light penetration).

In conclusion, we would like to dwell on the fact that the present study demonstrates the feasability of a photochemical treatment, through a grafting reaction, of a cellulosic substrate. It is obvious that the results cannot be easily extrapolated to the case of an industrial process of surface modification of the pulps. In fact, due to its low penetration, UV light can be reserved in surface treatment, during or after the making of paper. However, we are convinced that such a study is of interest for the understanding of processes such those occuring in the photopolymerization of di or trifunctionnal monomers in order to get coatings for paper or wood (14).

Aknowledgments - The work is supported in part by the Centre National de la Recherche Scientifique and the Centre Technique du Papier of Grenoble. Collaborators who are actively contributing to this program are : R. SALVIN and L. TRAN THAN.

Literature cited

1. GARNETT, J.L., ACS symposium series 1977, 48, 334
2. REINHARDT, R.M. ; HARRIS, J.A., Text. Res.J 1980, 1, 139
3. OGIWARA, Y ; HAYASE, K ; KUBOTA, H ; J. Appl. Polym. Sci. 1979, 23, 1
4. GUTHRIE, J.T., Prog. Polym. Sci., 1978, 6, 31
5. TAKAHASHI, A ; TAKAHASHI, S ; Kobunshi Ronbunshu 1980, 37, 151
6. HEBEISH, A ; GUTHRIE, J.T. "The chemistry and technology of cellulosic copolymers" Springer Verlag N.Y. 1981
7. STANNET, V.T. ; MEMETEAT, T J. Polym. Sci. Polym. Symposia 1978, 64, 57
8. MERLIN, A ; FOUASSIER, J.P. Angew. Makromol. Chem. 1980, 86, 109
9. HEROLD, R ; FOUASSIER, J.P. Angew. Makromol. Chem. 1980, 86, 123
10. HEROLD, R ; FOUASSIER, J.P. Starch 1981, 3, 90
11. HEINE, H.G. ; RUDOLPH, H ; ROSENKRANZ, H.J. J. Polym. Sci. Appl. Polym. Symp. 1975, 26, 157
12. MERLIN, A ; FOUASSIER, J.P. J. Chim. Phys. 1981, 78, 3
13. HON, N.S. J. Polym. Sci., Polym. Chem. Ed. 1976, 14, 14
14. LONG, T.T. ; SALVIN, R ; FOUASSIER, J.P. to be published
15. HEROLD, R. ; BENDAIKHA, T ; FOUASSIER, J.P. ; DECKER, C unpublished results.

RECEIVED December 17, 1981.

8

Photoinduced Grafting Reactions in Cellulose and Cellulose Derivatives

DAVID N.-S. HON

Virginia Polytechnic Institute and State University, Department of Forest Products, Blacksburg, VA 24061-0299

HOW-CHING CHAN

Mount Sinai School of Medicine, The City University of New York, New York, NY 10029

Photoinduced free radicals generated in cellulose and cellulose derivatives, namely, methyl cellulose, ethyl cellulose, acetyl cellulose, hydroxyethyl cellulose and carboxymethyl cellulose, were examined by electron spin resonance (ESR) spectroscopy, and their capability of initiating graft copolymerization was studied. ESR findings revealed that a high concentration of free radicals was generated in all samples irradiated with ultraviolet light of $\lambda > 254$ nm and $\lambda > 280$ nm. Free radicals generated in cellulose were due to chain scission, dehydrogenation, and dehydroxymethylation reactions, whereas free radicals generated in cellulose derivatives were due to cleavage of substituted side chains. Grafting reactions of cellulose took place at the main backbone, whereas grafting reactions of cellulose derivatives took place at the substituted side chains. High degrees of grafting and grafting efficiency were inevitably obtained from samples treated with ultraviolet light of $\lambda > 280$ nm in homogeneous media.

Cellulose, the most abundant renewable agricultural raw material, is transformed into multifarious products affecting every phase of our daily life. The presence of active hydroxyl groups in cellulose has been utilized in a variety of chemical reactions to produce commercially important cellulose derivatives, such as cellulose ethers and cellulose esters. Although the practical purpose of cellulose derivatization is by and large to improve various properties of the original cellulose, these cellulose derivatives are often not competitive with most of the petrochemically derived synthetic polymers. In order to provide a better market position for cellulose derivatives, there is little doubt that further chemical modification is required. Grafting of vinyl monomers onto cellulose and

0097-6156/82/0187-0101$6.00/0
© 1982 American Chemical Society

cellulose derivatives may improve the intrinsic properties of these polymers.

Several papers dealing with grafting reactions in cellulose derivatives have been reported recently. Bardhan et al. (1) and Mukhopadhyay et al. (2) grafted acrylamide and acrylic acid, respectively, onto methyl cellulose using potassium persulfate as the initiator. Hebeish et al. (3, 4) grafted acrylonitrile and methyl acrylate onto acetylated cotton and partially carboxymethylated cotton. Grafting of methyl methacrylate to cellulose nitrate by ceric ion initiation has been developed by Santappa and his co-workers (5, 6). The radiation induced grafting of acrylonitrile onto cyanoethylated cotton and cellulose acetate have been investigated by Arthur et al. (7) and by Stannett et al. (8), respectively. Grafting of styrene to carbonilated celluloses by γ-irradiation was done by Guthrie et al. (9). In spite of the common use of ultraviolet light for grafting vinyl monomers onto cellulose, there is as yet no literature available on photoinduced grafting reactions in cellulose derivatives. Research on grafting vinyl monomers onto cellulose derivatives by ultraviolet light is in progress in our laboratory. This paper will report the detection of photoinduced free radicals in cellulose and various cellulose derivatives, namely, methyl cellulose, ethyl cellulose, acetyl cellulose, hydroxyethyl cellulose and carboxymethyl cellulose, and their capability of initiating graft copolymerization. The electron spin resonance (ESR) studies reveal that most of the free radicals were generated at the substituted side chains of cellulose derivatives by ultraviolet light and these free radicals are responsible for grafting and homopolymerization reactions.

Experimental

Materials. Cellulose and cellulose derivatives used for this study are summarized in Table 1. Commercial products were used as received without further purifications. Methyl methacrylate was used as the monomer, which was purified by sodium hydroxide/sodium chloride extraction followed by distillation under reduced pressure (10).

Methods

Photoirradiation and ESR Measurements. The samples of cellulose and cellulose derivatives were packed uniformly into clear fused Suprasil quartz tubes (O.D. 4 mm), which did not produce any ESR signal during the irradiated sequences. The quartz tubes containing the samples were evacuated to a constant pressure (10^{-6} mm Hg) and sealed. The source of ultraviolet irradiation was a high pressure mercury-xenon compact arc lamp (Conrad Hanovia type 901 B0011, 200 W) which

TABLE I. Cellulose Derivatives Characteristics

Specimen	Manufacturer	Type
Cellulose	Buckeye Cellulose	95.8% α-cellulose from yellow poplar
Ethyl Cellulose	Hercules, Inc.	N-4
Carboxylmethylcellulose (Fibrous)[a]	---	D.S. = 2.2
Carboxylmethylcellulose (Powder)	Hercules, Inc.	12M8
Hydroxyethylcellulose	Hercules, Inc.	250GR
Acetyl cellulose (Fibrous)[b]	---	35% CH_3CO

[a]Preparation in laboratory. (Methods in Carbohydrate Chemistry, R. L. Whistler, ed., Vol. III. Academic Press, New York, 1963, p. 322.

[b]Preparation in laboratory. (Methods in Carbohydrate Chemistry, R. C. Whistler, ed., Vol. III. Academic Press, New York, 1963, p. 193.)

emitted wavelengths ranging from 254 nm to 2000 nm, with the strongest wavelength at 365 nm. For certain studies, a Corning glass filter (color specification number, 0-53; glass code number, 7440) was used to eliminate wavelengths shorter than 280 nm. In all cases of ultraviolet irradiation, the distance between light source and sample was about 30 cm, and the sample was kept in liquid nitrogen in a Dewar flask. ESR spectra of irradiated samples in the Dewar flask were recorded using a spectrometer (Varian E-12) operated at X-band with 100 KHz modulation. The spectra were recorded at 9.3 GHz at 3300 gauss. The g values were measured by comparison with the pitch sample provided by Varian Associates.

Graft Copolymerization. Graft copolymers of cellulose and cellulose derivatives were prepared in a quartz reactor containing 0.5 g oven dried sample, 10 ml monomer and 100 ml water at 45°C. The grafting mixtures or solutions were photo-irradiated with the ultraviolet light of $\lambda > 254$ nm and $\lambda > 280$ nm. Grafted products of cellulose were washed with distilled water and extracted with acetone for 72 hours to remove homopolymers; grafted products of cellulose derivatives were washed and extracted with benzene repeatedly, the final products were then freeze-dried. Degree of grafting and grafting efficiency were calculted according to the following formula:

$$\text{Degree of grafting (\%)} = \frac{A-B}{B} \times 100$$

$$\text{Grafting efficiency (\%)} = \frac{A-B}{(A-B) + C} \times 100$$

Where A is weight of grafted sample after extraction, B is weight of original sample (oven-dry), and C is weight of homopolymer.

Results and Discussion

Photoinduced Free Radicals in Cellulose and Cellulose Derivatives. Most cellulose and cellulose derivatives are susceptible to photochemical reaction in the presence of ultraviolet light below the wavelength of 340 nm (11) via a free radical process. These free radical intermediates may be capable of initiating graft copolymerization reaction if they are not decayed rapidly or undergo secondary termination reactions. The detection of free radicals generated in cellulose and cellulose derivatives by ultraviolet light was studied. ESR spectra of cellulose and cellulose derivatives used for this study are shown in Figures 1 and 2 for samples treated with ultraviolet light of $\lambda > 280$ nm and $\lambda > 254$ nm respectively. It is obvious that free radicals were generated in all samples irradiated with both light sources. The ESR

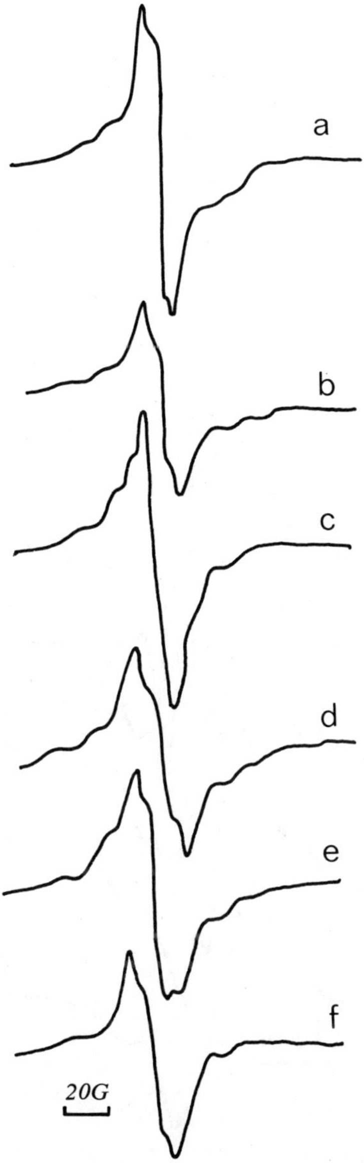

Figure 1. ESR spectra of cellulose and cellulose derivatives irradiated with UV light of λ > 280 nm for 60 min at 77 K. Spectra were recorded at 77 K. Key: a, cellulose; b, methylcellulose; c, ethylcellulose; d, acetylcellulose; e, hydroxyethylcellulose; f, carboxymethylcellulose.

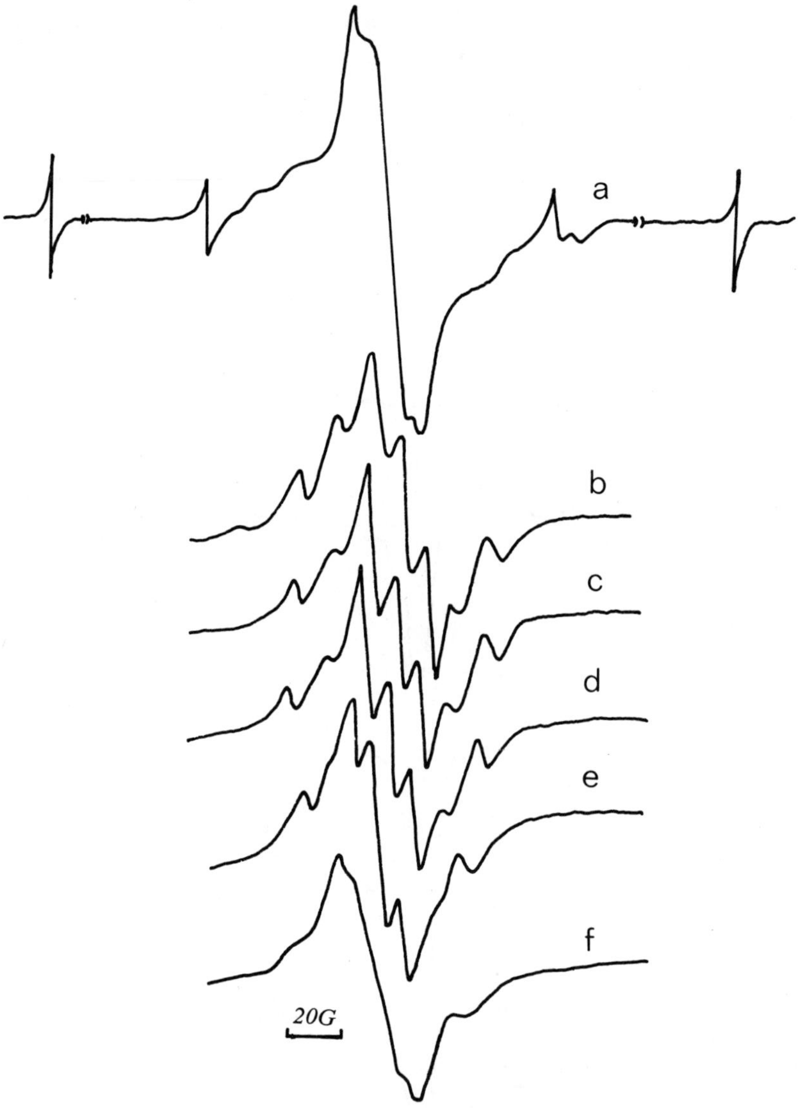

Figure 2. ESR spectra of cellulose and cellulose derivatives irradiated with UV light of λ > 254 nm for 60 min at 77 K. Spectra were recorded at 77 K. Key: a, cellulose; b, methylcellulose; c, ethylcellulose; d, acetylcellulose; e, hydroxyethylcellulose; f, carboxymethylcellulose.

signals derived from samples irradiated with light of $\lambda > 280$ nm exhibited ill-defined multiplet hyperfine splitting while prominent signals with better resolution were observed from those with light of $\lambda > 254$ nm because of higher energy involved. From the latter light source, an eleven-line signal with a g value of 2.003 was detected in cellulose, and seven-line signals with identical g values were detected from methyl cellulose, ethyl cellulose, acetyl cellulose, hydroxyethyl cellulose (weak) and carboxymethyl cellulose (weak).

Interpretation of the eleven-line signal of cellulose has been published elsewhere (12). It was the result of superimposing signals generated from six kinds of radical species, as a consequence of chain scission, dehydrogenation and dehydroxymethylation reactions in photoirradiated cellulose. Details of the mechanism of free radical formation were also discussed elsewhere (13, 14). A seven-line signal was observed from photoirradiated methyl cellulose (Figure 3). When this photoirradiated sample was warmed gradually from 77°K to ambient temperature for 1 minute and kept again at 77°K, a poorly resolved five-line spectrum was observed (Figure 3b). When it was warmed for 5 minutes, the signal was further decayed to a singlet signal with a line-width of 21 gauss (Figure 3c). By subtracting b from a in Figure 3, a quartet spectrum with intensity ratio 1:3:3:1 is obtained (Figure 3d). Since this four-line signal has a splitting constant of 22.8 gauss, we assigned this signal to methyl radicals. Consequently, it is obvious that demethylation took place in methyl cellulose, as shown in equation 1.

$$\text{Cellulose—OCH}_3 \xrightarrow{h\nu} \underset{\text{(singlet)}}{\text{Cellulose—O}\cdot} + \underset{\text{(quartet)}}{\cdot\text{CH}_3} \qquad (1)$$

The singlet signal is hence due to the alkoxy radical, which has a longer life time than carbon radicals at ambient temperature.

When ethyl cellulose was irradiated with light of $\lambda > 254$ nm for 60 minutes at 77°K, a complicated eleven-line signal was observed (Figure 4). When this sample was warmed to ambient temperature for 5 minutes, only a singlet signal with a line-width of 26 gauss was observed. Analyzing the splitting constants of the multi-intensity resonance peaks, it reveals that, in addition to the singlet component, the spectrum was superimposed with two signals: a strong six-line component with a hyperfine splitting constant of 25 gauss (∇ in Figure 4), and a weak fine-line component with a hyperfine splitting constant of 25 gauss (∘ in Figure 4). This implies that deethylation took place preferentially during photoirradiation to create a radical pair of six-line and singlet components. A hydrogen abstraction reaction also likely occurred generating the five-line component. Accordingly, the mechanisms are illustrated in equations 2, 3 and 4.

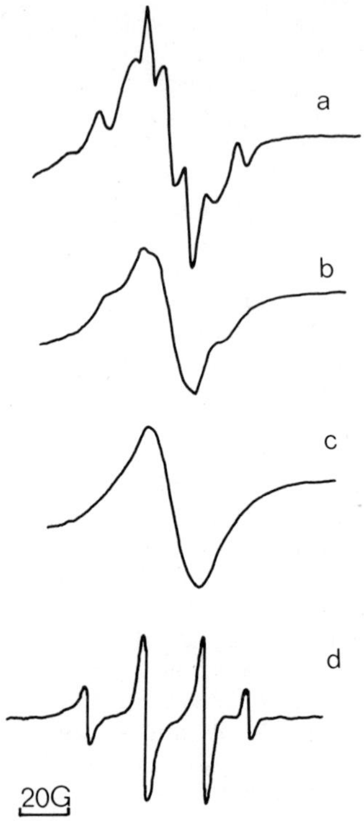

Figure 3. Changes in ESR spectra of methylcellulose. Key: a, initial spectrum observed at 77 K immediately after irradiation with light of λ > 254 nm at 77 K for 60 min; b, after sample was warmed to ambient temperature for 1 min; c, after sample was warmed to ambient temperature for 5 min; d, a quartet component which was obtained by subtracting b from a.

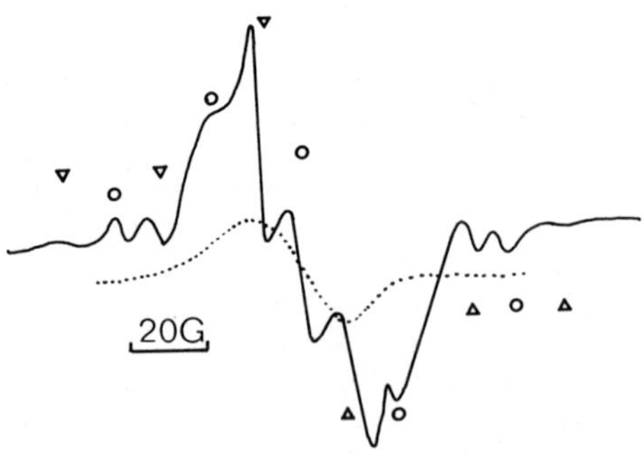

Figure 4. ESR spectra of ethylcellulose irradiated with UV light of λ > 254 nm at 77 K for 60 min. Spectra were recorded at 77 K. Dotted line signal was obtained after sample was warmed to ambient temperature for 5 min.

$$\text{Cellulose} - O - CH_2CH_3 \xrightarrow{h\nu} \underset{\text{(singlet)}}{\text{Cellulose} - O\cdot} \quad + \underset{\text{(6-line)}}{\cdot CH_2CH_3} \qquad (2)$$

$$\text{Cellulose} - O - CH_2CH_3 \xrightarrow{h\nu} \text{Cellulose} - O - \underset{\text{(5-line)}}{CH_2-CH_2\cdot} \qquad (3)$$

$$\text{Cellulose} - O - CH_2CH_3 \xrightarrow{h\nu} \text{Cellulose} - O - \underset{\text{(5-line)}}{\dot{C}H - CH_3} \qquad (4)$$

The hyperfine splitting constant of the six-line component is closed to the ethyl radicals of ethyl iodide (15).

A prominent seven-line signal was observed from acetyl cellulose when irradiated with light of $\lambda > 254$ nm for 60 minutes at 77°K (Figure 5). The warm-up treatment, as for methyl cellulose, revealed that a four-line signal with a hyperfine splitting constant of 22.8 gauss was generated, indicating deacetylation and demethylation took place, as illustrated in equations 5 and 6.

$$\text{Cellulose}-O\overset{O}{\overset{\|}{C}}-CH_3 \xrightarrow{h\nu} \text{Cellulose}-O\cdot \quad +\cdot\overset{O}{\overset{\|}{C}}-CH_3 \qquad (5)$$

$$\cdot\overset{O}{\overset{\|}{C}}-CH_3 \longrightarrow CO + \cdot CH_3$$

$$\text{Cellulose}-O-\overset{O}{\overset{\|}{C}}-CH_3 \xrightarrow{h\nu} \text{Cellulose}-O-\overset{O}{\overset{\|}{C}}\cdot + \cdot CH_3 \qquad (6)$$

Details of the free radical formation mechanism were discussed elsewhere (16).

When hydroxyethyl cellulose was irradiated with the same light, a seven-line signal was also observed (Figure 6). When this sample was warmed to ambient temperature, a five-line signal with a hyperfine splitting constant of 29 gauss disappeared, only a singlet signal with a line-width of 24 gauss was observed. It is clear that dehydroxyethylation took place for this sample. Reaction is shown in equation 7.

$$\text{Cellulose}-O-CH_2-CH_2OH \xrightarrow{h\nu} \underset{\text{(Singlet)}}{\text{Cellulose}-O\cdot} \; + \; \underset{\text{(5-line)}}{\cdot CH_2CH_2OH} \qquad (7)$$

A seven-line like signal was detected from carboxymethyl cellulose irradiated with ultraviolet light of 254 nm (Figure 7). When this sample was warmed to ambient temperature for 60 seconds, this multiplet spectrum was transformed rapidly into a prominent doublet signal with a hyperfine splitting constant of 20 gauss. This indicated that the primary free radicals

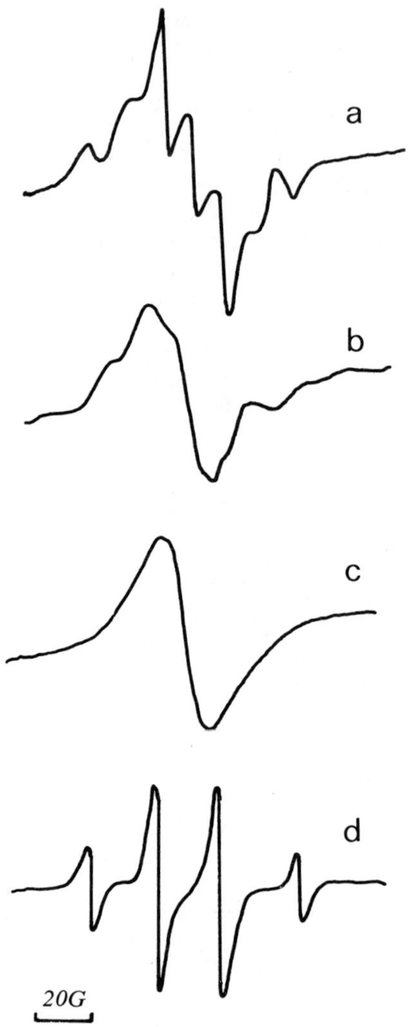

Figure 5. Changes in ESR spectra of acetylcellulose. Key: a, initial spectrum observed at 77 K immediately after irradiation with UV light of λ > 254 nm at 77 K for 60 min; b, after sample was warmed to ambient temperature for 1 min; c, after sample was warmed to ambient temperature for 5 min; d, a quartet component which was obtained by subtracting b from a.

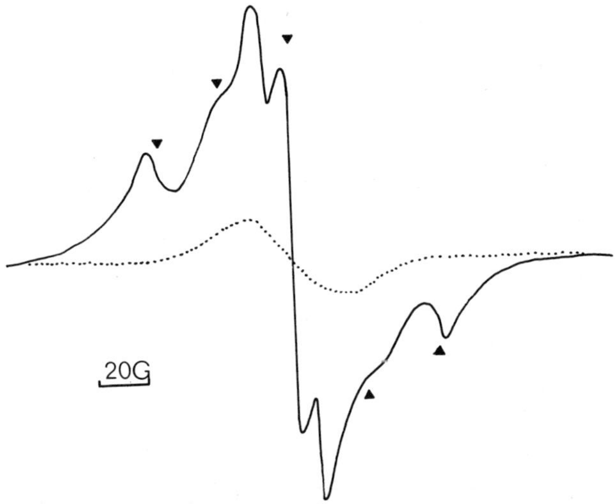

Figure 6. ESR spectra of hydroxyethylcellulose irradiated with UV light of λ > 254 nm at 77 K for 60 min. Spectra were recorded at 77 K. Dotted line signal was obtained after sample was warmed to ambient temperature for 5 min.

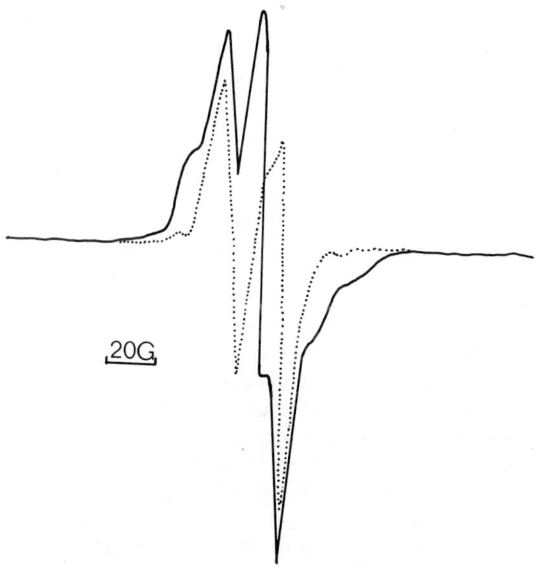

Figure 7. ESR spectra of carboxymethylcellulose irradiated with UV light of λ > 254 nm at 77 K for 60 min. Spectra were recorded at 77 K. Dotted line signal was obtained when sample was warmed to ambient temperature for 1 min.

generated were transformed into a radical site attributing the doublet signal. Based on the chemical structure of carboxymethyl cellulose, it is likely that the following reaction took place (equation 8):

$$\text{Cell—O—CH}_2\text{COOH} \xrightarrow{h\nu} \text{primary radicals} \xrightarrow{\text{heat}} \underset{\text{(doublet)}}{\text{Cell—O—}\dot{\text{C}}\text{H—COOH}} \quad (8)$$

From this ESR information, it is obvious that more free radicals were generated when samples were irradiated with ultraviolet light of $\lambda > 254$ nm than those with light of $\lambda > 280$ nm. Several types of free radicals were generated in cellulose due to chain scission, dehydrogenation and dehydroxymethylation, whereas free radicals generated in cellulose derivatives were by and large due to the cleavage of the substituted side chains.

Graftability of Photoinduced Cellulosic Free Radicals. Various free radicals are generated in cellulose and cellulose derivatives by ultraviolet light, which may be capable of initiating graft copolymerization reactions with vinyl monomers. The graftability of these photoinduced free radicals in homogeneous and heterogeneous media was studied.

The degree of grafting and grafting efficiency of grafting reactions induced with ultraviolet light of $\lambda > 280$ nm and $\lambda > 254$ nm were scrutinized.

The results obtained by using ultraviolet light of $\lambda > 280$ nm are shown in Figures 8 and 9 for degree of grafting and grafting efficiency respectively. It appears that the degree of grafting increased as the function of irradiation time. After four hours, the degree of grafting for various samples are in the order of fibrous carboxymethyl cellulose > cellulose > hydroxyethyl cellulose > ethyl cellulose > methyl cellulose > powdered carboyxmethyl cellulose > acetyl cellulose. Powdered carboxymethyl cellulose and acetyl cellulose showed a rather low degree of grafting. However, the grafting efficiency for these samples varied. After four hours of reaction, they are in the order of fibrous carboxymethyl cellulose > hydroxyethyl cellulose > methyl cellulose > powdered carboxymethyl cellulose > ethyl cellulose > cellulose > acetyl cellulose.

When carboxymethyl cellulose and hydroxyethyl cellulose were used as the substrates, the grafting efficiency was over 90%, indicating that fewer homopolymers were produced. It is of great interest to observe that ethyl cellulose, cellulose and acetyl cellulose, which were all dispersed in the grafting mixture, exhibited low grafting efficiency, whereas carboxymethyl cellulose, hydroxyethyl cellulose and methyl cellulose, which were dissolved in the grafting solution, exhibited high grafting efficiency. These findings suggest that grafting

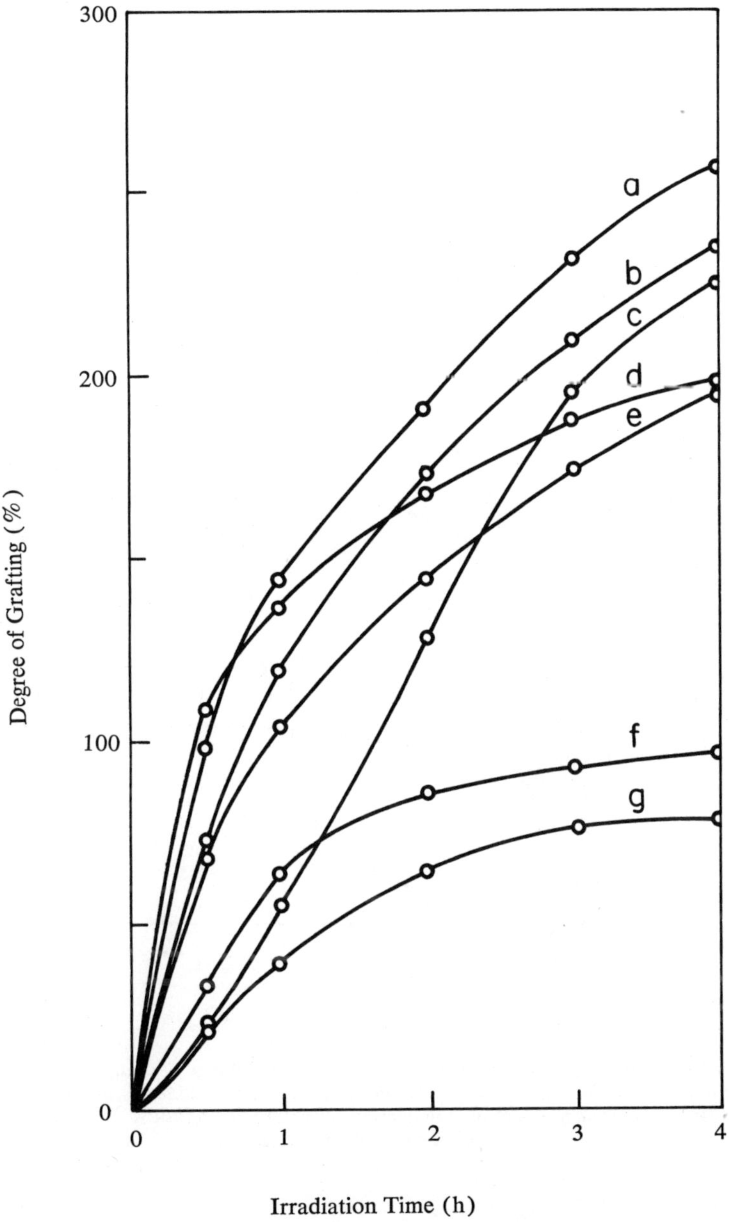

Figure 8. Degree of grafting for cellulose and cellulose derivatives initiated by UV light of λ > 280 nm. Key: a, fibrous carboxymethylcellulose; b, cellulose; c, hydroxyethylcellulose; d, ethylcellulose; e, methylcellulose; f, powdered carboxymethylcellulose; g, acetylcellulose.

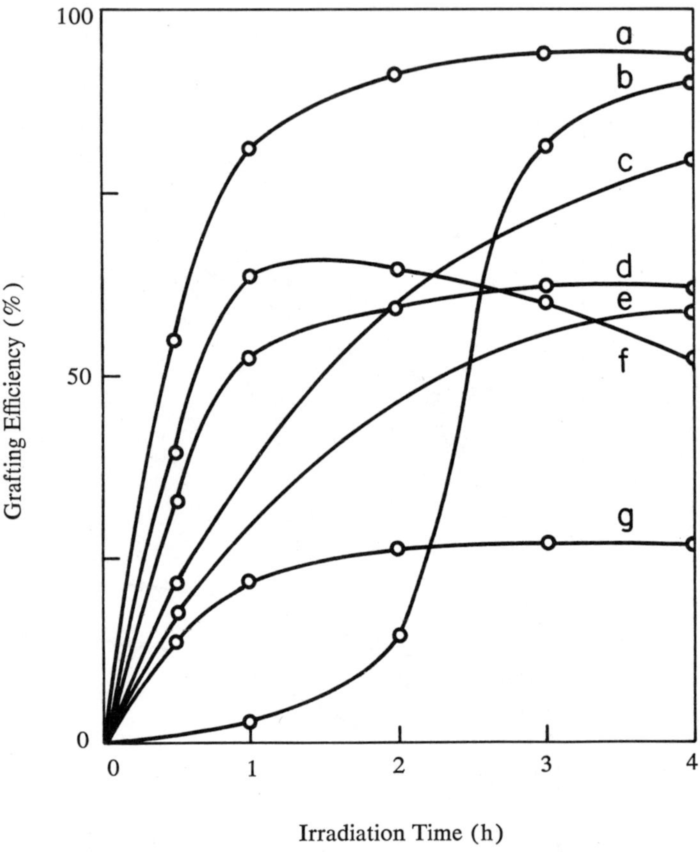

Figure 9. Grafting efficiency for cellulose and cellulose derivatives initiated by UV light of λ > 280 nm. Key: a, fibrous carboxymethylcellulose; b, hydroxyethylcellulose; c, methylcellulose; d, powdered carboxymethylcellulose; e, ethylcellulose; f, cellulose; g, acetylcellulose.

reaction carrying out in a homogeneous system achieves a high degree of grafting with less homopolymer formation. The grafting reactions are not favored in a heterogeneous system and a high degree of homopolymer formation is likely to occur. However, when the grafting reactions were carried out using ultraviolet light of λ > 254 nm, the degree of grafting (Figure 10) and grafting efficiency (Figure 11) were rather low for all samples, implying that a large amount of homopolymer was formed. This may be ascribed to the shorter wavelengths and higher energy involved, which created greater amounts of free radicals in cellulose and cellulose derivatives. Most of the low molecular free radicals generated from the samples are likely to undergo a homopolymerization reaction. As a result, the grafting efficiency is decreased.

Conclusions

Based on the ESR and graft copolymerization reactions studied on cellulose and particularly on cellulose derivatives, the following conclusions may be drawn:

1. Free radicals were generated in cellulose and cellulose derivatives irradiated with ultraviolet light of λ > 254 nm and λ > 280 nm. More free radicals were generated by using the former light due to the shorter wavelength, i.e., stronger energy of the light. Several types of free radicals were generated in cellulose due to chain scission, dehydrogenation and dehydroxymethylation reactions; whereas free radicals generated in cellulose derivatives, namely, carboxymethylcellulose, methyl cellulose, ethyl cellulose, hydroxymethyl cellulose and acetyl cellulose, were by and large due to the cleavage of the substituted side chains. Grafting reactions of cellulose took place at the main backbone, whereas grafting reactions of cellulose derivatives took place at the substituted side chains.

2. Free radicals in cellulose and cellulose derivatives by ultraviolet light of λ > 254 nm and λ > 280 nm were capable of initiating grafting reactions. A high degree of grafting efficiency, that is, low degree of homopolymer formation, was observed from samples using ultraviolet light of λ > 280 nm. Due to the shorter wavelength, high energy involved from the light of λ > 254 nm, although more free radicals were generated from the substrates, they were unfavorably participating in the homopolymer formation reaction.

3. Grafting reactions carried out in the homogeneous medium for carboxymethyl cellulose, methyl cellulose, and hydroxyethyl cellulose, always achieved higher degree of grafting efficiency than those samples such as cellulose, ethyl cellulose and acetyl cellose, irradiated in the heterogeneous medium.

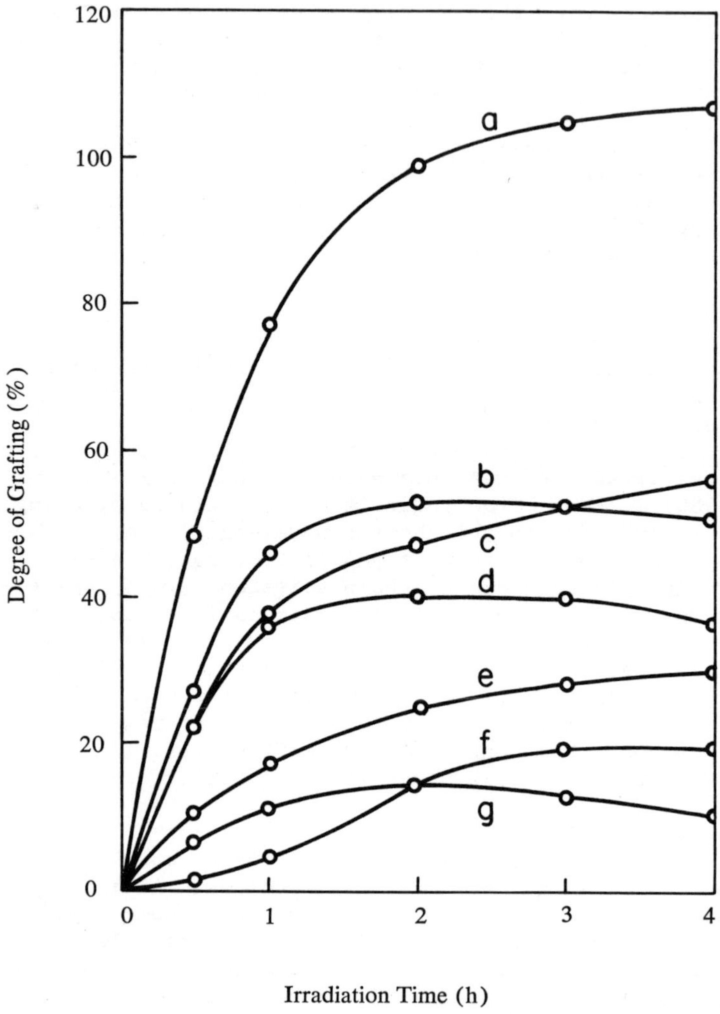

Figure 10. Degree of grafting for cellulose and cellulose derivatives initiated by UV light of λ > 254 nm. Key: a, fibrous carboxymethylcellulose; b, ethylcellulose; c, methylcellulose; d, cellulose; e, powdered carboxymethylcellulose; f, hydroxyethylcellulose; g, acetylcellulose.

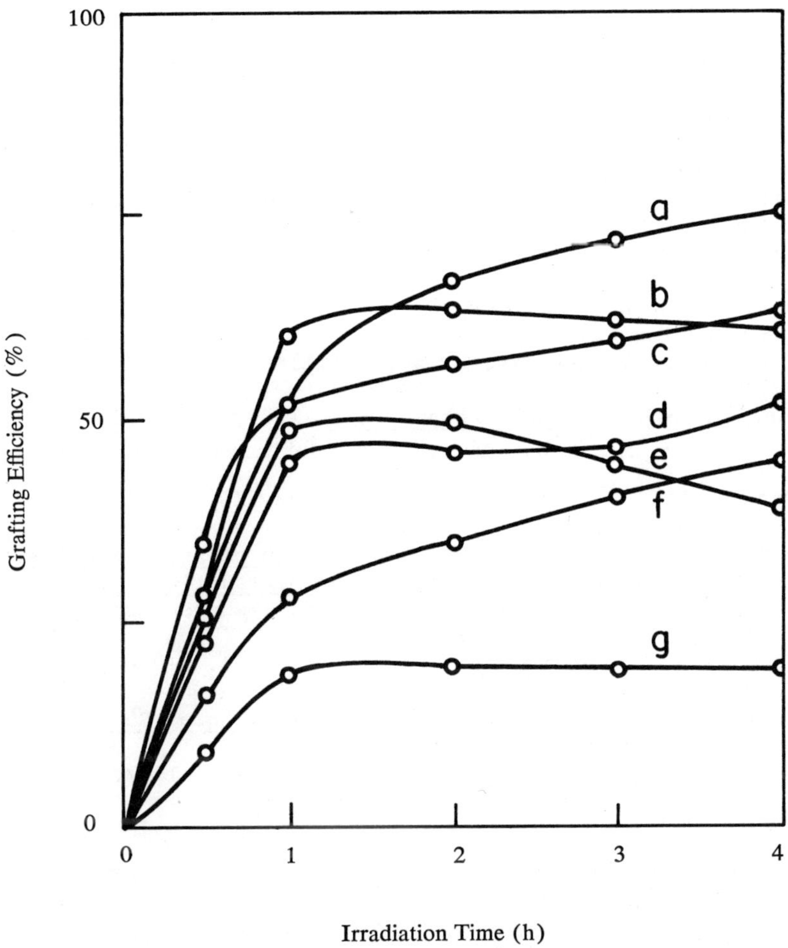

Figure 11. Grafting efficiency for cellulose and cellulose derivatives initiated by UV light of λ > 254 nm. Key: a, fibrous carboxymethylcellulose; b, cellulose; c, powdered carboxymethylcellulose; d, methylcellulose; e, ethylcellulose; f, hydroxyethylcellulose; g, acetylcellulose.

Literature Cited

1. Bardhan, K.; Mukhopadhyay, S.; Chatterjee, S. R. J. Polym. Sci. Polym. Chem. Ed. 1977, 15, 141.
2. Mukhopadhyay S.; Prasad J.; Chatterjee, S. R. Makromol Chem. 1975, 176, 1.
3. Hebeish, A.; Kantouch A.; El-Rafie, M. H. J. Appl. Polym. Sci. 1977, 15, 11.
4. Kantouch, A.; Hebeish, A.; El-Rafie, M. H. Europ. Polym. J. 1970, 6, 1575.
5. Sadhakar, D.; Srinivasan, K.S.V.; Joseph, R. T.; Santappa, M. J. Appl. Polym. Sci. 1979, 23, 2923.
6. Srinivasan, K.S.V.; Hon, D.N.-S.; Santappa, M. This volume, p.155.
7. Demint, R. J.; Arthur, J. C., Jr.; McSherry, F. Textile Res. J. 1961, 31, 821.
8. Wellons, J. D.; Williams, J. L.; Stannett, V. J. Polym. Sci. Part A, 1967, 5, 1341.
9. Guthrie, J. T.; Huglin, M. B.; Phillips, G. O. Europ. Polym. J. 1972, 8, 747.
10. Hon, D.N.-S. J. Appl. Polym. Sci. 1979, 23, 3591.
11. Hon, N.-S. J. Polym. Sci. Polym. Chem. Ed. 1975, 13, 1347.
12. Hon, N.-S. J. Polym. Sci. Polym. Chem. Ed. 1975, 13, 2653.
13. Hon, D.N.-S. Photochemical Degradation of Lignocellulosic Materials in Development in Polymer Degradation - 3 (N. grassie, ed.), Applied Science Publishers, Essex, England, 1981, pp. 229-281.
14. Hon, N.-S. J. Polym. Sci. Polym. Chem. Ed. 1976, 14, 2497.
15. Fenrick, H. W.; Willard, J. E. J. Amer. Chem. Soc. 1963, 85, 3731.
16. Hon, N.-S. J. Polym. Sci. Polym. Chem. Ed. 1977, 15, 725.

RECEIVED December 24, 1981.

9

Photoinduced Graft Copolymerization of Cellulose Derivatives

TAKAHASHI AKIRA and TAKAHASHI SHIRŌ

Kogakuin University, Department of Industrial Chemistry, Faculty of Engineering, Shinjuku-ku, Tokyo, 160, Japan

Photoinduced graft copolymerization of cellulose derivatives especially containing carbonyl groups in the absence of a photosensitizer, has been investigated. Grafting of methyl methacrylate onto dialdehyde and dicarboxyl celluloses proceeded more easily than onto monocarboxyl cellulose. The grafting efficiency of the former two samples attained to 95-100%. The observation shows that graft reaction starts with the photolysis of carbonyl groups and that the degree of grafting and the apparent number of grafted chains are varied with the species and the positions of carbonyl groups. The effects of wavelength, grafting activity of monomer and solvent were also examined. Grafting to nitrocellulose took place by the photolysis of nitro groups and the grafted chains were shortened because of the degradative chain transfers. Whereas grafting to acetylcellulose hardly occurred with higher acetic acid contents or increasing degree of polymerization. In case of regenerated cellulose, grafting to the end of cellulose proceeds more readily than to the medium.

Photoinduced graft copolymerization of vinyl monomer onto cellulose in the presence of such a photosensitizer as dyes (1), metal ions (2,3) or hydrogen peroxide (4) have been reported. In every case, the grafting efficiencies were usually low. Grafting onto cellulose containing photodecomposable groups was also carried out by a few researchers. Kadooka *et al*. (5) reported that acrylamide and acrylonitrile easily grafted to cellophane treated with methyl vinyl ketone. Accordingly, it is expected that by introducing functional groups into cellulose, it may accelerate the photo induced graft polymerization and increase the apparent number of grafted chains. The authors reported that the photoinduced graft copolymerization onto dialdehyde cellulose derivatives occurred easily and homopolymer scarcely produced (6-9).

0097-6156/82/0187-0119$6.50/0
© 1982 American Chemical Society

This paper describes the graft polymerization of various monomers onto cellulose derivatives containing carbonyl groups without photosensitizer. Especially the effects of functional group, wavelength, monomers, solvent and reaction temperature on the grafting, the chain length and the apparent number of grafted chains are investigated.

Experimental

Preparation of Cellulose Derivatives. The cellulose derivatives were prepared by the methods shown in Figure 1 using a purified cotton linter (C-) cellulose. The 2,3-dialdehyde (Al-) cellulose was prepared by treating the C-cellulose with sodium periodate (10). The 2,3-dicarboxyl (Cd-) cellulose was obtained by oxydizing the Al-cellulose with sodium chlorite (11). The 6-carboxyl (Cm-) cellulose was formed by oxidizing the C-cellulose with nitrogen dioxide gas (12). The content of oxidized groups, namely carbonyl and carboxyl groups, were determined by the methods using semicarbazide (13) and calcium acetate (14), respectively.

Carbonyl group contents of Al-cellulose increased with prolonging the sodium periodate oxidation time, with very little formation of carboxyl groups. Oxidizing Al-cellulose with sodium chlorite converted 90-97% of carbonyl groups into carboxyl groups (Cd-cellulose). The results indicate that the bond between C_2 and C_3 position of glucose unit is open.

Cm-cellulose mainly contained carboxyl groups at C_6 and also slight carbonyl groups (12). The Cm-cellulose having 90 mmol functional groups contained 0.6% nitrogen which was determined by the elementary analysis (Fig. 9). Treating Cm-cellulose with sodium borohydride reduced the carbonyl groups and oxidizing it with sodium chlorite decreased the carbonyl groups.

Nitro (N-) cellulose was prepared by nitrizing the C-cellulose with the mixtures of sulfuric and nitric acid solution with various concentration ratios.

Commercial acetyl (Ac-) cellulose (Daicel Co., Ltd.) was prepared by partial hydrolysis of triacetate obtained by acetylating the C-cellulose with a mixture of acetic anhydride, acetic acid and sulfuric acid.

Regenerated (R-) cellulose was obtained by hydrolyzing the Ac-cellulose with sodium hydroxide-acetone aqueous solution (15). According to the Alexander and Michell's report (16), the degree of polymerization of R-cellulose was determined viscometrically, after it was nitrated with the mixture of phosphoric anhydride and fumic nitric acid then dissolved in ethyl acetate.

Reagents. Using monomers, stylene (St) and methyl methacrylate (MMA) were distilled at reduced pressure, and acrylonitrile (AN), vinyl acetate (VAc) and methyl acrylate (MA) were distilled at ordinary pressure under nitrogen using a Widmer distilled

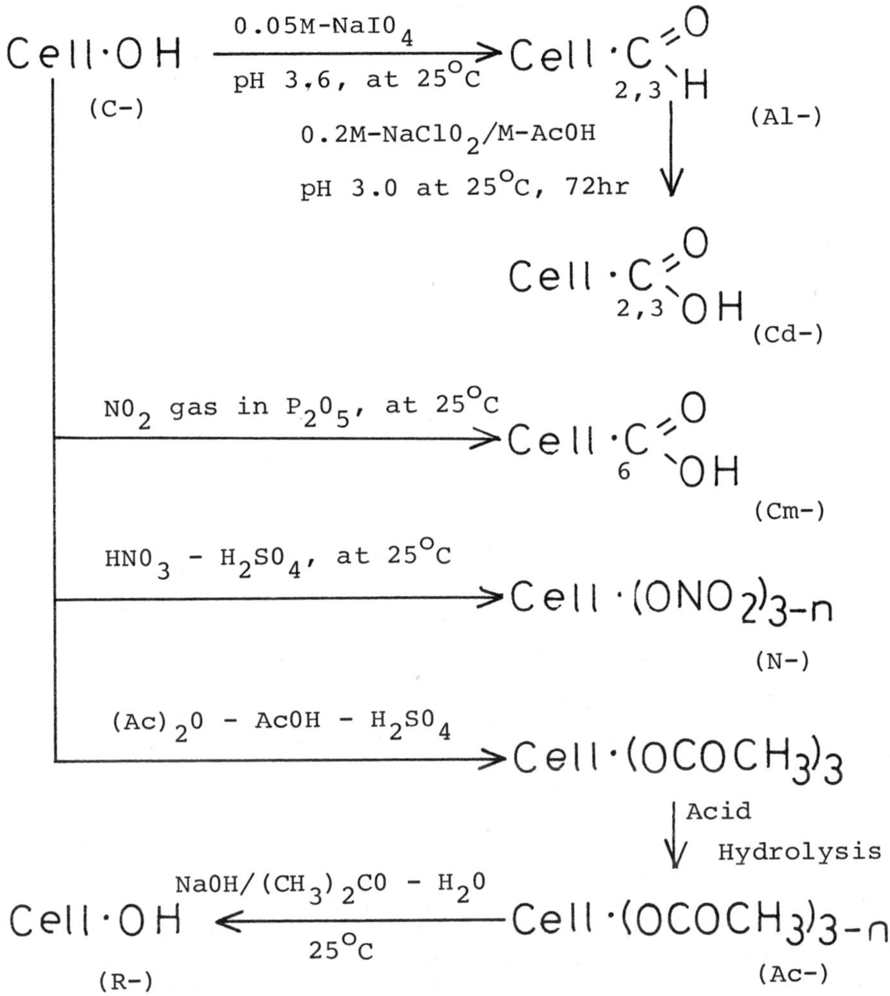

Figure 1. Preparation of cellulose derivatives.

apparatus. The fifteen species of organic solvents adding polymerization system were all purified by distillation.

Graft Polymerization. The cellulose was weighed into a quartz (Q) and a Pyrex (P) tube (15 mm in diameter and 110 mm long). Then water, organic solvent (not used in some cases) and monomer were placed in it flushed with nitrogen gas. The reaction vessel was fixed into a Riko rotary photochemical reactor and irradiated with ultraviolet light (160W low pressure mercury lamp). The energy distribution of the light source is shown in a previous paper (6). The polymerization tubes were shaken well every five minutes in addition to the stirring with the instrument. The polymerization product was precipitated into methanol or water containing hydroquinone. Then, the homopolymer was thoroughly extracted, in case of PMMA, PVAc and PMA with hot acetone, PSt with hot benzene, and PAN with hot dimethyl formamide. The total conversion, the degree of grafting and the grafting efficiency were calculated by the method reported previously (17).

Isolation of Grafted Chain and Measurement of Molecular Weight. The PMMA grafted was obtained by hydrolyzing the graft copolymer with 72% sulfuric acid (18) and purified by reprecipitation. The molecular weight was determined by the viscosity of acetone solution (19). The apparent number of grafted chains was calculated from the degree of grafting per 100g cellulose and the molecular weight of grafts.

Results and Discussion

Graft onto 2,3-Dialdehyde Cellulose. Fig. 2 shows the results of the graft polymerization of methyl methacrylate onto Al-cellulose having various carbonyl contents. The degree of grafting rapidly increased with increasing carbonyl groups, up to 50 mmol carbonyl groups and leveled off. The grafting efficiencies were as large as 95%. The presence of a small amount of carbonyl groups greatly shortened the chain length (P> Q-tube) and increased the apparent numbers of grafted chains (Q> P-tube). The calculated results show that using a Q-tube, one seventieth of the carbonyl groups becomes a radical site. Hardly any graft reaction occurred using the C-cellulose under the same conditions shown in Fig. 2 and in the system adding sodium periodate without light (20). These results suggest that the graft polymerization proceeds mainly via the photolysis of carbonyl groups and these groups participate not only the initiation reaction but also the termination with growing polymer radicals. Furthermore, since using a Q-tube caused the scission of the trank polymer besides the photolysis of carbonyl groups (6), the grafting to the chain end is also possible. Ogiwara *et al.* (21) observed that irradiation with ultraviolet light on Al-cellulose resulted acyl radicals.

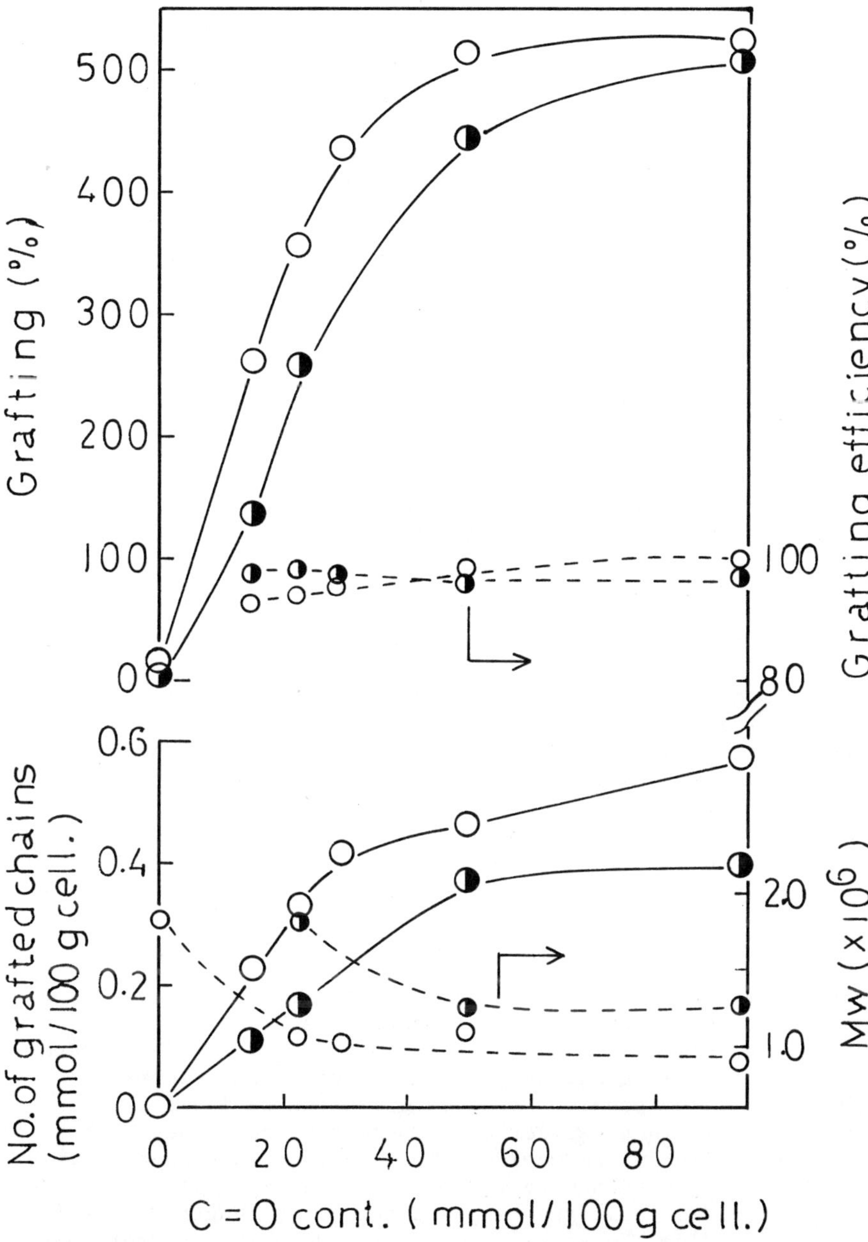

Figure 2. Effect of carbonyl group content on the yield of copolymer, molecular weight, and apparent number of grafted chains in the pMMA-grafted dialdehyde-cellulose. Conditions: cellulose, 0.3 g; H_2O, 10 mL; MMA, 2 mL at 50°C for 1 h. Key (reaction tube): ○, quartz; ◑, Pyrex.

Fig. 3 indicates the grafting activity of monomer onto the Al-cellulose containing 40.5 mmol carbonyl groups. The grafting rate decreased in the following order: methyl acrylate > methyl methacrylate > vinyl acetate > acrylonitrile > stylene. Apparently the monomer with high polarity grafts easily accept acrylonitrile. Furthermore the grafting efficiencies were very high, especially acrylonitrile and methyl methacrylate, except vinyl acetate.

Grafting reactions were also carried out in organic solvents. Fig. 4 shows the polymerization of methyl methacrylate in the fifteen species of organic solvents having various dielectric constants and water systems without cellulose. The numbers in the figure express the dielectric constants of the solvent in the order. As the organic solvent (No. 1-6) with the dielectric constants around 2 was added, the conversion became 0-11% and the value of dioxane (No. 2), which is soluble in water, was the highest. As the solvents with dielectric constants above 8 were used, the conversion increased, except cyclohexane (No. 11). These results reveal clearly that the polymerization is extremely affected by the dielectric constant of the solvent, and also the affinity with water and the mechanisms of primary radical formation by photolysis.

By taking into account those data, the graft polymerization of methyl methacrylate onto cellulose was done using the solvents (No. 2, 12, 13 and 14-) water system (Fig. 5). Increasing the solvent concentrations decreased the degree of grafting, but all of the grafting efficiencies were at high levels of 85-95%. These indicate that the graft polymerization took place between cellulose radicals and monomers. The chain length decreased with increasing the solvent concentrations and were markedly affected by solvent used.

From these results, the chain transfer constants to the solvent were determined by the Mayo's equation (22).

$$1 / Pn = 1 / Pno + Cs \times [S] / [M]$$

where [S] and [M] are the initial concentration of solvent and monomer, respectively. Cs is the chain transfer constant of the solvent. As demonstrated in Fig. 6, 1 / Pn vs. [S] / [M] have linear relationships and Cs are determined by the slopes of these lines. The results obtained are shown in Table I. Cs values decrease in the order of carbone tetrachloride > i-butanol > acetone > dimethyl formamide. These values show good agreement with the values obtained by the homogeneous system (23). Consequently, the degree of grafting and the chain length can be controlled by adding the solvent with various chain transfer constants.

Graft onto 2,3-Dicarboxyl Cellulose. Cd-celluloses with different carboxyl group contents were grafted with methyl metha-

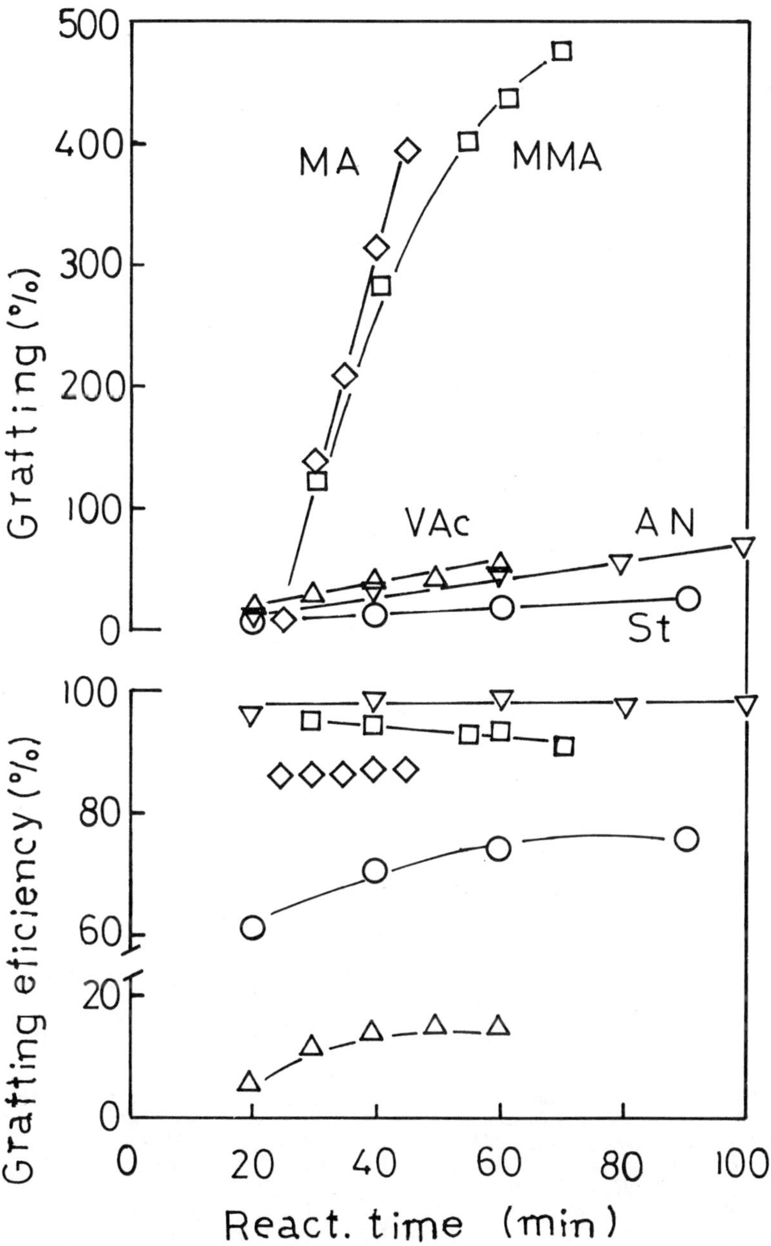

Figure 3. Effect of reaction time on the yield of copolymer in the various vinyl polymer-grafted dialdehydecellulose using a quartz tube. Conditions: cellulose (C = 0, 40.5 mmol/100 g), 0.3 g; monomer, 1.5 M*; H_2O, (12 − monomer) mL; 50°C. Key: ◇, MA; □, MMA; △, VAc; ▽, AN; ○, St.*

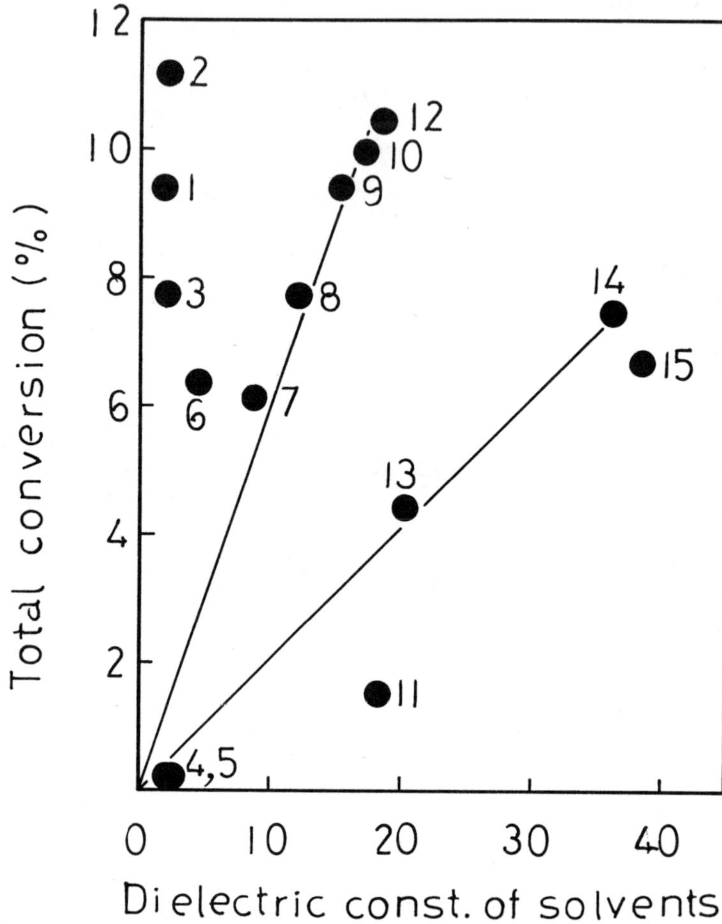

Figure 4. Effect of dielectric constant of solvents on the total conversion in the polymerization of MMA in solvent–water systems using a quartz tube. Conditions: H_2O, 8 mL; solvent, 2 mL; MMA, 2 mL; at 50°C for 3 hr. The solvents (and dielectric constants) are 1, C_6H_{12} (2.02); 2, $C_4H_8O_2$ (2.21); 3, CCl_4 (2.23); 4, C_6H_6 (2.27); 5, $C_6H_5CH_3$ (2.38); 6, $CHCl_3$ (4.70); 7, CH_3Cl (8.9); 8, $(CH_3)_3COH$ (12.2); 9, $C_2H_5COCH_3$ (15.45); 10, $CH_3(CH_2)_2CH_2OH$ (17.7); 11, $C_6H_{10}O$ (18.3); 12, $(CH_3)_2CHCH_2OH$ (18.7); 13, $(CH_3)_2CO$ (20.5); 14, $HCON(CH_3)_2$ (36.71); 15, CH_3CN (37.5).

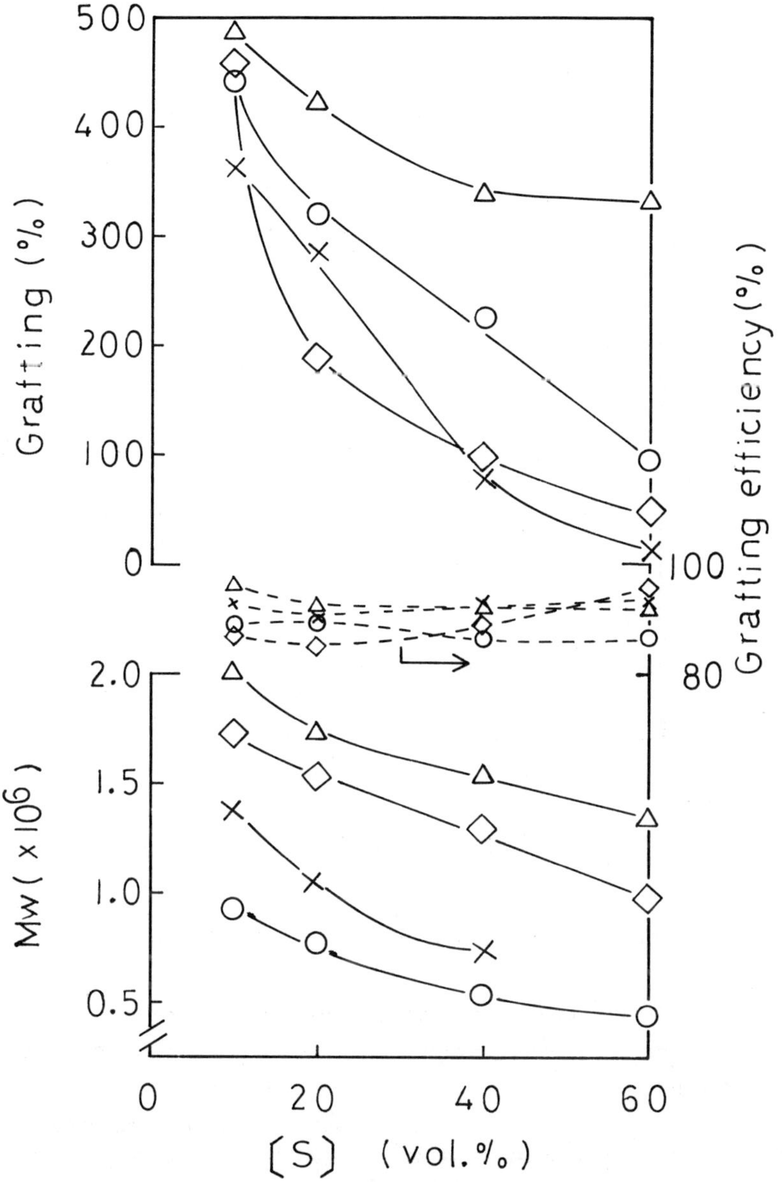

Figure 5. Effect of organic solvent concentration on the yield of copolymer and molecular weight of grafted chains in the pMMA-grafted dialdehydecellulose using a quartz tube. Conditions: cellulose (C = 0, 29.8 mmol/100 g), 0.3 g; H_2O + solvent, 10 mL; MMA, 2 mL; 50°C. Key to solvent: ○, CCl_4, 1.5 h; ×, $(CH_3)_2CHCH_2OH$, 1.25 h; △, $HCON(CH_3)_2$, 1 h; ◇, $(CH_3)_2CO$, 3 h.

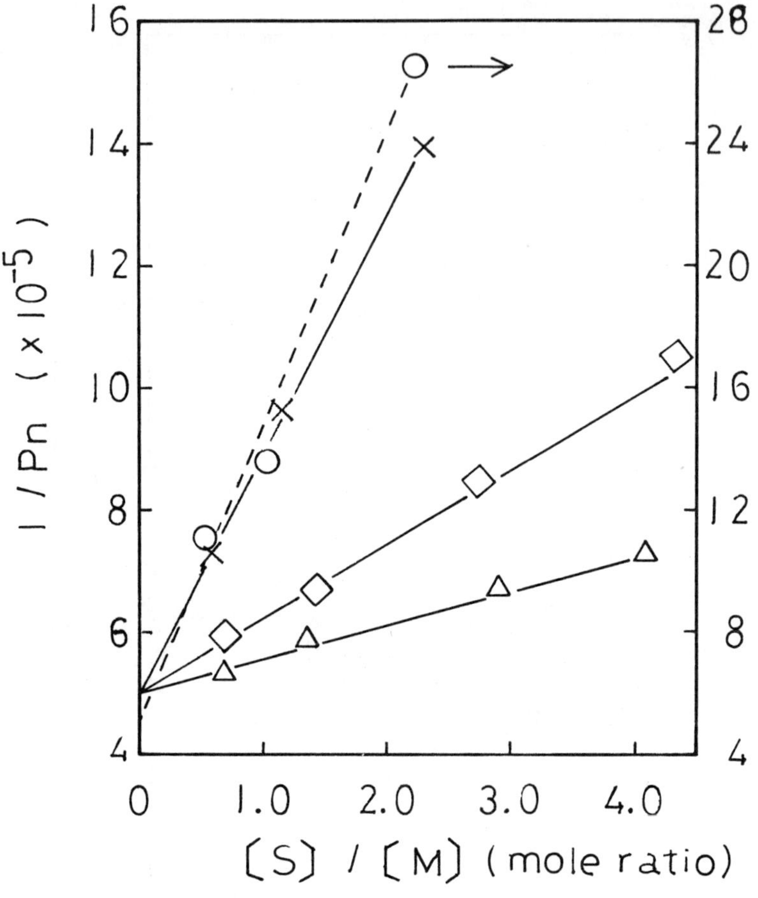

Figure 6. Plots of 1/Pn vs. [S]/[M]. Key to solvent: ○, CCl_4, 1.5 h; ×, $(CH_3)_2$-$CHCH_2OH$, 1.25 h; △, $HCON(CH_3)_2$, 1 h; ◇, $(CH_3)_2CO$, 3 h.

Table I. Chain transfer constant of solvents

Solvent	Cs ($\times 10^{-5}$) Experimental	Referencial[23]
Carbon tetrachloride	9.3	9.25
i-butanol	3.8	1.0
Acetone	1.3	1.95
Dimethyl formamide	0.7	

crylate (Fig. 7). As compared with Fig. 2, the degrees of grafting and the apparent numbers of grafted chains rapidly increased with a slight increase in carboxyl groups and became constant after 10 mmol. In this case, different from the Al-cellulose, one out of twenty-five carboxyl groups becomes a radical site (in case of quartz tube). This suggests that the initiation mechanisms of the cellulose having carboxyl groups are different from those of the cellulose having carbonyl groups. The triangle in Fig. 7 shows the results of the grafting of methyl methacrylate onto Cd-cellulose treated with sodium borohydride. The degree of grafting and the apparent number of grafted chains were lower than those of Cd-cellulose. Since carboxyl groups are not reduced with sodium borohydride, these differences are attributed to the reduction of the chain ends. Consequently, it is concluded that the grafting takes place more easily by the photolysis of dicarboxylic groups and also at the end of cellulose molecules.

Fig. 8 shows the effect of polymerization temperature on the grafting. The degree of grafting increased as the temperature and carboxyl group contents increased. The grafting efficiencies were 90-95%. The apparent activation energies of the graft copolymerization calculated by Arrhenius' plots are shown in Table II. These values are extremely smaller than that of the original radical polymerization. These support that the grafting starts with the photolysis of carboxyl groups.

Graft onto Mono-Carboxyl Cellulose. The graft polymerization of Cm-cellulose and its derivatives with methyl methacrylate are shown in Fig. 9. The degree of grafting of Cm-cellulose decreased with increasing the functional group contents. However, as the Cm-cellulose was reduced or oxidized, the degree of grafting increased, and decreased if the cellulose samples (contained 0.6 N%) had 60 and 80 mmoles functional groups. As compared with the results of Cd-cellulose (Fig. 7), the degree of grafting, the grafting efficiency and the apparent number of grafted chains were considerably low. From these results, the graft polymerization is extremely affected by the substituted position of the functional groups and also by the radical formed. Elmolenko *et al*. (24) reported that irradiation with ultraviolet on Cm- and Cd-cellulose resulted in dehydration. However the precise mechanisms are not clear, this must be studied in the future.

As described above, since the degree of grafting of Cm-cellulose containing nitro groups decreased, the grafting of nitrocellulose was investigated. The degree of grafting increased with increases of nitro group contents (Q > P-tube) and the grafting efficiencies were as high as 85-95%, as shown in Fig. 10. However, by comparing with the results of Al- and Cd-celluloses, the degree of grafting, the chain length and the numbers of grafts were low and rather near the values of Cm-cellulose. The graft reaction of C-cellulose did not occur in the same conditions and the infrared spectrum showed the decrease in nitro group

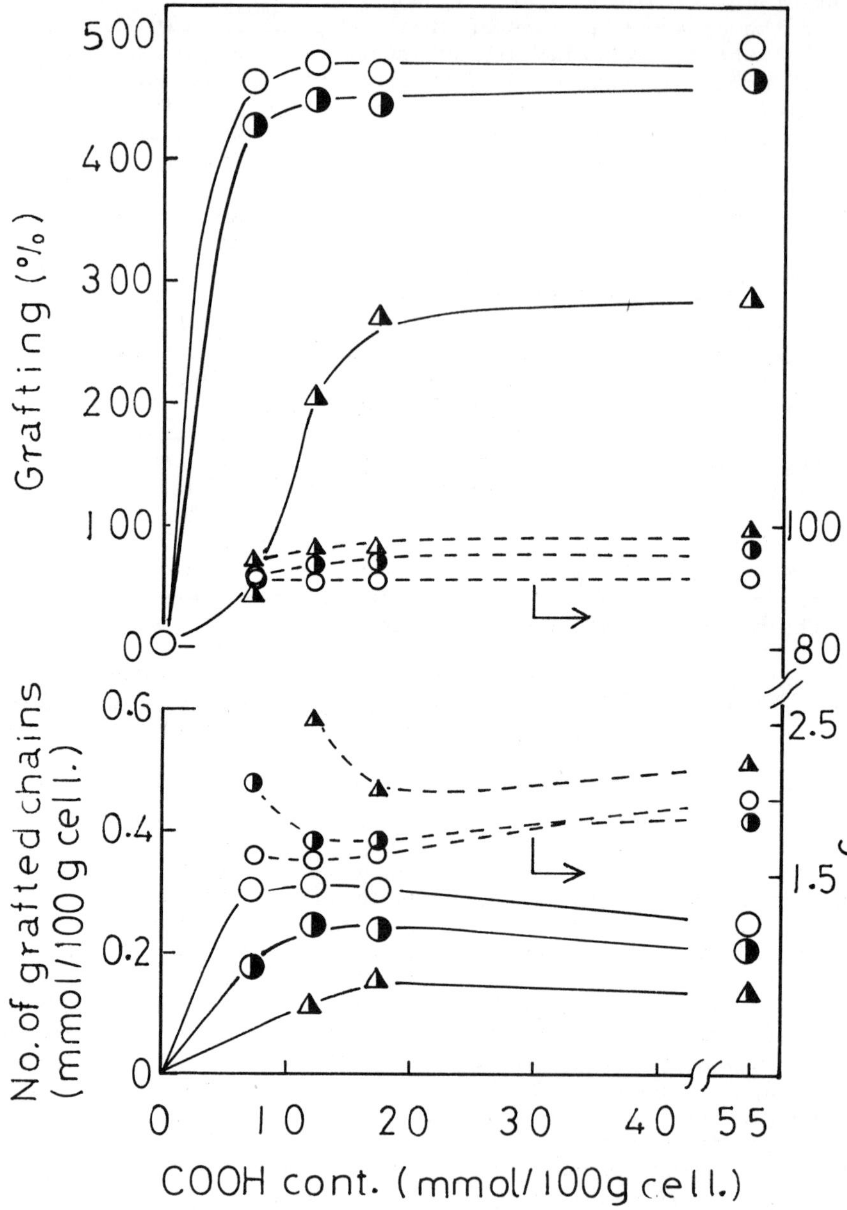

Figure 7. Effect of carboxyl group content on the yield of copolymer, molecular weight, and apparent number of grafted chains in the pMMA-grafted dicarboxyl-cellulose. Conditions: cellulose, 0.3 g; H_2O, 10 mL; MMA, 2 mL. Key (reaction tube): open mark, quartz (50°C, 1 h); half closed mark, Pyrex (40°C, 1 h).

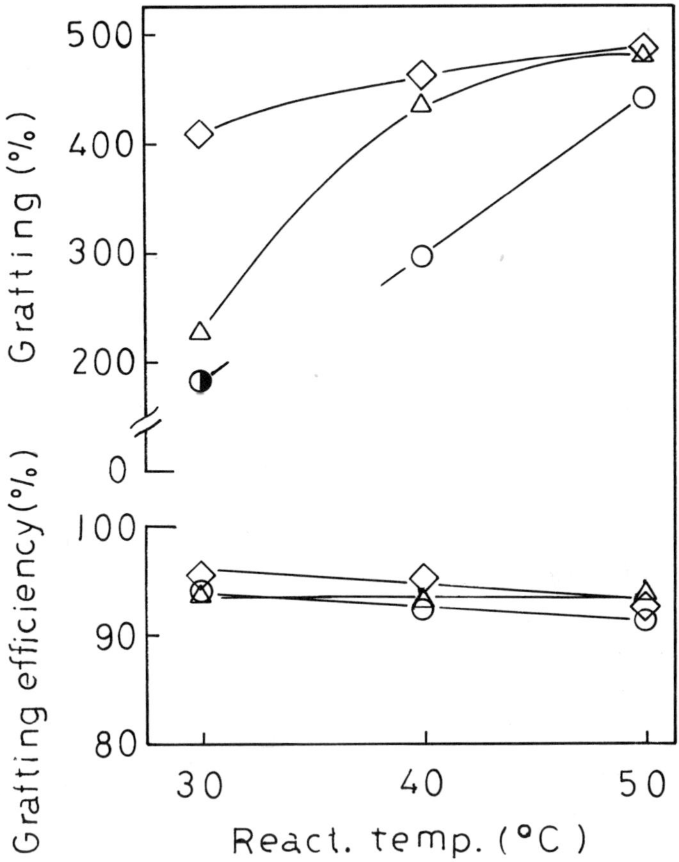

Figure 8. Effect of reaction temperature on the yield of copolymer in the pMMA-grafted dicarboxylcellulose using a Pyrex tube. Conditions: cellulose, 0.3 g; H_2O, 10 mL. Key [cellulose COOH content (mmol/100 g)]: ○, 6.8; △, 18.1; ◇, 55.0. Key (reaction time): open mark, 1 h; half closed mark, 1.4 h.

Table II. Apparent activation energies of grafting

Cellulose sample	COOH content (mmol/100g cell.)	Ea (kcal/mol)
	6.8	11.4
Dicarboxyl-	18.1	7.3
	55.0	1.1

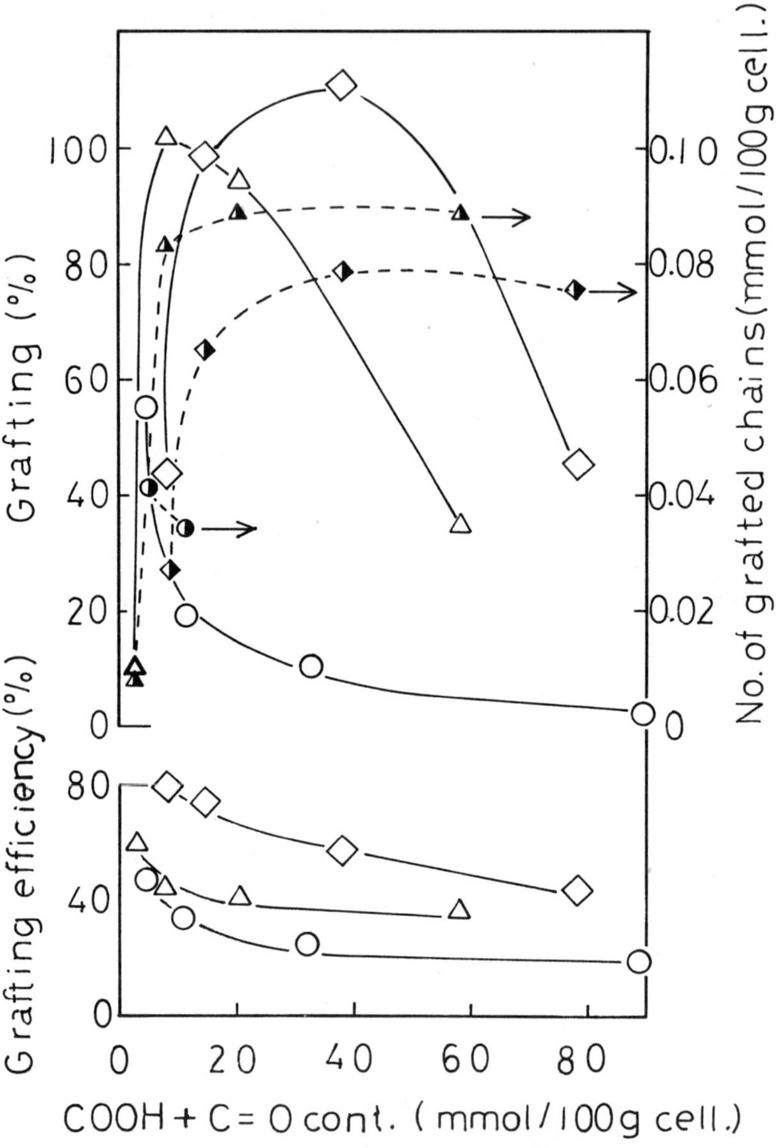

Figure 9. Effect of carboxyl and carbonyl group content on the yield of copolymer and apparent number of grafted chains in the pMMA-grafted monocarboxylcellulose derivatives using a quartz tube. Key (cellulose treatment): ○, original; △, treated with $NaBH_4$; ◇, treated with $NaClO_2$.

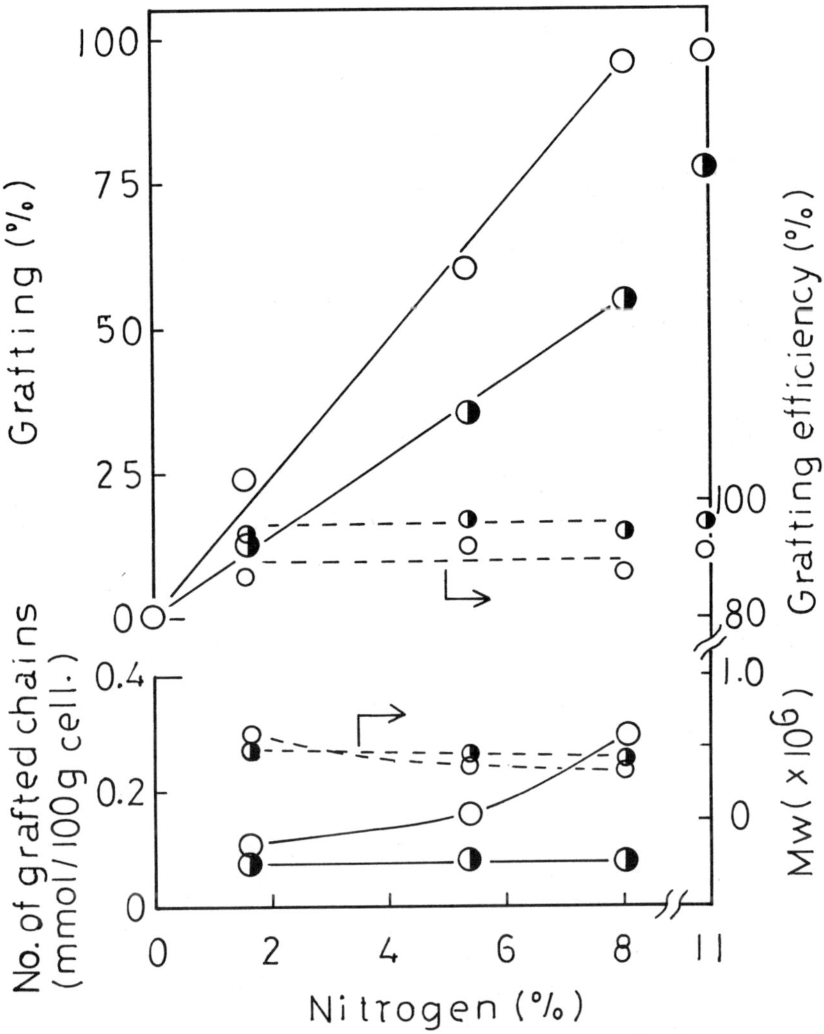

Figure 10. Effect of nitrogen content on the yield of copolymer, molecular weight, and apparent number of grafted chains in the pMMA-grafted nitrocellulose. Conditions: cellulose, 0.3 g; H_2O, 10 mL; MMA, 2 mL; at 50°C for 1 h; Key (reaction tube): ○, quartz; ◑, Pyrex.

intensities at 1640, 1280 and 840 cm^{-1} after irradiation with ultraviolet light (Fig. 11). From these results, it is apparent that the grafting of N-cellulose starts by the photolysis of nitro groups. Furthermore, the chain lengths of Al- (Fig. 2) and Cd- (Fig. 7) cellulose were 1-1.8 x 10^6 and 1.6-2.5 x 10^6, respectively. But the values of N-cellulose were extremely small, 0.3-0.6 x 10^6. These suggest that the degradative chain transfer to nitro groups may occur. The precise mechanism is now studied. Chapiro et al. (25) observed that the graft copolymerization of vinyl acetate and styrene by the radioactive methods occurred by the chain transfer to nitro groups.

Graft onto Acetylcellulose. The graft copolymerization of methyl methacrylate onto the Ac-cellulose obtained by partial hydrolysis of triacetate was studied (Fig. 12). At 30-50% acetic acid contents, the degree of grafting decreased with increasing acetyl groups or the degree of polymerization, and above 50% acetic acid content, the degree of grafting was almost zero. The degree of grafting and the apparent number of grafted chains using a Q-tube were greater than those of using a P-tube. These indicate that the graft polymerization does not proceed by the photolysis of acetyl groups, but by the scission of cellulose. Kudrna (26) reported that on irradiation with 248 nm light, the glucosidic bond in triacetylcellulose was ruptured; on irradiation with longer 248 nm light, the acetate group was split off and vinyl polymers were grafted without chain scission.

Therefore, the hydrolyzed cellulose acetate was used to examine the graft which took place at the end of celulose. Fig. 13 presents the results. In case of degree of polymerization 95 and 145 R-cellulose, increasing monomer concentration resulted in increases of the degree of grafting and the apparent number of grafted chains. But the values for degree of polymerization 162 R-cellulose were constant. The grafting efficiency was high at a low degree of polymerization.

The chain transfer constant to the polymer was determined by the following equation (27).

$$1 / Pn = 1 / Pno + Cp \times [P] / [M]$$

where [P] and [M] are the initial concentration of the trank polymer and the monomer, respectively. Cp is the chain transfer constant to the trank polymer. As shown in Fig. 14 (top), the plots of 1 / Pn and [P] / [M] show linearity. Cp are obtained by the declining of the slope, then the results are shown in Table III. The Cp values increase with decreasing the degree of polymerization of R-cellulose. These values are almost equal to the value (1.0 x 10^{-4}) obtained by the homogeneous copolymerization between the cellulose model sample and acrylonitrile (28).

Since Cp is the value summarized all the bonds in the trank polymer, the chain transfer constant to the end and the medium

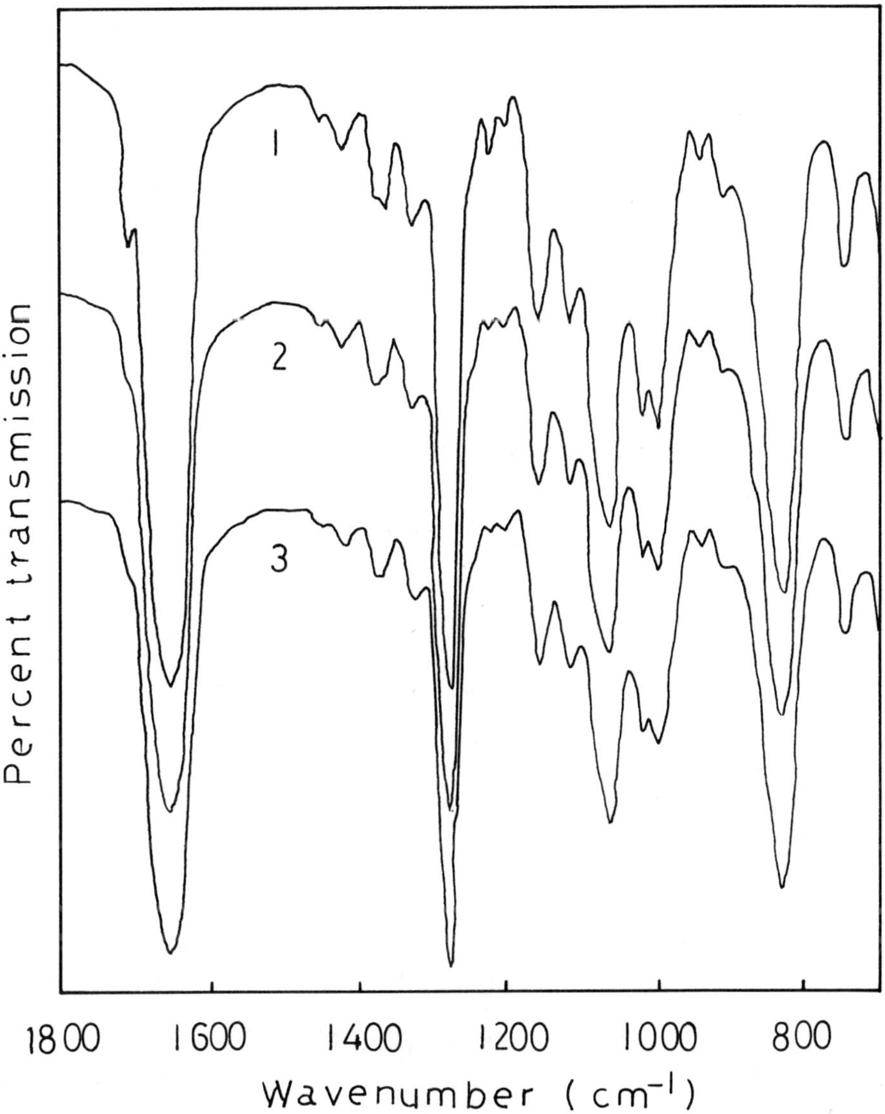

Figure 11. IR spectra of nitrocellulose by film method. Original sample 1, without hν; 2, with hν for 1 h; 3, with hν for 2 h.

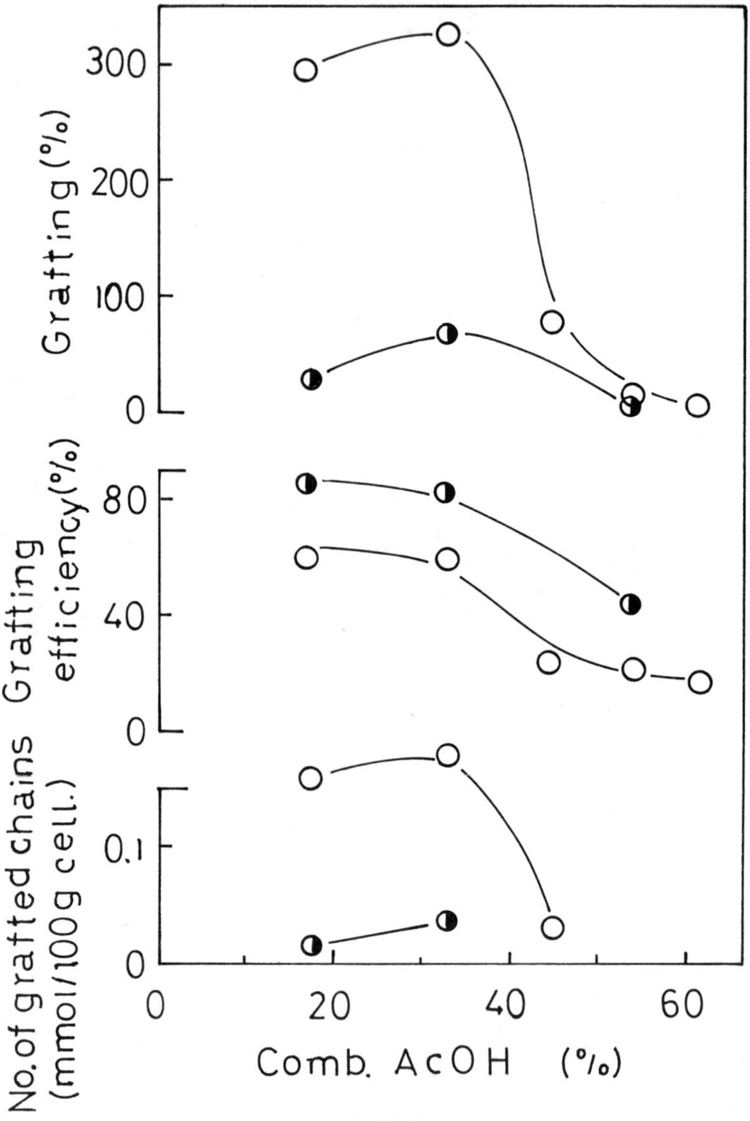

Figure 12. Effect of combined acetic acid content on the yield of copolymer and apparent number of grafted chains in the pMMA-grafted acetylcellulose. Conditions: cellulose, 0.3 g; H_2O, 10 mL; MMA, 2 mL; at 60°C for 4 h. Key (reaction tube): ○, quartz; ◑, Pyrex.

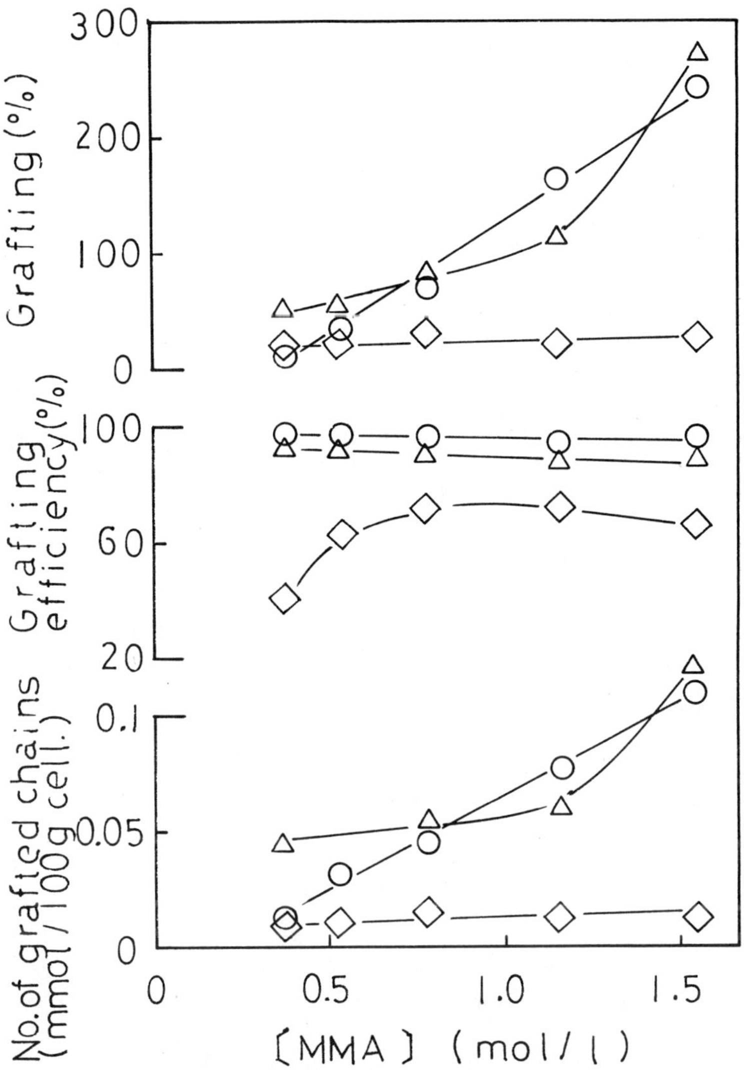

Figure 13. Effect of monomer concentration on the yield of copolymer and apparent number of grafted chains in the pMMA-grafted regenerated cellulose using a quartz tube. Conditions: cellulose, 0.3 g; H_2O, (12 — monomer) mL; at 50°C for 6.25 h. Degree of cellulose polymerization: ○, 95; △, 145; and ◇, 162.

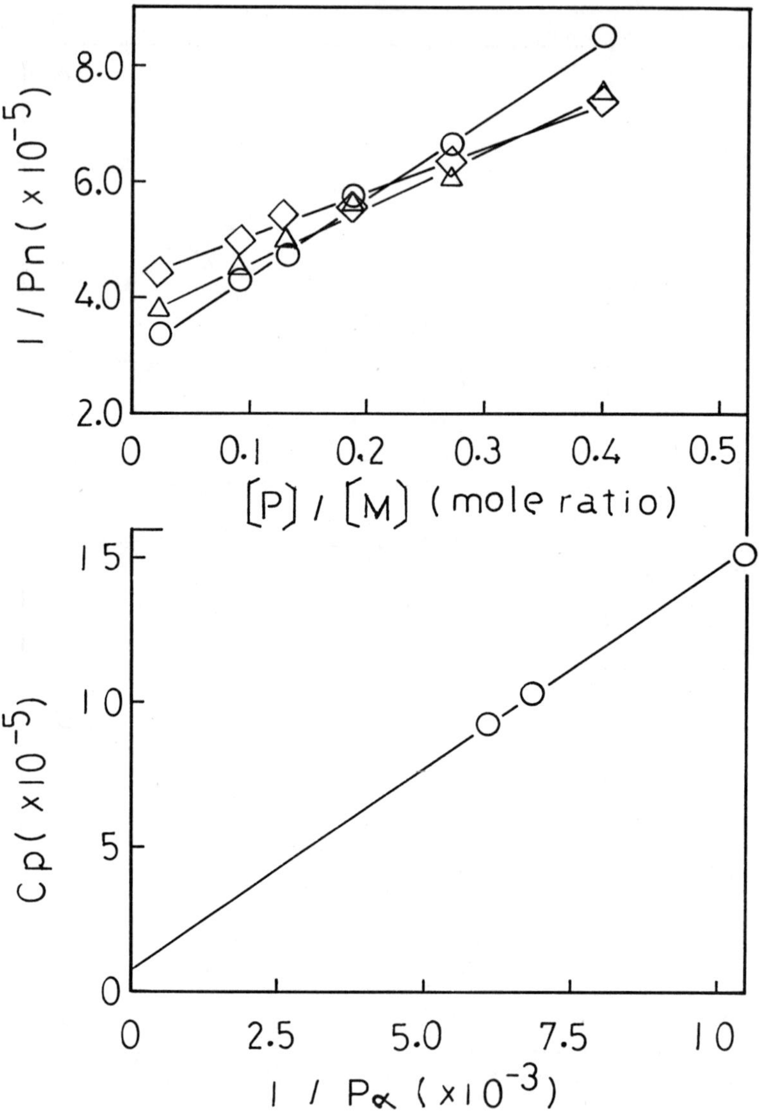

Figure 14. Top: Plots of 1/Pn vs. [P]/[M]. Bottom: Plots of Cp vs. 1/Pα. Degree of cellulose polymerization: ○, 95; △, 145; and ◇, 162.

Table III. Chain transfer constant of trank polymers

Cellulose sample	Degree of polymerization	Cp ($\times 10^{-4}$)	C_{PE} ($\times 10^{-2}$)	C_{PM} ($\times 10^{-5}$)
	95	1.5		
Regenerated-	145	1.0	1.4	0.8
	162	0.9		

also were examined to calculate by the following equation (27).

$$Cp = C_{PM} + C_{PE} \times 1 / P\alpha$$

where C_{PE} and C_{PM} are the chain transfer constant to the end and the medium of the trank polymer. Pα is the degree of polymerization of the trank polymer. As shown in Fig. 14 (bottom), the relation of Cp to 1 / Pα shows a straight line. C_{PE} and C_{PM} are determined by the declining of the line and the intersection of Y axis, respectively. These results are given in Table III. It is concluded that using a R-cellulose with a low degree of polymerization, the grafting more easily proceeds to the end than the medium of the trank polymer.

Literature Cited

1. Geacintov, N.; Stannett, V.: Abrahamson, E. W.; Hermans, J. J. Appl. Polym. Sci. 1960, 3, 54.
2. Kubota, H.: Ogiwara, Y. Ibid. 1972, 16, 337.
3. Ogiwara, Y.: Kubota, H.: Murata, Y. Ibid. 1974, 18, 3455.
4. Kubota, H.: Ogiwara, Y. Ibid. 1970, 14, 2879.
5. Kadooka, M.: Tanaka, M.: Murata, J. Soc. Polym. Sci. Jap. 1965, 22, 107.
6. Takahashi, A.: Takahashi, S. J. Soc. Fiber Sci. & Tech. Jap. 977, 33, T508.
7. Takahashi, A.: Omata, T.; Takahashi, S. Ibid. 1979, 35, T196.
8. Takahashi, A.: Takahashi, S. Soc. Polym. Sci., Jap. 1980, 37, 151.
9. Takahashi, A.: Takahashi, S. J. Soc. Fiber Sci. & Tech. Jap. 1980, 36, T397.
10. Whistler, R. L., Ed. "Methods in Carbohydrate Chemistry"; Academic: New York, 1963; Vol. III; p165.
11. Whistler, R. L., Ed. Ibid. p182.
12. Whistler, R. L., Ed. Ibid. p168.

13. Yamamoto, J.: Miyata, T. J. Chem. Soc. Jap., Ind. Chem. Sect. 1959, 62, 109.
14. Davidson, G. F.; Nevell, T. P. J. Textile Inst. 1948, 39, T102.
15. Whistler, R. L., Ed. "Methods in Carbohydrate Chemistry; Academic: New York, 1963; Vol. III; p103.
16. Alexander, W. J.; Mitchell, R. L. Anal. Chem. 1949, 21, 1498.
17. Takahashi, A.; Takahashi, S. J. Chem. Soc. Jap., Ind. Chem. Sect. 1971, 74, 2541.
18. Ide, F. Ibid. 1961, 64, 1489.
19. Chinai, S.; Matlack, J.; Resnik, A. J. Polym. Sci. 1955, 17, 39.
20. Sugahara, Y.; Takahashi, A.; Takahashi, S. J. Soc. Fiber Sci. & Tech. Jap. 1981, 37, T187.
21. Ogiwara, Y.; Hayase, K.; Kubota, H. J. Appl. Polym. Sci. 1979, 23, 1.
22. Otsu, T. "Radical Polymerization"; Kagaku Dozin: Tokyo, 1971; Vol. I; p126.
23. Otsu, T. Ibid. p128.
24. Ermolenko, I. N.; Savastenko, G. N. Vysokomol. Soedin., Ser B. 1975, 17, 538.
25. Chapiro, A.; Fowx, J.; Jendrychowska-Bonamour, A. M. Eur. Polym. J. 1977, 13, 637.
26. Kudrna, Stanislav. Chem. Abstr. 1972, 76, 5049b.
27. Otsu, T. "Radical Polymerization"; Kagaku Dozin; Tokyo, 1971; Vol. I; p149.
28. Toda, T. J. Polym. Sci. 1962, 58, 411.

RECEIVED December 24, 1981.

10

Acid Effect in UV- and Radiation-Induced Grafting of Styrene to Cellulose

CHYE H. ANG, JOHN L. GARNETT and STAN V. JANKIEWICZ
The University of New South Wales, Kensington, N.S.W., Australia 2033

DAVID SANGSTER
Australian Atomic Energy Commission, Lucas Heights, N.S.W., Australia 2033

Further work on the mechanism of the acid enhancement in both UV and gamma ray grafting is reported. The copolymerisation of styrene in methanol and dioxan to cellulose is used as model system. Acid enhances grafting and homopolymer formation in both radiation systems. Analysis of the homopolymer from the grafting runs indicates that acid reduces the chain length, but increases the numbers, of grafted chains. Variables affecting the homopolymer yields in acid have been examined and include radiation dose, dose-rate and concentration of monomer in solvent. Water is shown to have a different effect to acid. $\bar{M}_n$ values on all homopolymers from the above runs are lower in acid than in neutral solution. The significance of the molecular weight data in grafting and homopolymerisation is discussed especially with respect to possible commercial applications of the graft copolymers of cellulose.

UV and ionising radiation are useful initiators for the grafting of a variety of monomers to cellulose and its derivatives (1 - 11). Methods for enhancing the yield of copolymerisation are important especially with cellulose which is a backbone polymer sensitive to radiation decomposition (1,12). Thus at 10^6 rads, irradiated cellulose exhibits carbonyl group formation, carboxyl group formation and chain cleavage in the approximate ratio of 20:1:1. At slightly above 10^8 rads cellulose becomes completely soluble in 1% NaOH. It is therefore preferable to keep the total radiation dose to a minimum when radiation grafting to cellulose and any additives which will accelerate the copolymerisation would be useful.

Recently inclusion of mineral acid in the grafting solution was found to enhance the grafting yields for a typical system such as the copolymerisation of styrene in methanol to cellulose in the presence of gamma rays (13). The acid effect now appears

0097-6156/82/0187-0141$6.00/0
© 1982 American Chemical Society

to be a relatively general phenomenon in these reactions, being applicable in the grafting of a wide range of monomers to a variety of natural and synthetic trunk polymers (14,15,16). Analogous acid enhancement effects have also been observed in UV grafting reactions (15,17), especially those processes which are accelerated by sensitisers. However, the mechanism of the acid effect, particularly for the gamma ray grafting of monomers, such as methyl methacrylate, to polyolefins (18,19) is controversial. The initial suggestion was that the acid enhancement effect was due to increased G(H) yields from reaction between the increased H^+ concentration in the solution with thermalised, secondary electrons from the primary radiolysis act (13,14). Other workers have proposed that dissolved oxygen and even traces of water may be contributing factors in the phenomenon (18).

In the present work, further mechanistic data for the acid effect in both UV and gamma radiation systems are reported, particularly molecular weight studies on the species present in the grafting solutions. The significance of these acid effects in possible commercial applications of the resulting copolymers of cellulose is also briefly mentioned.

Experimental

For the grafting experiments, methanol (Allied Corp., acetone-free, ACS reagent grade) and 1,4-dioxan (analytical grade) were used as solvents. The mineral acids were analytical reagent grade whilst styrene was donated by Monsanto (Aust.) Ltd. Inhibitor was removed from monomer by washing with dilute sodium hydroxide. Two methods for subsequently purifying the styrene were adopted. In the first, styrene was distilled under reduced pressure, the middle 80% fraction collected, stored at -2°C on fused calcium chloride, then redistilled under reduced pressure immediately prior to use. In the second procedure, styrene was purified by passing down an alumina column (4). Monomer obtained from both methods was checked for purity by gas chromatography. Because of the possible problem from trace impurities present in methanol, particularly, this solvent was fractionated twice in the presence of 2,4-dinitrophenylhydrazine and concentrated sulfuric acid, the fractionation being performed in a dry atmosphere as previously outlined (20).

In the radiation grafting work, experiments were performed in spent fuel element and cobalt-60 facilities of the Australian Atomic Energy Commission, dose rates being determined by ferrous sulfate dosimetry (G(Fe)= 15.6). For the specific irradiations, which were carried out in quadruplicate, small strips (5 x 4cm) of cellulose (Whatman No. 41, W.R. Balston Pty. Ltd., double acid-washed chromatography grade) were equilibrated at 65% relative humidity for 24 h at 23°C, folded into test tubes (16 x 1.2 cm), monomer solution (6ml) added to the tubes which were then lightly stoppered. After irradiation, the strips were washed with cold methanol to remove residual acid and monomers. The strips were then extracted in benzene for 100 h dried, humidified to 65% relative humidity at 23°C for 24 h and weighed. Removal of acid in cold solvent prior to extraction is essential otherwise any residual acid present attacks the cellulose during hot solvent treatment leading to decomposition of the grafted polymer.

For the UV grafting experiments, a Philips 90W high pressure mercury vapour lamp fitted with a quartz envelope was utilised. Sample preparation was similar to the ionising radiation method except that tubes with larger volumes (25ml) of grafting solution containing sensitiser were used. Tubes were held in a rotating rack and were generally irradiated at a distance of 12cm from the source. Actinometry was performed by using the uranyl nitrate-oxalic acid system.

For radiolysis studies of the grafting solutions without cellulose, i.e. homopolymer runs, tubes were either lightly stoppered or sealed off at 10^{-3} Torr after three freeze-thaw cycles. For the molecular weight determinations on the homopolymers, oligomer solution was poured into excess methanol, the precipitated polystyrene collected, dried and analysed by g.p.c. on a Waters Associates Model ALG/GPS201 instrument.

Results

Typical acid enhancement effects in the grafting of styrene to cellulose in the presence of gamma rays are shown in Figure 1. Methanol which wets and swells the backbone polymer is a good solvent for the polymerisation whereas the longer chain alcohol, *n*-octanol, is a poor grafting solvent for cellulose. This finding is consistent with previous experimental observations (4,14). Inclusion of mineral acid in the methanol monomer solution leads to a marked increase in grafting yield at all styrene concentrations studied, both at the Trommsdorff peak and at concentrations well removed from this region. If *n*-octanol is added to methanol, the mixed solvent (50/50 v/v) gives improved grafting yields when compared to either of the two individual alcohols. Addition of acid to the mixed solvent systems enhances the polymerisation up to 35% monomer concentration at the radiation dose and dose-rate studied.

In like manner, inclusion of mineral acid in the photosensitised grafting of styrene in methanol to cellulose enhances the polymerisation yields significantly (Figure 2). The results obtained with phosphoric and sulfuric acids were consistent with earlier data (13,14). Dioxan is also a useful solvent for these reactions when uranyl ion is used as sensitiser. However dioxan is not as efficient as methanol for the grafting reaction. The grafting in dioxan exhibits a Trommsdorff peak at 60% monomer concentration. The results in Figure 2 also show that the unsensitised process does occur in dioxan as in other solvents, but

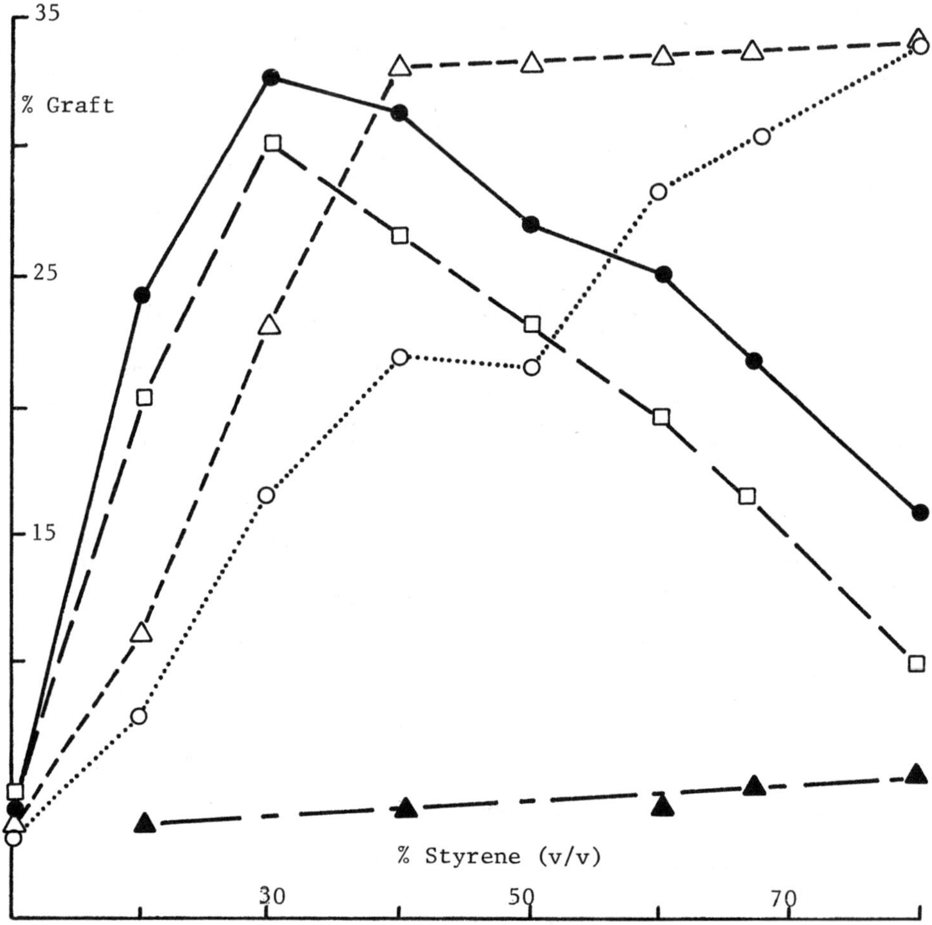

Figure 1. Gamma ray grafting of styrene in alcohols to cellulose in neutral and acid solutions. Dose, 0.2 × 10^6 rad at 4 × 10^4 rad/h. Key: ▲, octanol; ○, methanol; △, methanol/1-octanol(1:1); ●, methanol + 1% H_2SO_4 (v/v); □, methanol/octanol (1:1) + 1% H_2SO_4.

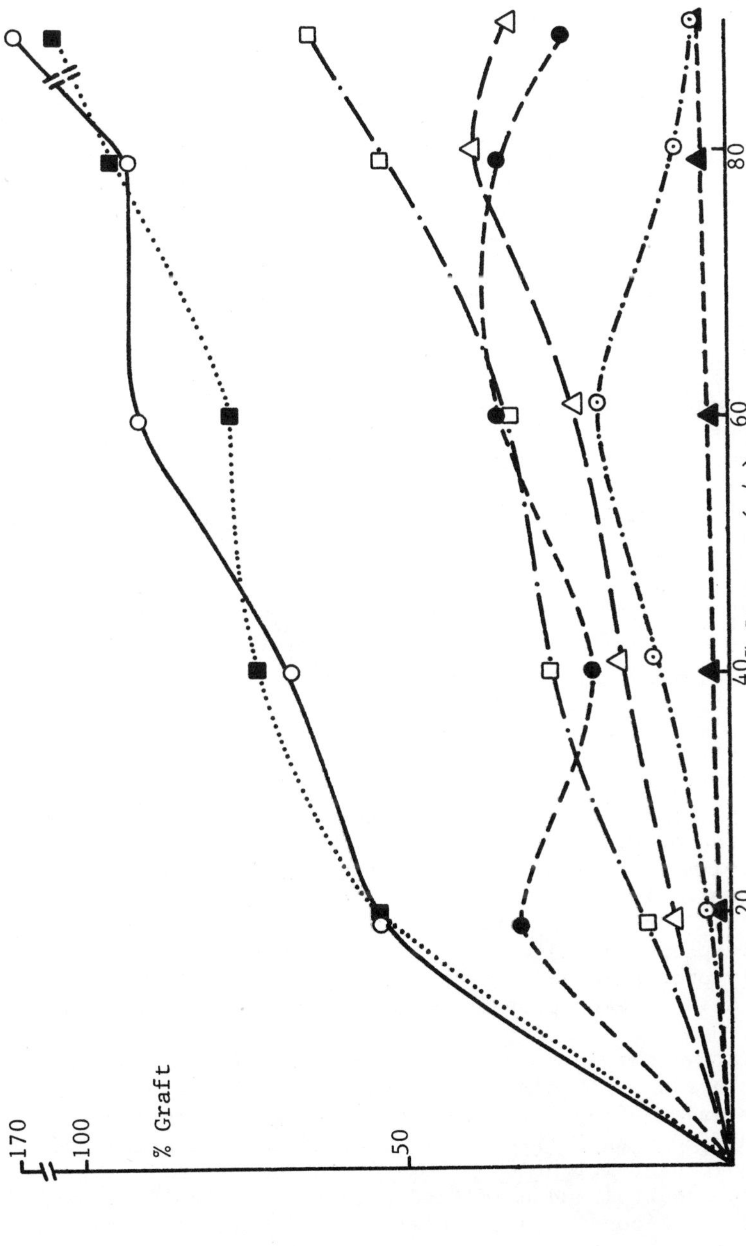

Figure 2. Effect of uranyl nitrate (1% w/v) as sensitizer on UV grafting of styrene to cellulose in neutral and acid media. Time 24 h at 24 cm from 90 W high pressure UV lamp. Key: ○, *graft in methanol + sensitizer +* H_3PO_4 *(1% w/v);* ■, *graft in methanol + sensitizer +* H_2SC_4 *(1% w/v);* □, *graft in methanol + sensitizer;* △, *homopolymer in methanol + sensitizer;* ●, *homopolymer in methanol + sensitizer +* H_3PO_4 *(1% w/v);* ▲, *graft in dioxane, no sensitizer;* ⊙, *graft in dioxane + sensitizer.*

is markedly less efficient unless uranyl ions or organic sensitisers are present.

Homopolymer Formation. The data in Figure 2 show that in the UV system homopolymer formation is competitive with grafting. A similar situation exists when ionising radiation is the initiator and acid enhances the homopolymer yields as well as the grafting (15). Furthermore, in preliminary studies (15), it has been found that in the presence of a trunk polymer such as cellulose, i.e. grafting conditions, the yield and molecular weight of the homopolymer formed is unaffected under the experimental radiation conditions used. Since homopolymer yields are independent of the presence of cellulose, the homopolymer reaction can be studied by following the simple radiation polymerisation of styrene in an appropriate solvent. Acid effects can also be examined in such homopolymer systems and the significance of the data extrapolated to analogous grafting reactions. For the present work, dioxan is a suitable solvent since (i) it solubilises polystyrene and the monomer solution remains homogeneous during irradiation and (ii) it is a useful solvent for grafting styrene to cellulose. It is thus possible to investigate the conditions under which common intermediates may exist in both grafting and homopolymerisation processes. The results in the following Tables I - VIII thus concern acid effects in the radiation homopolymerisation of styrene in dioxan.

Variables Affecting Conversion to Polystyrene. The conversion of styrene in dioxan (30% w/w) reaches a maximum at approximately 0.1M H_2SO_4 (Table 1). Beyond this acidity, the yield of hompolymer is not significantly increased under the radiation conditions used. At acid concentrations in excess of 0.7M there is a fall off in yield of polystyrene which may be attributed to the lower solubility of polystyrene at higher acid levels so that continued polymerisation is hampered by increasing immiscibility. The data in the same table show that the effect of water on the homopolymerisation is different from that of acid. The polymer conversion is also favored at the lower monomer concentrations (Table II) and is enhanced by the presence of acid at each monomer concentration studied. The conversion is also directly proportional to the total radiation dose (Table III) and inversely proportional to the dose rate (Table IV). The addition of acid in each case enhanced the resulting yields of homopolymer.

Molecular weight data ($\overline{M}_n$ values) for the above runs are particularly useful. Thus $\overline{M}_n$ values decrease with increasing acidity, the major effect of acid inclusion again occurring in the 0 - 0.1M H_2SO_4 range (Table V). Inclusion of acid does not change the molecular weight profiles of these samples. Polymers collected from neutral and acidified solution all exhibit normal Gaussian distributions in their molecular weights. In the

TABLE I. Effect of Acid and Water on Conversion of Styrene in Dioxan to Polystyrene[a]

Reagent Conc. (M)	Conversion (% w/w) H_2O	H_2SO_4
0	37	37
0.1	-	50
0.3	43	52
0.5	-	51
0.7	-	50
1.0	-	43

[a] Styrene in dioxan (30% w/w). Total dose 6.5×10^6 rad at 4.0×10^4 rad/hr.

TABLE II. Effect of Styrene Concentration on the Conversion of Monomer to Polymer in Presence of Sulfuric Acid[a]

Styrene Concentration (% v/v)	Conversion (% w/w) Neutral	0.1M H_2SO_4
20	10.9	14.9
40	10.5	14.8
60	9.7	14.1
80	8.9	12.3
90	7.8	11.0

[a] Dioxan solution. Total dose of 1.6×10^6 rad at dose rate of 3.3×10^4 rad/hr at 23°C.

TABLE III. Effect of Total Dose on the Conversion of Styrene Monomer to Polymer in Presence of Sulfuric Acid[a]

Total Dose (10^6 rad)	Conversion (% w/w) Neutral	0.1M H_2SO_4
0.18	1.4	2.1
0.94	7.8	9.8
1.14	9.5	12.4
1.90	14.5	18.8
2.86	19.9	27.0
3.82	25.8	34.2
4.78	33.0	42.0

[a] Styrene in dioxan (30% w/w). Dose rate of 4.0×10^4 rad/hr at 23°C.

TABLE IV. Effect of Dose Rate on Conversion of Monomer to Polymer in Presence of Sulfuric Acid.[a]

Dose Rate (10^3 rad/hr)	Conversion (% w/w) Neutral	$0.1\ H_2SO_4$
90	12.4	16.9
52	16.2	21.1
33	19.8	27.7
24	24.2	33
18.5	35.6	46.4
9.1	44	53.2

[a] Styrene in dioxan (30% w/w). Dose of 2.15 x 10^6 rad at 26°C.

TABLE V. Effect of Acid and Water on $\overline{M}_n$ Molecular Weight of Polymer Formed from Styrene in Dioxan.[a]

H_2SO_4 (M)	$\overline{M}_n$	H_2O (M)	$\overline{M}_n$
0.0	15,300	0.0	15,300
0.1	11,300	-	-
0.3	10,700	0.3	14,500
0.5	10,600	-	-
0.7	10,100	-	-
1.0	8,000	-	-

[a] Styrene in dioxan (30% w/w). Total dose of 6.5 x 10^6 rad at 4.0 x 10^4 rad/hr.

TABLE VI. Effect of Styrene Concentration on $\overline{M}_n$ of Polymer Formed from Styrene in Presence of Sulfuric Acid.[a]

Styrene Concentration (% v/v).	$\overline{M}_n$ Neutral	$0.1\ M\ H_2SO_4$
20	17,900	13,050
40	37,450	29,550
60	58,600	52,350
80	84,450	78,200
90	112,550	92,000

[a] Dioxan solution. Total dose of 1.6 x 10^6 rad at dose rate of 3.3 x 10^4 rad/hr at 23°C.

presence of equimolar amounts of water, the $\bar{M}_n$ values do not decrease to the same extent as with acid, suggesting different mechanistic pathways for the effect of the two additives. As the concentration of styrene in the solution increases, the $\bar{M}_n$ values increase (Table VI). With the total dose, $\bar{M}_n$ reaches a maximum at 2 x 10^6 rad (Table VII) and is inversely proportional to dose-rate (Table VIII). Inclusion of acid for each of the runs in Tables VI to VIII leads to a lowering of the corresponding $\bar{M}_n$ value. The increased yields of polymer of lower molecular weight has a pronounced effect on the rate of viscosity increase for acidified samples compared with neutral solutions. After exposure to 7.5 x 10^6 rad delivered at 4 x 10^4 rad/hr samples containing H_2SO_4 in the range 0.3-0.7M increased 800% on starting viscosity whereas the neutral reference increased by only 450%.

Discussion

Mechanism of Acid Effect in Radiation Grafting. In earlier studies (13,14,15,21) on the mechanism of the acid effect on radiation grafting of styrene (MH) in dioxan or methanol (SH) to cellulose(Cell-OH)the sequence of events depicted in Equations (1) - (9) was proposed to explain the phenomenon. Thus, cellulose can undergo bond

$$\text{Cell-OH} \longrightarrow \text{Cell-O}^{\bullet} + \text{H} \quad (1)$$
$$\text{SH} \longrightarrow \text{S}^{\bullet} + \text{H} \quad (2)$$
$$\text{Cell-OH} + \text{H} \longrightarrow \text{Cell-O}^{\bullet} + \text{H}_2 \quad (3)$$
$$\text{Cell-OH} + \text{S}^{\bullet} \longrightarrow \text{Cell-O}^{\bullet} + \text{HS} \quad (4)$$
$$\text{Cell-OH} + \text{H}^{+} \longrightarrow \text{Cell-OH}_2^{+} \quad (5)$$
$$\text{Cell-OH}_2^{+} + e \longrightarrow (\text{Cell-O}\overset{}{\text{H}})^{*} + \text{H} \quad (6)$$
$$(\text{Cell-OH})^{*} \longrightarrow \text{Cell-O}^{\bullet} + \text{H} \quad (7)$$
$$\text{SH} + \text{H} \longrightarrow \text{SH}_2^{+} \quad (8)$$
$$\text{SH}_2^{+} + e \longrightarrow \text{SH} + \text{H} \quad (9)$$

rupture by direct irradiation effects yielding sites for grafting (Equation 1). Hydrogen abstraction reactions involving radiolytic fragments from the solvent can also activate the backbone polymer (Equations 2 - 4). Hydrogen ions from the added acid can equilibrate with the cellulose to form ionic species which can capture thermalised electrons from the primary radiation act to give excited cellulose molecules and H atoms (equations 5 and 6). Decomposition of the excited cellulose molecules yields radical sites for grafting and additional H atoms (Equation 7). In the presence of methanol or dioxan solvent, additional hydrogen atoms can be formed by reactions depicted in Equations 8 and 9. Overall the acid effect was therefore attributed to increased G(H) yields in acidified solvent upon irradiation, leading to enhanced H atom scavenging and the creation of more grafting sites in the trunk polymer by abstraction reactions. Radical scavenger styrene can then react with the above species leading to the observed acid enhancement in both grafting and homopolymerisation processes.

TABLE VII. Effect of Total Dose on $\overline{M}_n$ Molecular Weight of Polystyrene Produced in Presence of Sulfuric Acid.[a]

Total Dose (10^6 rad)	$\overline{M}_n$ Neutral	0.1 M H_2SO_4
0.18	22,900	19,800
0.94	23,500	19,500
1.14	23.400	19,550
1,90	24,900	18,100
2.86	25,100	18,200
3,82	22,150	16,640
4.78	21,000	16,400

[a] Styrene in dioxan (30% v/v). Dose rate of 4.0×10^4 rad/hr at 23°C.

TABLE VIII. Effect of Dose Rate on $\overline{M}_n$ Molecular Weight of Polystyrene Produced in Presence of Sulfuric Acid.[a]

Dose Rate (10^3 rad/hr)	$\overline{M}_n$ Neutral	0.1M H_2SO_4
90	17,800	13,600
52	21,500	17,360
33	25,500	22,900
24	30,200	26,300
18.5	33,800	24,800
9.1	41,100	33,300

[a] Styrene in dioxan (30% v/v). Total dose of 2.15×10^6 rad at 26°C.

In more recent fundamental studies on the radiolysis of binary mixtures of aromatic compounds with methanol (22), an increase in scavenging products was observed in the presence of added acid. Extrapolated to the styrene - dioxan (or methanol) system, this would mean an increase in concentration of styrene-dioxan (or methanol) intermediates (MS•) in acid solutions. Combined with the increase in G(H) in the presence of acid, leading to more grafting sites by abstraction reactions, the increase in MS• species should lead to an enhancement in both graft and homopolymer at certain monomer concentrations. In terms of the charge-transfer model for grafting (13,14), the binary mixture study also suggests that acid may affect the stability of the intermediate complex formed in the copolymerisation leading to possible changes in the subsequent mechanistic pathway for the formation of copolymer and homopolymer, especially involving reactions with solvent fragments.

Significance of Present Homopolymerisation Data. The results from the current homopolymerisation experiments can now be used to refine the above mechanisms for homopolymerisation and, by extrapolation, to grafting processes also. The fact that inclusion of cellulose (or other trunk polymer) in the monomer solution (i.e. grafting conditions) does not alter the $\overline{M}_n$ values can be used to advantage since there is no need to incorporate cellulose in the subsequent experiments involving studies of the variables affecting the molecular weights of the oligomers in the grafting solution. It is particularly important from the homopolymer runs in the tables that acid not only consistently increases the conversion to polymer, it also reduces significantly the $\overline{M}_n$ values in all experiments studied in Tables V-VIII. The oligomer formed in acid solution therefore contains shorter chains and, to explain the higher conversions, must be present in greater numbers. Such a conclusion would be consistent with an enhancement in G(H) yields with inclusion of acid (Equations 6,7 and 9), leading ultimately to much higher radical concentrations in the reaction solution. Higher radical concentrations would not only directly increase the extent to which polymerisation occurs but would also achieve a greater incidence of radical termination by combination. The end result is significantly greater numbers of shorter chains. The interplay between these two factors is responsible for the observed enhancement in rate of viscosity increase in the monomer solution containing acid during irradiation.

This explanation for the acid effect on the radiation polymerisation of styrene solutions has a direct application in a grafting context in that higher G(H) yields or lowered effective concentrations of radical scavenging impurities will enhance existing hydrogen abstraction reactions from the substrate to create potential grafting sites as well as increasing the concentration of oligomer chains in the bulk of the solution. Because there is a dynamic equilibrium between bulk monomer solution and monomer solution absorbed by trunk polymer, any change in properties of the oligomer in the bulk solution must affect the properties and nature of the graft. Thus although the oligomer chains are shorter in acid grafting solutions, the numbers of these chains are higher and, because of the smaller size of these chains, the radical intermediates associated with them can diffuse more readily into the swollen trunk polymer to give enhanced yields owing to higher concentration and higher mobilities associated with their smaller size. The lower $\bar{M}_n$ and larger numbers of shorter chains in the presence of acid are also consistent with the induced and enhanced Trommsdorff peaks observed in radiation grafting (13,14), since the lower molecular weights lead to a high solubility of oligomer radicals in monomer solution, resulting in an increase in viscosity at the Trommsdorff peak and a reduction in chain termination, leading to a marked acceleration in grafting. This interpretation would also explain why the gel effect in radiation polymerisation is strongly dependent on structure of solvent, those solvents that are strong hydrogen atom donors in a radiation chemistry context being the most efficient in accelerated grafting. Finally in the current styrene-dioxan system, the data show that the roles of water and acid are significantly different, an important conclusion in view of the analogous controversy in the methyl methacrylate - methanol grafting solution (14,18,19,23).

Mechanism of the Acid Effect in UV Grafting and Homopolymerisation. It is of interest that acid also enhances both the grafting and homopolymer yields under certain experimental conditions for the analogous photosensitised copolymerisation of styrene in methanol and/or dioxan to cellulose. Preliminary $\bar{M}_n$ experiments (15) on these UV homopolymers, similar to the data in Table V, indicate that acid affects the $\bar{M}_n$ values significantly, in some instances reducing them to even lower values than in the corresponding ionising radiation work. However the acid effect in UV grafting appears to be more complicated than in the gamma ray system. Thus acid itself can affect cellulose in two ways (i) producing both intercrystalline and intracrystalline swelling which loosens the "order-disorder" structure of the cellulose making it more accessible to reagents and (ii) acting as a catalyst in the hydrolysis of the cellulose leading to uncoiling of the chains and improved monomer accessibility. Such hydrolysis reactions involve an intermediate complex between the glycosidic oxygen and the proton. In the presence of UV, breakage of the glycosidic bond of the complex is facilitated leading to additional grafting sites.

In the sensitised UV copolymerisation, radical pathways similar to the ionising radiation process are available for grafting (Equations (10-12,4,UO_2^{2+} being representative sensitiser).

Thus grafting sites can be formed by

$$UO_2^{2+} + h\nu \longrightarrow (UO_2^{2+})^* \quad (10)$$

$$(UO_2^{2+})^* + Cell\text{-}OH \longrightarrow UO_2^{+} + H^+ + Cell\text{-}O^\bullet \quad (11)$$

$$(UO_2^{2+})^* + SH \longrightarrow S^\bullet + H^+ + UO_2^{+} \quad (12)$$

intermolecular hydrogen abstraction of the cellulose with excited sensitiser molecules or solvent radicals. Because of the residual water in cellulose (~7% at ambient conditions) acid enhancement effects in photosensitised grafting may be observed by processes similar to the sulfuric acid catalysed photopolymerisation of monomers in aqueous solution (Equations 13 and 14) where in the presence of acid additional radicals are formed.

$$SO_4^{2-} + H_2O \xrightarrow{h\nu} SO_4^{\bar{\cdot}} + H^\bullet \quad (13)$$

$$SO_4^{\bar{\cdot}} + H_3O^+ \longrightarrow H_2SO_4 + OH^\bullet \quad (14)$$

Importance of Acid Effects in Radiation and UV Induced Grafting and Homopolymerisation. In preparative terms, it is significant that in both the simple polymerisation of monomers such as styrene and the grafting to cellulose by the same monomers in the presence of gamma rays or UV, acid can enhance yields of products. Thus, in the presence of acid, lower total radiation doses are required to achieve a particular percentage graft, therefore the trunk polymers are exposed to lower radiation doses resulting in less degradation. Molecular weight studies show that the oligomer chains are shorter in acid grafting solutions, but the number of chains is high. This leads to lower $\bar{M}_n$ values for homopolymers and would be expected to affect the length of the grafted chains in cellulose. This point is being clarified at present by digesting the grafted celluloses and analysing the synthetic copolymer component. The inclusion of acid in both homopolymer and grafting systems thus possesses the additional advantage of yielding homopolymers and copolymers with potentially different structures and properties to the corresponding materials from neutral solution. The acid effect thus increases the spectrum of polymers and copolymers available for industrial applications due to structural differences.

Acknowledgement

The author wishes to thank the Australian Institute of Nuclear Science and Engineering, the Australian Atomic Energy Commission and the Australian Research Grants Committee for the support of this research.

Literature Cited

1. Chapiro, A., "Radiation Chemistry of Polymeric Systems", Interscience: New York, 1962.
2. Krassig, H.A.; Stannett, V.T. Advan. Polymer Sci., 1965, 4, 111.
3. Arthur, J.C. Jr. Advan. Chem. Ser., 1971, 99, 321.
4. Dilli, S., Garnett, J.L.; Martin, E.C., Phuoc, D.H. J. Polymer Sci. C, 1972, 37, 57.
5. Guthrie, J.T.; Haq, Z. Polymer, 1974, 15, 133.
6. Geacintov, N.; Stannett, V.T.; Abrahamson, E.W.; Hermans, J.J. J. Appl. Polymer Sci., 1960, 3, 54.
7. Arthur, J.C.Jr. Polymer Preprints, 1975, 16, 419.
8. Oster, G.; Yang, N.L. Chem. Rev., 1968, 68, 125.
9. Kubota, H.; Murata, T.; Ogiwara, T. J. Polym. Sci., 1973, 11, 485.
10. Tazuke, S.; Matoba, T.; Kimura, H.; Okada, T. ACS Symp. Ser. 121 1980, 217.
11. Davis, N.P.; Garnett, J.L.; Urquhart, R. J. Polymer Sci Polymer Lett Ed., 1976, 14, 537.
12. Bosworth, R.C.L.; Ernst, I.T.; Garnett, J.L. Int. J. Appl. Rad. Isotopes, 1961, 11, 152.
13. Dilli, S.; Garnett, J.L.; Phuoc, D.H. Polymer Lett., 1973, 11, 711.
14. Garnett, J.L. J. Rad. Phys. Chem., 1979, 14, 847.
15. Garnett, J.L.; Jankiewicz, S.V.; Sangster, D.F. J. Rad. Phys. Chem., in press.
16. Hoffman, A.S.; Ratner, B.D. J. Rad. Phys. Chem., 1979, 14, 831.
17. Garnett, J.L.; Leeder, J.D. ACS Symp. Ser. 121, 1980, 243.
18. Pinkerton, D.M.; Stacewicz, R.H. J. Polym. Sci. Polymer Lett. Ed., 1976, 14, 287.
19. Chappas, W.J.; Silverman, J. J. Rad. Phys. Chem., 1979, 14, 847.
20. Ekstrom, A.; Garnett, J.L. J. Chem. Soc. A., 1968, 2416.
21. Garnett, J.L.; Phuoc, D.H.; Airey, P.L.; Sangster, D.F. Aust. J. Chem., 1976, 29, 1459.
22. Fletcher, G.; Garnett, J.L. unpublished work.
23. Garnett, J.L.; Yen, N.T. J. Polymer Sci. Polymer Lett Ed., 1974, 12, 225.

RECEIVED January 28, 1982.

11

Graft Copolymer of Cellulose Nitrate

K. S. V. SRINIVASAN and D. N.-S. HON
Virginia Polytechnic Institute and State University, Department of Forest Products, Blacksburg, VA 24061-0299
M. SANTAPPA
Central Leather Research Institute, Adyar, Madras, India

Grafting of vinyl monomers onto cellulose nitrate in heterogeneous and homogeneous media initiated by ceric ions, and benzoyl peroxide and azobisisobutyronitrile, respectively, have been studied. Isolation of graft copolymers from homopolymers and unreacted cellulose nitrate was conducted by using selective solvent extraction technique. The graft copolymers were characterized by infrared spectroscopy, gel permeation chromatography, viscosity measurements, and nuclear magnetic resonance. The effects of various grafting parameters, viz., initiator concentration, monomer concentration, reaction time and the substrate concentration on grafting efficiency and degree of grafting were examined. Probable grafting reaction mechanisms for heterogeneous and homogeneous media are proposed.

Methods for graft copolymerization of vinyl monomers onto cellulose initiated by ultraviolet irradiation (1-3), high energy radiation (4-9), redox and oxidation reactions (10-12), and mechanoreactions (13-15) have been extensively investigated in different parts of the world. However, grafting of vinyl monomers onto cellulose derivatives as a means of altering properties of the base polymer has attracted little attention until recently. The graft copolymerization of methyl methacrylate and acrylonitrile onto methyl cellulose in the absence of radical inhibitors (16) and by the persulphate ion (17), and of styrene onto cellulose acetate (18), has been reported. Modification of cellulose nitrate has not been reported until recently in the literature despite its versatile application in surface coating and leather industries (19-22).

Cellulose nitrate film cast from solution is highly brittle and is made flexible for surface coating applications by the addition of plasticizers like dibutylphthalate, dioctylphthalate,

0097-6156/82/0187-0155$7.00/0
© 1982 American Chemical Society

etc. (23,24). The drawback with these plasticizers is that they tend to migrate on aging, resulting in brittleness of the film. We have envisaged that one of the best means of altering the properties of cellulose nitrate could be by grafting of vinyl monomers onto cellulose nitrate. By proper choice of vinyl monomers, the desired flexibility of the cellulose nitrate film could be achieved and migration of molecules could be prevented because the copolymers are chemically grafted to the cellulose nitrate. In addition to this property, grafting of proper vinyl monomers could impart high gloss, water repellency, and degradation from UV light to cellulose nitrate film. This paper deals with the grafting of monomers to cellulose nitrate initiated by ceric ammonium nitrate in heterogeneous medium and by benzoyl peroxide and azobisisobutyronitrile in homogeneous media and the characterization of the grafted products.

Experimental

Materials. Commercially available cellulose nitrate (½ sec)(CN) with 11.8-12.2% nitrogen content supplied by Ashahi Chemicals and Hercules, Inc., was purified by washing with water and drying under vacuum. Methyl methacrylate monomer supplied by Rohm and Haas was freed from inhibitor by washing with sodium hydroxide solution, drying over anhydrous sodium sulphate, and distilling under vacuum. Ceric ammonium nitrate (CAN) reagent grade was used without purification. A stock solution of 0.1N ceric solution in 1N nitric acid was prepared and stored in a refrigerator prior to the experiment. Benzoyl peroxide (BPO), azobisisobutyronitrile (AIBN), methanol, benzene, petroleum ether, and methyl isobutyl ketone were A.R. grade and used without purification.

Methods

Ceric Ion-Initiated Graft Copolymerization. The requisite amount of cellulose nitrate was dispersed in 300 ml of water in a Waring blender after allowing the polymer to swell for 30 minutes. The heterodispersed solution was then transferred to a three-necked flask equipped with a glass stirrer and a nitrogen inlet. The required quantity of monomer and the initiator were added, and then the purified nitrogen was purged into the solution for 30 minutes. The polymerization was allowed to proceed for a specified length of time at 30°C. Hydroquinone was added at the end of the reaction time to arrest the polymerization. The grafted products were separated by filtration and were washed with distilled water repeatedly.

Benzoyl Peroxide and Azobisisobutyronitrile-Initiated Graft Copolymerization. Cellulose nitrate (CN) (10 g) was dissolved in methyl isobutyl ketone (90 ml) in a three-necked flask and to this solution methyl methacrylate (10 ml) and benzoyl peroxide (1 g) were added; dry nitrogen was passed through the solution for 30 minutes. The nitrogen was then stopped and the reaction vessel was immersed in a thermostatic bath maintained at 70°C, and the polymerization was carried out with gentle stirring. After 3 hours, the contents were poured into an excess of petroleum ether (b.p. 60-80°C) to precipitate the copolymer.

Viscosity. The viscosity measurements of the PMMA graft chains obtained by hydrolysis of graft copolymer were carried out in benzene at 30°C using an Ubbelhode viscometer, and the number average molecular weight of the grafted chain was calculated using the equation(25):

$$[\eta] = 8.96 \times 10^{-5} M_n^{0.76}$$

The definitions of grafting parameters are summarized as follows:

$$\text{Percent Grafting} = \frac{\text{Wt. of (CN-g-PMMA)} - \text{wt. of CN}}{\text{Wt. of NC}} \times 100$$

Grafting Efficiency (%) =

$$\frac{\text{Wt. of PMMA grafted}}{\text{Wt. of PMMA grafted} + \text{Wt. of homopolymer (PMMA)}} \times 100$$

Number of Grafting Sites =

$$\frac{\text{Wt. of PMMA grafted/Mol. wt. of grafted PMMA}}{\text{Wt. of CN used/Mol. wt. of CN}}$$

Infrared Spectra (IR) - Infrared spectra of cellulose nitrate, poly (methyl methacrylate) and cellulose nitrate-g-poly(methyl methacrylate) films were recorded using a Perkin-Elmer 337 grating IR spectrophotometer.

Nuclear Magnetic Resonance (NMR) - NMR spectra of cellulose nitrate-g-poly(methyl methacrylate) and cellulose nitrate were recorded in deuterated acetone using Perkin-Elmer-R-32 (90 MHz).

Gel Permeation Chromatography (GPC): GPC of all samples were recorded with a Waters Associates Model 244 high-pressure liquid chromatograph using a dilute solution (0.02% in tetrahydrofuran and a flow rate of 5 ml/min). The chromatograph was connected with four microstyragel columns in series (pore size 10^6, 10^5, 10^4, 10^3 Å).

Results and Discussion

An initial trial run of experiments was carried out to graft various monomers (styrene, vinyl acetate, acrylic acid, acrylamide, methyl methacrylate, methyl acrylate) onto cellulose nitrate initiated by ceric ions. No graft copolymerization was found to occur with styrene, vinyl acetate, acrylamide and acrylic acid, though grafting did occur with methyl methacrylate and methyl acrylate. Between these monomers, higher grafting percentage and grafting efficiency were observed for methyl methacrylate than methyl acrylate. Brauer and Termini (26) have also observed the absence of grafting with many monomers onto collagen (protein) when ceric ions were used as the initiator. This could be due to the specific reactivity of the monomer towards the specific backbone.

Isolation and Characterization of Grafted Products. The product obtained by the above copolymerization contains the backbone polymer (cellulose nitrate), graft copolymer, and homopolymer. Solvent extraction procedure was employed for isolation of the graft copolymer. The products separated by filtration were soxhlet extracted for 72 hours with benzene to remove poly(methyl methacrylate) homopolymer. The remaining products were again soxhlet extracted with methanol for 72 hours to remove unreacted cellulose nitrate. The remaining product is hence a true graft copolymer. This is one of the few instances in graft copolymerization where complete separation of the individual polymer can be achieved.

The infrared spectra of cellulose nitrate and the cellulose nitrate-g-poly(methyl methacrylate) are shown in Figure 1. Figure 1a shows the infrared spectrum of cellulose nitrate characterized by a broad medium band at 3400 cm^{-1} due to the hydroxyl group of cellulose nitrate and perhaps a trace amount of moisture in the sample, a weak absorption at 2900 cm^{-1} indicating a small number of C-H linkages, and 1280 cm^{-1} and 1655 cm^{-1} characteristic of covalent $-ONO_2$ asymetric and symmetric stretching vibrations, respectively. Figure 1b shows the IR spectrum of pure graft copolymer completely separated from poly(methyl methcrylate) and cellulose nitrate backbone. It shows a small peak for the >C=O group at 1740 cm^{-1} indicating that attachment of poly(methyl methacrylate) homopolymer to nitrocellulose backbone took place. This graft copolymer was hydrolyzed with 6N HCl at 115°C for 48 hrs to separate poly(methyl methacrylate)-grafted chains from cellulose nitrate. The IR spectrum of this poly(methyl methacrylate) gave a more pronounced >C=O peak at 1740 cm^{-1} (Figure 2). A physical blend of cellulose nitrate and poly(methyl methacrylate) after soxhlet extraction with appropriate solvent for poly(methyl methacrylate), however, shows no peak at 1740 cm^{-1} due to >C=O group. This further substantiates the formation of a true graft copolymer.

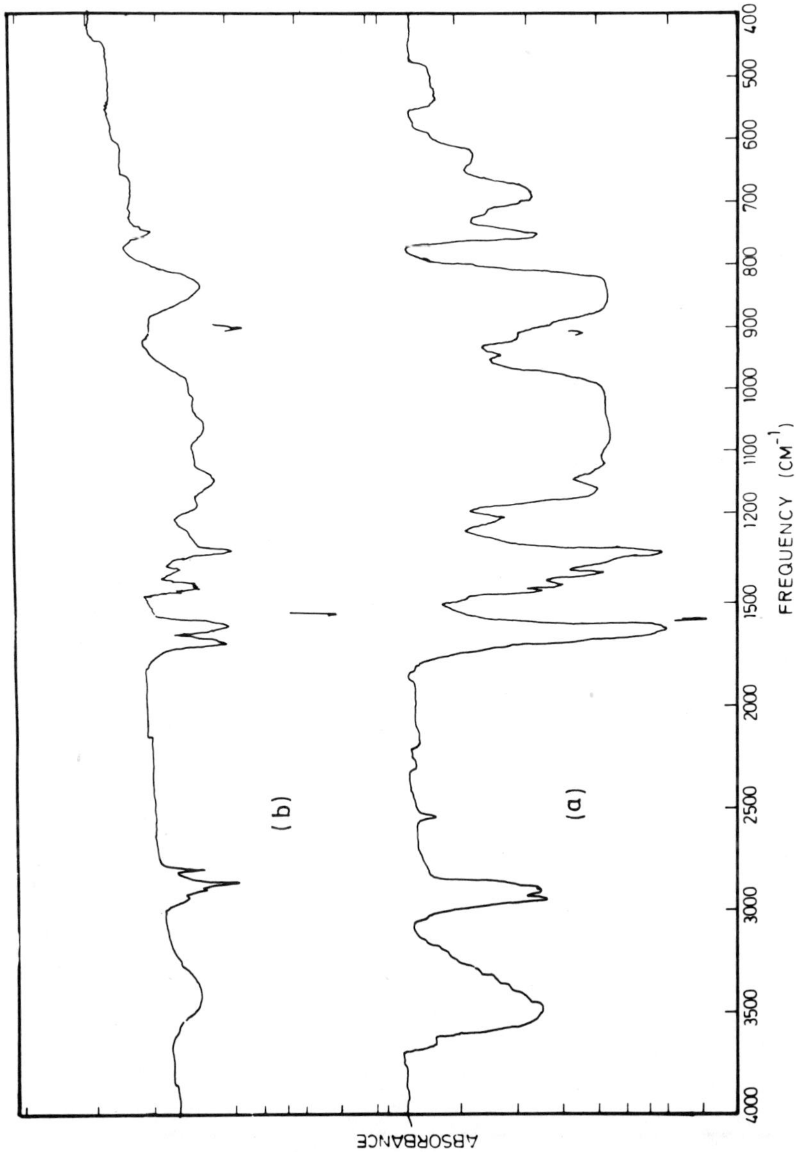

Figure 1. IR spectra of a, cellulose nitrate and b, cellulose nitrate-g-poly(methyl methacrylate).

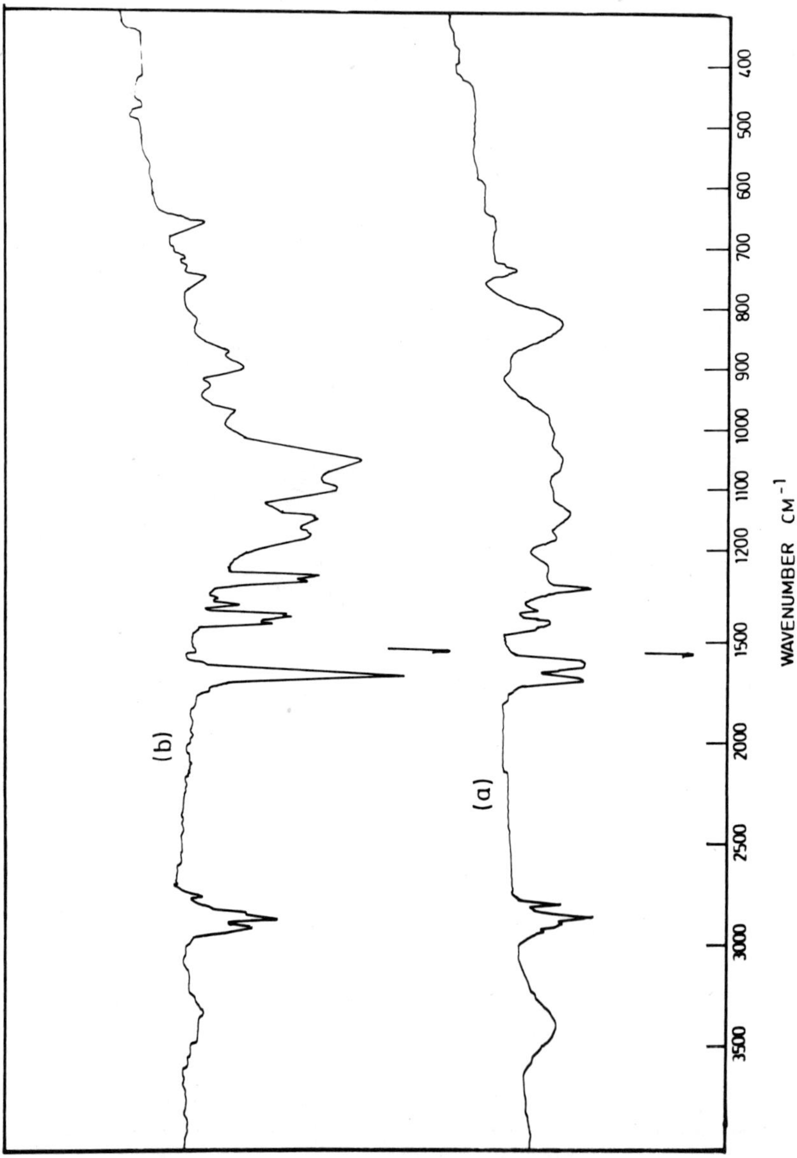

Figure 2. IR spectra of a, cellulose nitrate-g-poly(methyl methacrylate) and b, cellulose nitrate-g-poly(methyl methacrylate) after hydrolysis. (Reprinted, with permission, from Ref. 20. Copyright 1978, Wiley.)

The nuclear magnetic resonance spectra of cellulose nitrate and of cellulose-g-poly(methyl methacrylate) after separation from cellulose nitrate and poly(methyl methacrylate), are shown in Figures 3 and 4, respectively. The graft copolymer shows the presence of αCH_3 proton at 1.0δ , indicating the presence of poly(methyl methacrylate) chains attached to cellulose nitrate.

Gel permeation chromatography (GPC) of poly(methyl methacrylate) and cellulose nitrate showed elution volume peaks at 62.5 ml for PMMA and at 87.5 for cellulose nitrate (Figure 5), due to their difference in molecular weight. A mixture of poly(methyl methacrylate) and cellulose nitrate of the same ratio as that of the graft copolymer was recorded and two peaks in elution volume at almost identical positions were observed. This shows that the constituent homopolymers retain their identity in a physical mixture. The isolated graft copolymer showed a single peak in elution volume at 80.0 ml. The second peak in elution volume is absent in spite of poly(methyl methacrylate) attached to cellulose nitrate as revealed by infrared spectrum. Hence, these results indicate that GPC can be used as a technique to differentiate between homopolymer, physical mixture, and graft copolymer.

Grafting Parameters

Effect of Variables on Grafting Initiated by Ceric Ions

Effect of Grafting Time. The effect of grafting time on molecular weight of the grafted chain and grafting percentage is given in Table I and Figure 6. In the stages of graft copolymerization (up to 75 min), the ceric ion initiates a large number of growing PMMA branches. This is reflected by an increase in the grafting percentage up to 75 min; with grafting time greater than this, the grafting percentage remains more or less constant. The number average molecular weight of the grafted chains also increased up to 75 min and then remained constant.

Effect of Initiator Concentration. The effect of initiator concentration on the grafting percentage and on the number average molecular weight of the grafted chains is shown in Table II and Figure 7. A significant increase in grafting percentage was achieved when CAN concentration was 1.7 x 10^{-3} mol/1 and at higher concentration the degree of grafting decreased slightly. The number average molecular weight of the grafted chains also increased up to this concentration and further increase in concentration caused a decrease in number average molecular weight of the grafted chains. The probable reason for this is that a greater number of grafting sites were created by increase in the ceric ion concentration.

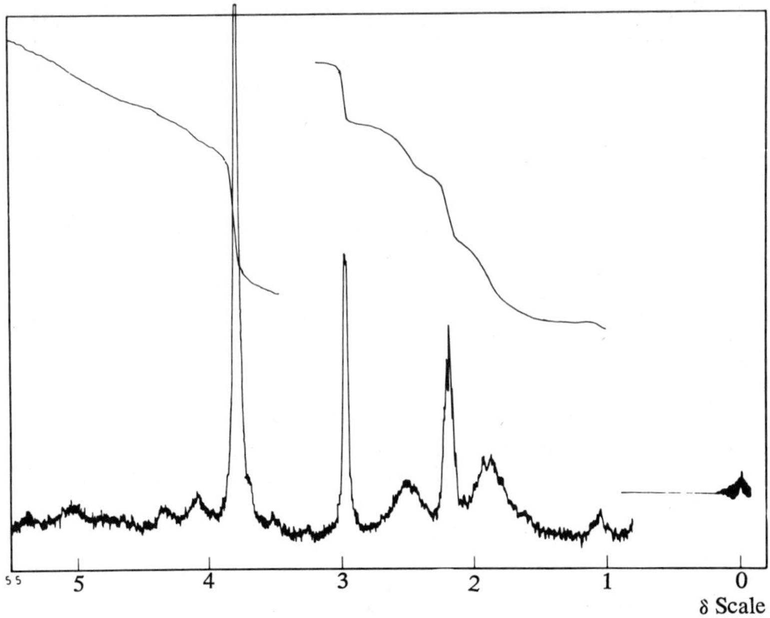

Figure 3. NMR spectrum of cellulose nitrate.

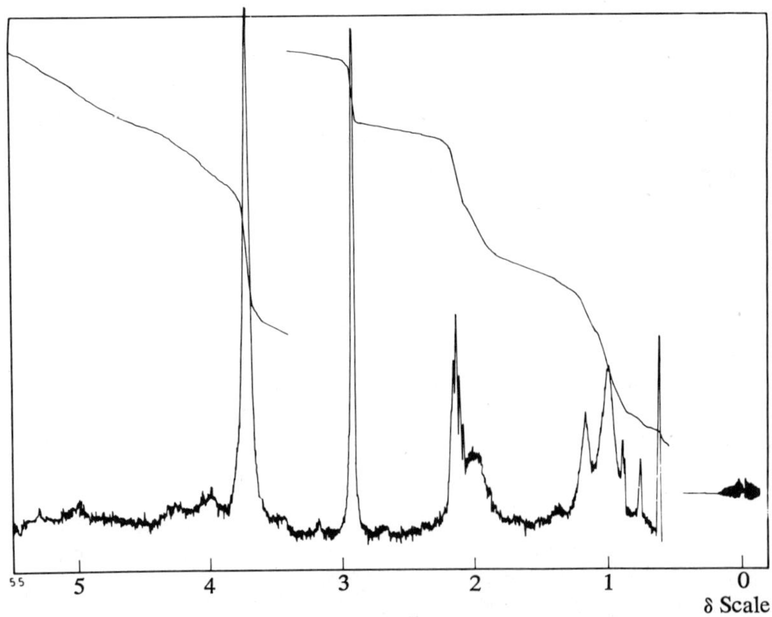

Figure 4. NMR spectrum of cellulose nitrate-g-poly(methyl methacrylate).

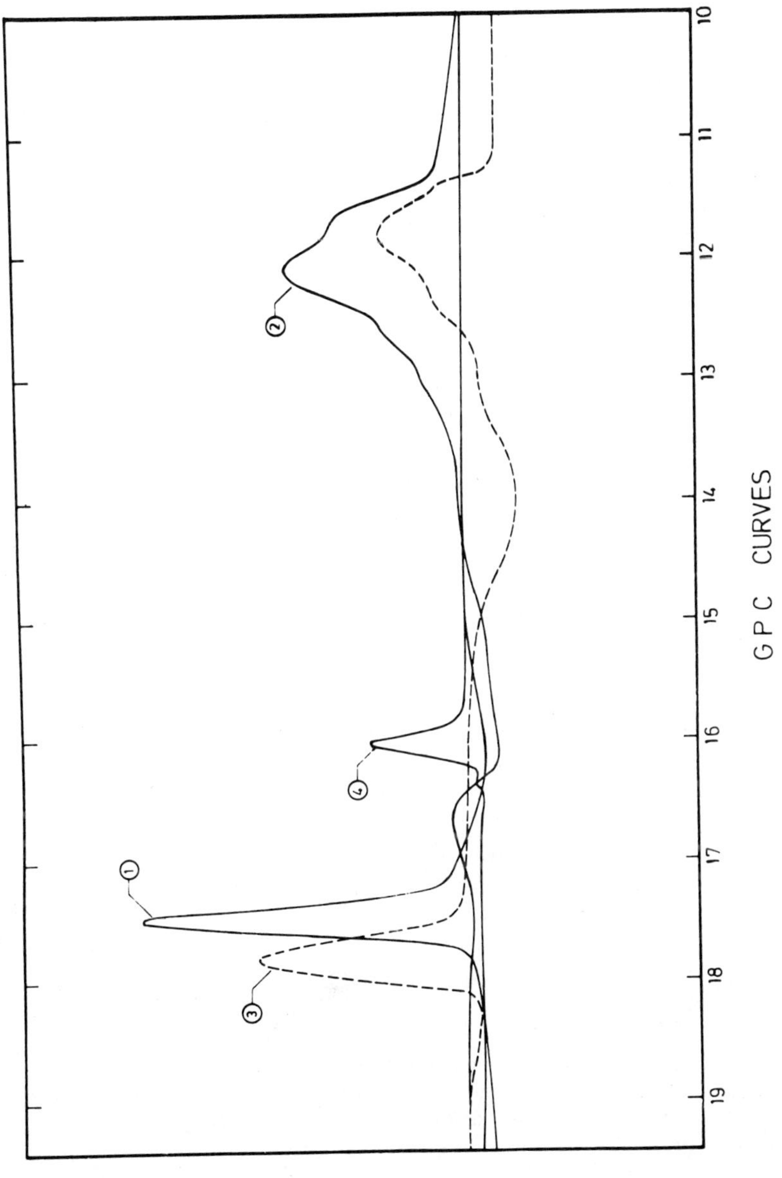

Figure 5. GPC curves of 1, cellulose nitrate; 2, poly(methyl methacrylate); 3, physical mixture of cellulose nitrate and poly(methyl methacrylate); and 4, graft copolymer of cellulose nitrate-g-poly-(methyl methacrylate). (Reprinted, with permission, from Ref. 21.

Copyright 1980, Springer-Verlag.)

Table I. Effect of Grafting Time on Percentage Grafting and Molecular Weight

Time	Grafting %	Mol. Wt. of Grafted Chains $M_n \times 10^5$	No. of Grafting Sites mole/mole
30	10.0	0.7	0.019
75	25.0	1.5	0.021
120	21.7	1.7	0.017
180	22.0	1.7	0.017
300	25.0	1.8	0.017

Grafting Conditions: Cellulose nitrate - 6 g; [MMA] - 0.25 mol/l; [CAN] - 1.7 x 10^{-3} mol/l; Temperature - 30°C; Total volume - 300 ml. (Reprinted with permission from reference 19.)

Table II. Effect of Initiator Concentration on Percent Grafting and Molecular Weight

No.	[CAN] mol/l x 10^{-3}	Grafting %	Molecular Weight $M_n \times 10^{-5}$	No. of Grafting Sites mole/mole
1	1.0	10.0	0.8	0.016
2	1.7	25.0	1.5	0.021
3	2.7	23.3	1.3	0.023
4	3.3	23.3	1.2	0.025

Grafting Conditions: Cellulose nitrate - 6 g; Time - 75 min.; [MMA] - 0.25 mol/l; Total volume - 300 ml; Grafting temperature - 30°C. (Reprinted with permission from reference 19.)

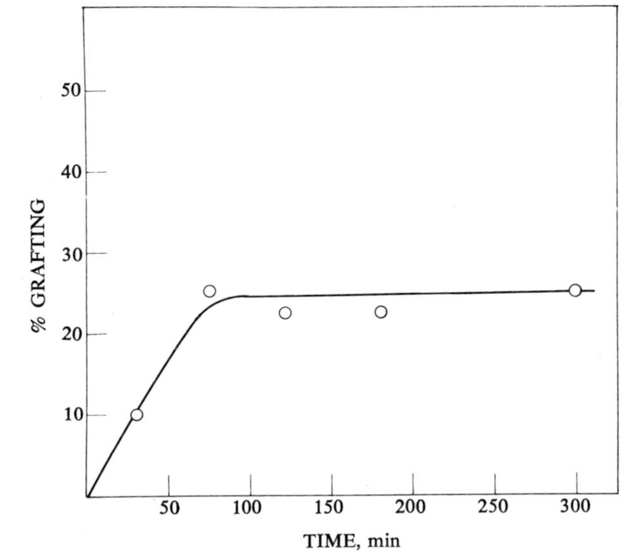

Figure 6. Effect of grafting time on the graft copolymerization of methyl methacrylate onto cellulose nitrate with ceric ammonium nitrate as initiator. (Reprinted, with permission, from Ref. 19. Copyright 1979, Wiley.)

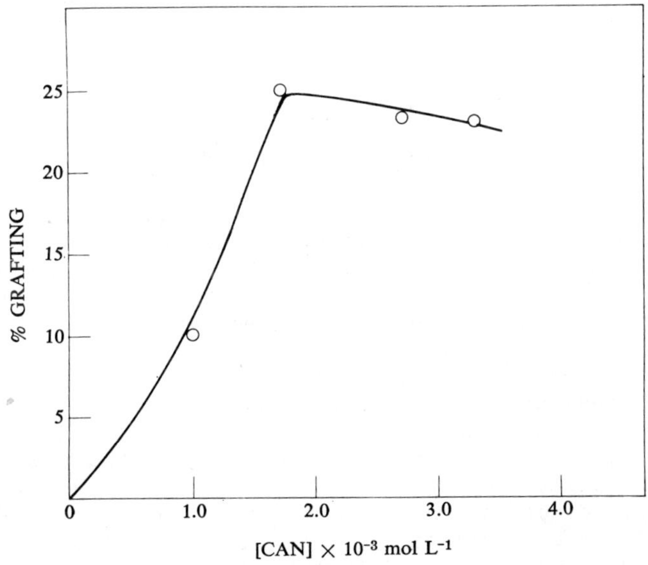

Figure 7. Effect of ceric ammonium nitrate concentration on the graft copolymerization of methyl methacrylate onto cellulose nitrate. (Reprinted, with permission, from Ref. 19. Copyright 1979, Wiley.)

Effect of Variation of Monomer and Cellulose Nitrate. The effect of varying monomer concentration and cellulose nitrate on the grafting percentage is shown in Figures 8 and 9 and Tables III and IV. As the monomer concentration increased, the grafting percentage and the number average molecular weight increased. It reached a maximum at a concentration of 0.22 mol/l and then decreased. The addition of higher monomer content caused agglomeration of cellulose nitrate resulting in the formation of lumps. The grafting percentage increased up to 6 g of cellulose nitrate and then decreased further with increase in cellulose nitrate. This is due to the increase in the relative ratio of cellulose nitrate to ceric ion.

Grafting of Vinyl Monomers to Cellulose Nitrate in Homogeneous Phase Initiated by Benzoyl Peroxide and Azobisisobutyronitrile. One of the critical problems in grafting reactions is the formation of homopolymer and its subsequent effects on grafting efficiency. If the homopolymer formed is insoluble in monomer solution or medium in which the grafting reaction is carried out, turbidity occurs and the grafting becomes erratic or even causes premature termination. Hence, the yield of homopolymer, which is always a competing reaction to grafting, is an extremely important factor in the grafting process. For industrial applications, if the homopolymer formed is in solution and does not interfere with the properties of the grafted product, it can be used as such in surface coating and other applications. Hence, the grafting of vinyl monomers in homogeneous medium is preferred to the heterogeneous phase to achieve greater efficiency of grafting and for direct applications.

Effect of Variables on Grafting Reaction in Homogeneous Medium. Tables V and VI and Figure 10 show the effect of benzoyl peroxide (BPO) and azobisisobutyronitrile (AIBN) concentrations on the percent grafting, grafting efficiency, and number average molecular weights of the grafted PMMA chains. The maximum in percent grafting was obtained at BPO = 3.46×10^{-2} mol/l and then decreased with further increase in concentration. It is presumed that up to this critical initiator concentration, all the initiator radical produced by decomposition of the initiator is utilized in producing radicals on the backbone polymer and/or homopolymer radicals. After the excess concentration of BPO = 1.77×10^{-2} mol/l and AIBN = 5.11×10^{-2} mol/l, initiator radicals are mostly involved in accelerating homopolymer formation. Consequently, the molecular weights of the grafted chains decrease with increase in the initiator concentration.

Tables VII and VIII and Figure 11 show the effect of reaction time on the percent grafting, grafting efficiency, and molecular weight of the grafted chains for BPO and AIBN systems.

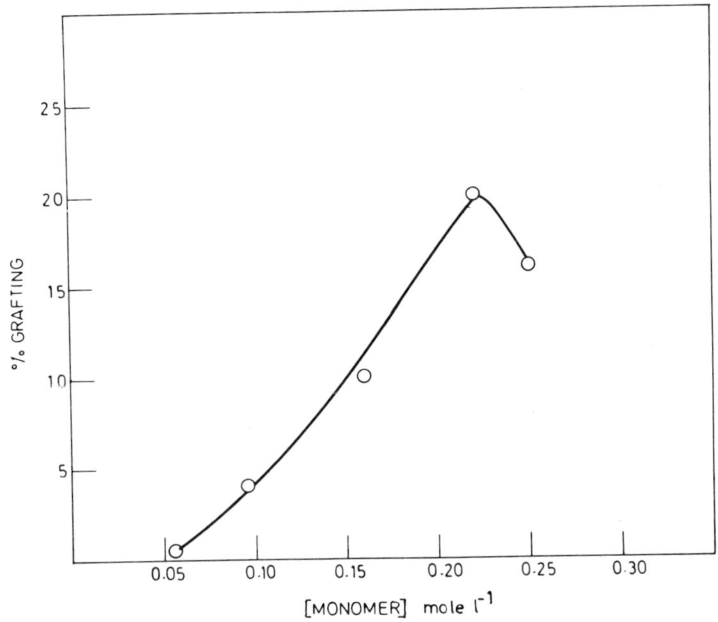

Figure 8. Effect of monomer concentration on the graft copolymerization of methyl methacrylate onto cellulose nitrate with ceric ammonium nitrate as initiator. (Reprinted, with permission, from Ref. 19. *Copyright 1979, Wiley.)*

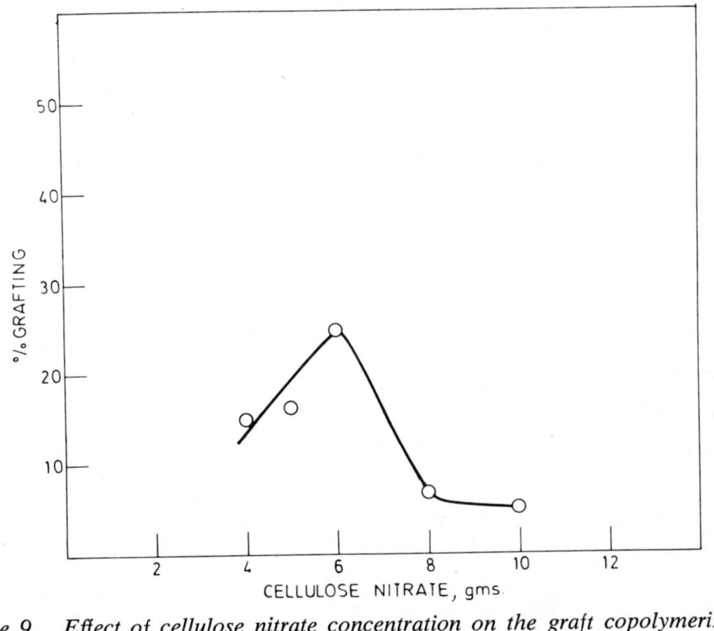

Figure 9. Effect of cellulose nitrate concentration on the graft copolymerization of methyl methacrylate onto cellulose nitrate with ceric ammonium nitrate as initiator. (Reprinted, with permission, from Ref. 19. *Copyright 1979, Wiley.)*

Table III. Effect of Monomer Concentration on Percentage Grafting and Molecular Weight

No.	[MMA] mol/l	Grafting %	Molecular Weight $M_n \times 10^5$
1	0.06	Trace	---
2	0.09	4.0	---
3	0.16	10.0	1.1
4	0.22	20.0	2.1
5	0.25	16.0	0.8

Grafting conditions: Cellulose nitrate - 5 g; Temperature - 30°C; Time - 75 min; [CAN] - 1.7×10^{-3} mol/l; Total volume - 300 ml. (Reprinted with permission from reference 19.)

Table IV. Effect of Cellulose Nitrate on Percentage Grafting and Molecular Weight

No.	Cellulose Nitrate (g)	Grafting %	Molecular Weight $M_n \times 10^5$
1	4	15.0	---
2	5	16.0	0.8
3	6	25.0	1.5
4	8	6.3	1.9
5	10	5.0	---

Grafting conditions: [MMA] - 0.25 mol/l; Temperature - 30°C; [CAN] - 1.7×10^{-3} mol/l; Total volume - 300 ml. (Reprinted with permission from reference 19.)

Table V. Effect of Initiator Concentration on Grafting Reaction with BPO as Initiator

No.	[BPO] mol/l x 10^2	Percent Grafting	Grafting Efficiency (%)	Molecular Weight $M_n \times 10^4$
1	0.692	19.2	63.6	1.43
2	1.385	27.2	71.0	1.28
3	2.078	38.7	72.9	1.21
4	2.771	42.7	77.3	1.18
5	3.464	46.8	81.5	1.17
6	4.157	42.8	73.5	1.17
7	4.850	41.4	83.2	1.18

Grafting conditions: Cellulose nitrate - 1.0 g; [MMA] - 0.626 mol/l; Time - 3 hrs; Temperature - 70 ± 0.1°C; Total volume - 15.0 ml.

Table VI. Effect of Initiator Concentration on Grafting Reaction with AIBN as Initiator

No.	[AIBN] mol/l x 10^{-2}	Percent Grafting	Grafting Efficiency %	Molecular Weight $M_n \times 10^4$
1	1.023	31.8	75.0	0.97
2	2.044	40.0	74.1	0.86
3	3.068	52.0	80.8	0.82
4	4.088	56.9	85.6	0.77
5	5.115	64.0	91.4	0.72
6	6.132	29.1	39.7	0.66
7	8.176	28.5	48.6	0.64

Grafting conditions: Cellulose nitrate - 1.0 g; [MMA] - 0.626 mol/l; Time - 3 hrs; Temperature - 70 ± 0.1°C; Total volume - 15 ml.

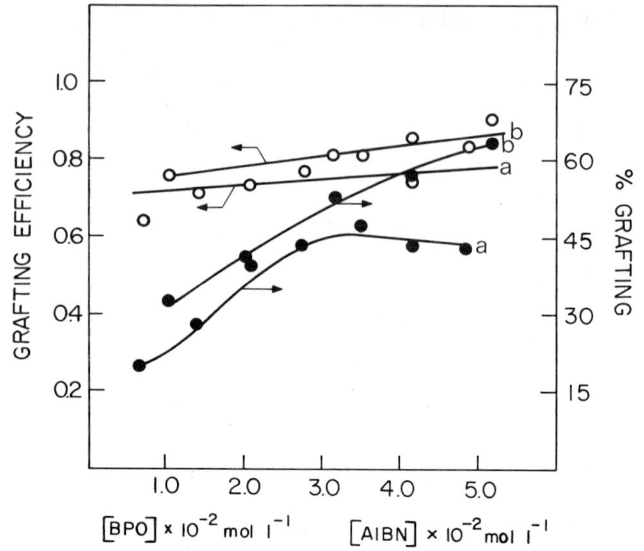

Figure 10. Effect of initiator concentration on the graft copolymerization of methyl methacrylate onto cellulose nitrate initiated by a, benzoyl peroxide and b, azobisisobutyronitrile. Key: ○, grafting efficiency; ●, percent grafting.

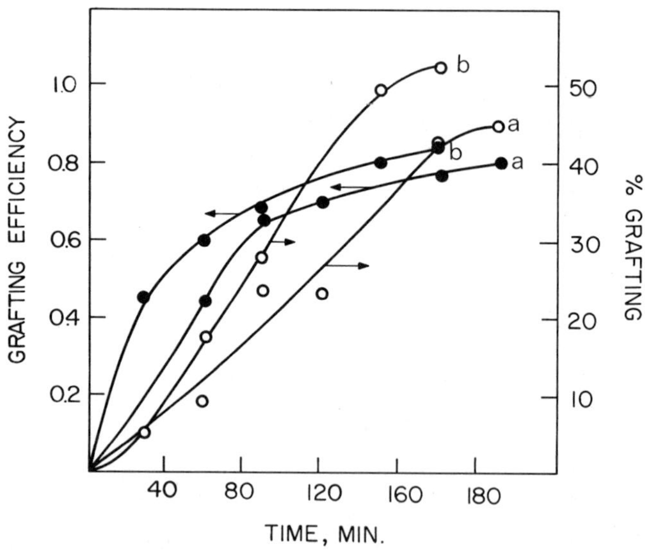

Figure 11. Effect of time on the graft copolymerization of methyl methacrylate onto cellulose nitrate initiated by a, benzoyl peroxide and b, azobisisobutyronitrile. Key: ●, grafting efficiency; ○, percent grafting.

Table VII. Effect of Reaction Time on Grafting Reaction with BPO as Initiator

No.	Time (min)	Percent Grafting	Grafting Efficiency %	Molecular Weight $M_n \times 10^4$
1	30	--	--	--
2	60	9.0	44.1	1.00
3	90	23.2	64.9	1.06
4	120	23.2	69.5	1.21
5	180	42.7	77.3	1.18
6	210	45.6	80.6	1.20

Grafting conditions: Cellulose nitrate - 1.0 g; [MMA] - 0.626 mol/1; Temperature - 70 ± 0.1°C; [BPO] - 2.77 x 10^{-2} mol/1; Total volume - 15.0 ml.

Table VIII. Effect of Reaction Time on Grafting Reaction with AIBN as Initiator

No.	Time (min)	Percent Grafting	Grafting Efficiency %	Molecular Weight $M_n \times 10^4$
1	30	4.8	44.6	0.58
2	60	17.2	60.1	0.54
3	90	28.1	67.7	0.59
4	150	50.3	80.6	0.60
5	180	56.9	85.6	0.72

Grafting conditions: Cellulose nitrate - 1.0 g; Temperature - 70 ± 0.1°C; [AIBN] - 4.08 x 10^{-2} mol/1; [MMA] - 0.626 mol/1; Total volume - 15.0 ml.

It can be seen that both percent grafting and grafting efficiency gradually increase with increase in grafting time. High molecular weight chains are formed in short reaction times, and the molecular weight increases with prolonged times.

Tables IX and X and Figure 12 show the dependence of percent grafting and grafting efficiency on methyl methacrylate concentration. It can be seen that the percent grafting increased with the monomer concentration in the initial stages and after reaching a maximum remained almost constant. In BPO-initiated reaction, grafting efficiency increased initially and then decreased with increase in monomer concentration. With AIBN as initiator, high values of grafting efficiency were obtained even at very low monomer concentrations, and then the grafting efficiency decreased with the increase of monomer concentration. This shows that at low monomer concentrations, most of the azo-radicals formed are utilized in grafting reactions rather than initiating homopolymerization of MMA. The molecular weights of the isolated PMMA grafts were found to increase with increase in monomer concentration with both initiators.

Denitration and Unsaturation of Cellulose Nitrate

Estimation of Nitrogen. Nitrogen analysis in cellulose nitrate, graft copolymer and cellulose nitrate as treated in grafting experiments but in the absence of monomer was attempted with the view that it would help inevaluating the mechanism of grafting reactions. The conventional nitrometer and Kjeldahl methods are not suitable for the estimation of nitrogen on cellulose nitrate since the graft copolymer is not completely soluble in sulfuric acid. Hence, volumetric (27) and colorimetric (28) methods developed for the determination of percent nitrogen in cellulose nitrate were used. The results obtained by these methods are given in Table XI. A difference of 4.9% in nitrogen content by volumetric study was found between the ungrafted samples and grafted cellulose nitrate. Cellulose nitrate as treated in the grafting reaction conditions in the absence of monomer also showed a loss of nitrogen (2.3%). A reduction in the intensity of $-ONO_2$ peak in the IR spectra of grafted cellulose compared to cellulose nitrate also supports the above result of the reduction of nitrogen content in the grafted samples (Figures 1 and 2).

Unsaturation. The cellulose molecule is a saturated one. Cellulose nitrate as treated in the graft copolymerization reaction conditions in the absence of monomer was found to be unsaturated by bromination method (Table XI). The amount of unsaturation was estimated as 0.14 per anhydroglucose unit.

Kinetic Aspects of AIBN-Initiated Graft Copolymerization. Benzoyl peroxide is much more successful in giving

Table IX. Effect of Monomer Concentration on Grafting Reaction with BPO as Initiator

No.	[MMA] mol/l	Percent Grafting	Grafting Efficiency %	Molecular Weight $M_n \times 10^4$
1	0.313	10.8	42.8	0.36
2	0.626	42.7	77.3	1.18
3	0.939	68.9	82.5	1.54
4	1.253	58.4	52.7	2.23
5	1.566	57.0	36.4	3.72
6	1.879	59.9	34.2	--

Grafting conditions: Cellulose nitrate - 1.0 g; [BPO] - 2.77 x 10^{-2} mol/1^{-1}; Total volume - 15.0 ml; Temperature - 70°C; Time - 180 min.

Table X. Effect of Monomer Concentration on Grafting Reaction with AIBN as Initiator

No.	[MMA] mol/l	Percent Grafting	Grafting Efficiency %	Molecular Weight $M_n \times 10^4$
1	0.313	37.4	96.3	0.43
2	0.626	62.4	87.4	0.77
3	0.939	69.6	66.5	1.07
4	1.253	70.2	49.9	1.66
5	1.566	69.5	41.7	--
6	1.879	72.6	38.6	3.10

Grafting conditions: Cellulose nitrate - 1.0 g; [AIBN] - 4.19 x 10^{-2} mol/l; Total time - 15.0 ml; Temperature - 70°C; Time - 180 min.

Table XI. Determination of Nitrogen Content and Unsaturation

Sample	% Nitrogen A[a]	% Nitrogen B[b]	Unsaturation[c]
Cellulose nitrate	11.3	11.6	No
Cellulose nitrate treated[d]	9.0	9.5	Yes
Cellulose nitrate-g-PMMA	6.4	5.2	--

[a] Obtained from colorimeter study.

[b] Obtained from volumetry study.

[c] Obtained from bromination study.

[d] Cellulose nitrate treated under identical grafting conditions without monomer.

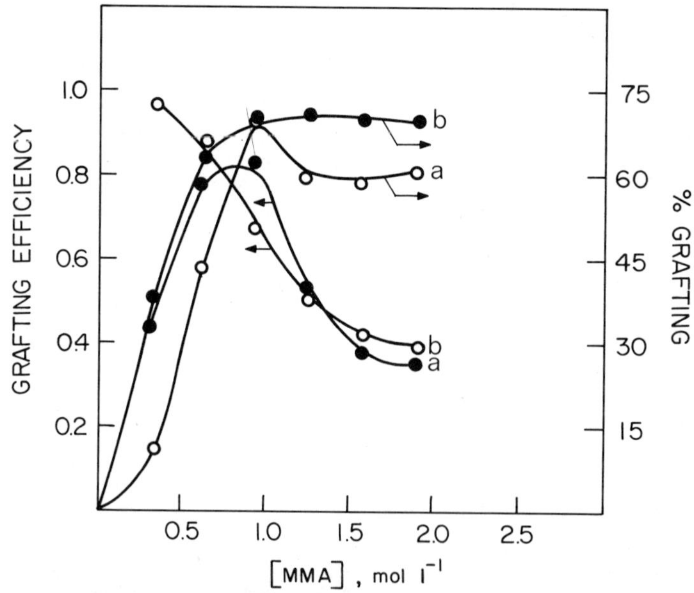

Figure 12. Effect of monomer concentration on the graft copolymerization of methyl methacrylate onto cellulose nitrate initiated by a, benzoyl peroxide and b, azobisisobutyronitrile. Key: ○, grafting efficiency; ●, percent grafting.

graft copolymers by the chain transfer mechanism than is AIBN, which forms radicals with weak hydrogen-abstracting properties (29). Although some authors aver that AIBN produces no grafting while others maintain that it produces a small percentage of grafting (30-32), in this study the very fact that AIBN produces a greater amount of graft copolymer of vinyl monomers onto cellulose nitrate rules out the possibility of a chain transfer mechanism for the grafting of vinyl monomers onto cellulose nitrate in homogeneous medium. The activation of energy for spontaneous decomposition of AIBN is about 30 kcal $mole^{-1}$. Rao et al. (33) have reported the activation energy of 10.7 kcal $mole^{-1}$ in the case of grafting of acrylic acid onto polyester fibers and the probable mechanism in this case is that the initiating species obtained for grafting appears to be only the growing homopolymer abstracting hydrogen from the backbone. The energy of activation for the AIBN-and BPO-initiated graft copolymerization of MMA onto cellulose nitrate was found to be 18 kcal $mole^{-1}$ and 16 kcal $mole^{-1}$, respectively. The higher activation energy obtained in our study may rule out the possibility of a chain transfer mechanism.

Probable Mechanism of Grafting. Cellulose nitrate molecules do not have a labile atom (as in the case of polyvinyl chloride or chlorinated rubber) which can be abstracted to create a grafting site on it by free radical initiators such as BPO or AIBN. The possibility of grafting sites on the cellulose nitrate molecule may therefore be either (a) at the free hydroxyl group, (b) at the nitro group, or (c) by some rearrangement in anhydroglucose unit (AGU). It is well known that the hydroxyl group in the cellulose molecule can form reactive sites only with redox initiators like ceric, manganic, etc., but have been ineffective with initiators like BPO or AIBN. We have observed no polymerization of MMA or grafting of MMA onto cellulose nitrate in the absence of initiators. Hence, the elimination of nitro group may not be causing the radical site to initiate grafting. The elimination of the NO_2 group and the formation of unsaturation on the polysaccharide may be a reactive site for the initiator to initiate the polymerization. Under mild conditions of oxidation of cellulose, it has been found that a keto group is formed at the C_2 without C_2-C_3 bond cleavage (34). McGee et al. concluded that the presence of a keto group in AGU resulting from the oxidation of celluronic acid by nitrogen dioxide would enolize to an unsaturated structure (35). Hence, the probable mechanism for grafting of vinyl monomers to cellulose nitrate may be due to the radical of initiator attacking the unsaturated position being formed in the cellulose nitrate (after removal of the NO_2 group and enolization) to create a radical site on the substrate to initiate a graft copolymer (Figure 13).

In the case of cellulose nitrate grafting initiated by ceric ammonium nitrate, grafting may probably be due to a complex

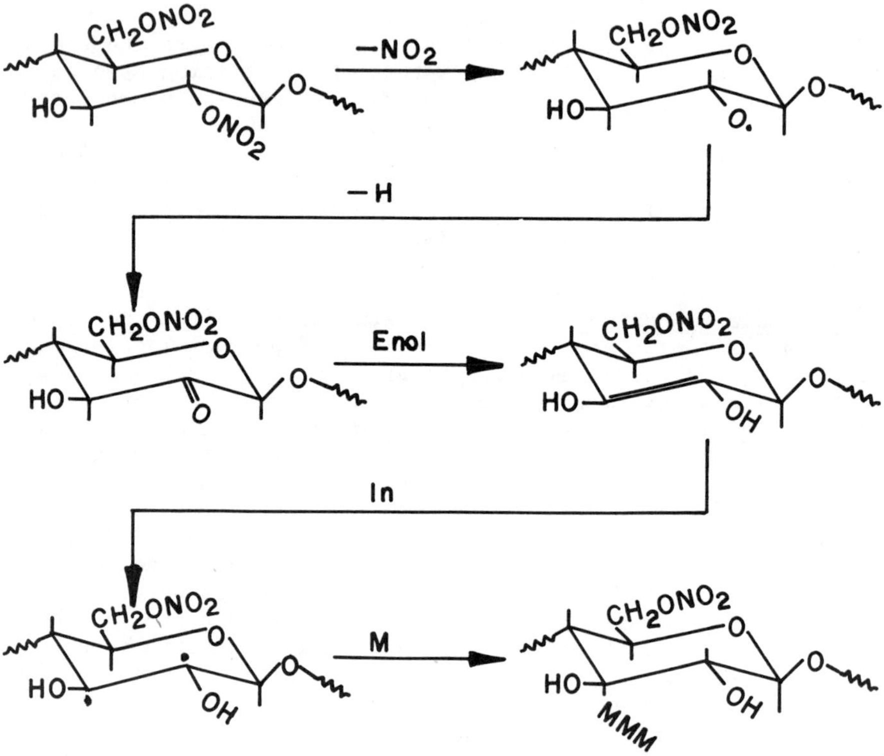

Figure 13. Probable mechanism of grafting of methyl methacrylate onto cellulose nitrate. Key: Enol, enolization; In, initiator; M, monomer.

formation between the ceric ion and the NO_2 group and subsequent formation of a radical on the backbone, resulting in grafting. We have not had any experimental evidence yet for the validity of this mechanism.

Conclusions

The present investigation revealed that the grafting of vinyl monomers onto cellulose nitrate can be successfully carried out using ceric ammonium nitrate in heterogeneous aqueous media and benzoyl peroxide and azobisisobutyronitrile in homogeneous organic media. Separation of graft copolymer from poly(methyl methacrylate) homopolymer and cellulose nitrate backbone has been carried out by selective solvent extration. IR, NMR, and GPC studies have substantiated the presence of true graft copolymer. The effects of various grafting parameters, viz., initiator concentration, monomer concentration, reaction time and the substrate concentration on grafting percentage, grafting efficiency, and the number average molecular weight of the grafted chains were examined. The grafting of vinyl monomers onto cellulose nitrate in homogeneous media gives rise to higher percentage of grafting than that in heterogeneous media. This is due to the precipitation and interference of homopolymer during the reaction in the heterogeneous medium. A probable mechanism for grafting of vinyl monomers onto cellulose nitrate has been proposed. It is likely that initiator radicals attack unsaturated groups formed in cellulose nitrate, by denitration and enolization, to initiate graft copolymerization reactions.

Acknowledgements

We (K.S.V.S. and M.S.) would like to express our gratitude to Dr. D. Sudhakar and Dr. K. Thomas Joseph for their contributions which helped make this paper possible.

Literature Cited

1. Arthur, J. C., Jr. "Advances in Macromolecular Chemistry"; Wallace M. Pasika, Ed.; Academic Press: New York, 1970; Vol. 2, pp 1-87.
2. Davis, N. P.; Garnett, J. L. J. Polym. Sci. 1976, 55, 287.
3. Davis, N. P.; Garnett, J. L. "Modified Cellulosics"; Rowell, R. M. and Young, R. A., Eds.; Academic Press: New York, 1978, p. 197.
4. Krassig, H. A.; Stannett, V. Adv. Polym. Sci. 1965, 4, 111.
5. Moore, P. W. Rev. Pure Appl. Chem. 1970, 20, 139.
6. Arthur, J. C.; Blouin, F. A. Paper presented at the International Symposium on Radiation-Induced Polymerization and Graft Polymerization, Battelle Memorial Institute, 1962, Preprints, p 319.

7. Dilli, S.; Garnett, J. L.; Martin, E. C.; Phuoc, D. H. J. Polym. Sci. 1972, 37, 57.
8. Guthrie, J. T. Progr. Polym. Sci. 1978, 6, 31.
9. Gurhrie, J. T. Polymer 1975, 16, 135.
10. Mino, G.; Kaizerman, J. J. Polym. Sci. 1958, 31, 242.
11. Ogiwara, Y.; Kubota, H. J. Appl. Polym. Sci. 1968, 12, 2575.
12. Ranby, B.; "Modified Cellulosics"; Rowell, R. M. and Young, R. A., Eds; Academic Press: New York, 1978, p 171.
13. Hon, David N.-S. J. Polym. Sci. 1980, 18, 1857.
14. Hon, David N.-S. J. Polym. Sci. 1979, 23, 3591.
15. Casale, A.; Porter, R. S. "Polymer Stress Reactions"; Academic Press: New York, 1978, p 161.
16. Simionescu, C. I.; Feldman, D.; Eiubotariu, M. J. Polym. Sci. C 1972, 37, 119.
17. Bardhon, K.; Mukpopadhyay, S.; Chatterjee, S. R. J. Polym. Sci. 1977, 15, 1977.
18. Wellons, J. D.; Williams, J. L. Stannett, V. J. Polym. Sci. 1967, 5, 1341.
19. Sudhakar, D.; Srinivasan, K. S. V.; Joseph, K. T.; Santappa, M. J. Appl. Polym. Sci. 1979, 23, 2923.
20. Sudhakar, D.; Srinivasan, K. S. V.; Joseph, K. T.; Santappa, M. J. Polym. Sci. Polym. Letters Ed., 1978, 16, 457.
21. Srinivasan, K. S. V.; Sudhakar, D. Polym. Bull. 1980, 2, 321.
22. Sudhakar, D.; Srinivasan, K. S. V.; Joseph, K. T.; Santappa, M. Polymer (London), in press.
23. Gordon, P. L.; Dolgin, G. J. "Surface Coatings and Finishes" Chem. Publ. Co., Inc.: New York, 1954.
24. Payne, H. F. "Organic Coating Technology," Vol. 1, J. Wiley & Sons, Inc.: New York, 1964.
25. Fox, T. G.; Kissinger, J. B. Mason, H. F.; Schule, E. M. Polymer 1962, 3, 71.
26. Brauer, G. M.; Termini, D. J. J. Appl. Polym. Sci. 1973, 17, 2557.
27. Stalcup, H.; Williams, R. W. Anal. Chem. 1955, 27, 543.
28. Kappelmeir, C. P. A., Ed. "Chemical Analysis of Resin-Based Coating Materials"; Interscience: New York, 1959.
29. Lewis, F. M.; Matheson, M. S. J. Am. Chem. Soc. 1949, 71, 747.
30. Minoura, Y.; Mori, Y.; Imoto, N. Makromol. Chem. 1957, 24, 205.
31. Mori, Y.; Minoura, Y. Imoto, N. Makromol. Chem. 1958, 25,1.
32. Dieyes, K.; Schuster, H. Makromol. Chem. 1967, 101, 200.
33. Rao, K. N.; Rao, M. H.; Lokhande, H. T.; Mody, N. R.; Jog, A. C. J. Appl. Polym. Sci. 1979, 23, 2133.
34. Heibeish, A. Mehta, P. C. J. Appl. Polym. Sci. 1968, 12, 1625.
35. McGee, P. A.; Fowler, W. F.; Unruh, C. C.; Kenyan, W. W. J. Am. Chem. Soc. 1948, 70, 2700.

RECEIVED January 12, 1982.

12

Computer Simulations of Grafting Reactions of Styrene onto Pre-irradiated Cellulose Acetate

TAMIO YASUKAWA and KENKICHI MURAKAMI

Tohoku University, Chemical Research Institute of Nonaqueous Solutions, 980 Sendai, Miyagi, Japan

Grafting of styrene onto gamma-ray pre-irradiated cellulose acetate from styrene/MeOH mixture solutions at 50°C were simulated with several models on the basis of kinetic parameters obtained by the bromine labelling method:average growth rate,average termination probability and initiation rate. In order to check the validity of models,the molecular weight distributions of graft chains were calculated and compared with observed ones.Those models where all graft radicals were assumed to have equal reactivities gave much narrow distributions relative to observations,whereas fair agreements were obtained with some models where growth rates of graft radicals were assumed to be diverged largely. Agreements were improved by assuming that local recombination rates are somewhat correlated with local growth rates. These results suggest that reactions proceed non-uniformly: graft radicals around the surface of micro-crystals grow and terminate more rapidly than those located inner parts of crystals , which may result in uneven distributions of graft components in final products.

One of the most fundamental factors to characterize graft copolymers is the distribution of graft chains. It is generally observed that the molecular weight distribution of graft chains in copolymers prepared by heterogeneous reactions is considerably different from that of homopolymers obtained by homogeneous reactions under similar conditions. Further , graft chains are not always distributed uniformly throughout materials. Heterogeneous reactions are generally difficult to study by ordinary analytical procedures owing to their complicated natures and computer simulation seems to be helpful. In a preceeding paper(1), we reported some preliminary results obtained by the simulation of grafting

0097-6156/82/0187-0179$6.00/0
© 1982 American Chemical Society

reactions of styrene onto pre-irradiated cellulose acetate films. In that study,however, the effects of heterogeneity of trunk polymer have not been discussed fully.In the present study , we shall study this effect in terms of the molecular weight distribution of graft chains MWD , as well as the spacial distribution of grafted components in copolymeric materials , by employing more elaborate simulation models.

Kinetic Parameters

Cellulose acetate films CA (DP=380 , acetic acid content 60% , thickness 0.05 mm) were irradiated with gamma-ray from a ^{60}Co source at dryice temperature to a total dose 4 x 10^5 rad. Grafting was performed at 50°C from styrene/MeOH mixtures under nitrogen atmosphere and kinetic parameters were determined by the bromine-labelling method (2). In Figure 1 are reproduced the kinetic parameters for the styrene/MeOH=1/1 system where GR denotes average growth rate of graft radicals ; PR ,average probability of termination of one graft radical and DA , rate of initiation (i.e. the number of CA radicals which react for the first time with styrene monomer to initiate grafting reaction at time t).

Simulation Models

Primary CA radicals produced by gamma-ray in amorphous regions decay out rapidly by recombinations or reactions with impurities. Hence, effective active sites for grafting may principally be located within or near the surface of micro-crystals as was shown for pre-irradiated polyethylene (3). Figure 2 shows a schematic illustration of reaction sites. Growth rates of graft radicals located within crystalline parts (region I) may be much smaller than those located on the surface of micro-fibrils (region VII).

In order to take into consideration the effects of diverged reactivities of graft radicals , we divided reacting sites into several parts, and kinetic parameters for respective regions were determined as follows. Over-all grafting rate is given by

$$\text{rate} = GR(t) \sum na_i(t) = \sum gr_i(t) \cdot na_i(t) \qquad (1)$$

where $na_i(t)$ and $gr_i(t)$ denote the number of graft radicals and local growth rate in the i region at time t . For convenience' sake , ratios of gr_i are assumed to be constant in the course of reactions and gr_2 was tentatively chosen as the reference to define a parameter β_i

$$\beta_i = gr_i(t)/gr_2(t) \qquad (2)$$

Local rate of initiation in respective regions are given by

$$da_i(t) = \gamma_i \cdot DA(t) \qquad (3)$$

with $\sum \gamma_i = 1$;i.e., γ_i is the distribution coefficient of initiating CA radicals in each region.

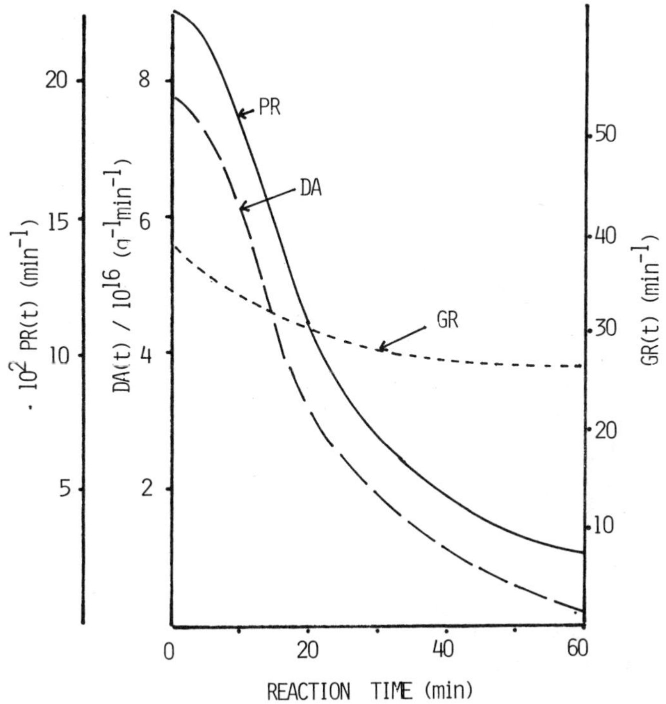

Figure 1. Kinetic parameters for grafting of styrene onto pre-irradiated cellulose acetate films at 50°C from a styrene/MeOH (1/1) solution. Key: GR, average growth rate of graft radical; PR, average termination probability; DA, initiation rate of grafting. (Adapted from Figure 6 of Ref. 2.)

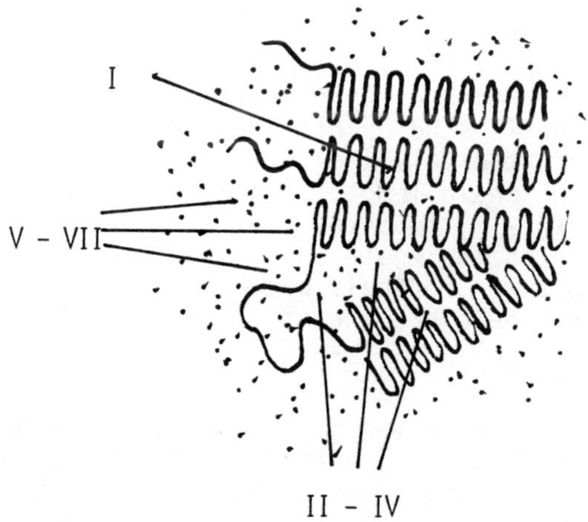

Figure 2. A schematic illustration of reaction sites.

At the first stage of grafting reaction , following relations hold

$$GR(1) \sum da_i(1) = \sum da_i(1) \cdot gr_i(t) \quad (4)$$

$$\therefore \quad \sum \beta_i \cdot \gamma_i = 1 \quad (5)$$

We might call the plot of γ_i against β_i the initiation rate spectrum for short. In Figure 3 are shown some examples of initiation rate spectra employed in the present study.

Grafting reactions are assumed to proceed independently in respective regions:i.e., graft radicals in each region have the same growth rate , chain terminations by recombination and disproportionation take place princiapally between radicals belonging to the same region and a chain transfer to polymer reproduces an active site in the same region. Hence , the time variation of the number of graft radicals with length L at time t in the region i is given by (1)

$$n_i(L,t) = n_i(L-gr_i(t-1),t-1) \cdot (1-p_{c,i}(L,t) - p_{tr})$$

$$\cdots\cdots\cdots\cdots\cdots\cdots \quad (6)$$

$$n_i(1,t) = da_i(t) + p_{trp} \sum_L n_i(L,t-1)$$

where $p_{c,i}(L,t)$ and p_{tr} denote the probabilities of chain termination by recombination and chain transfers , which are determined by following relations

$$PR(t) \sum na_i(t) = \sum pr_i(t) \cdot na_i(t) \quad (7)$$

$$pr_i(t) \sum_L n_i(L,t) = \sum_L (p_{c,i}(L,t) + p_{tr}) \cdot n_i(L,t)$$

$$p_{tr} = k_{trM} [M] + k_{trS} [S] + k_{trP} [P]$$

with ordinary abbreviations for rate constants and reaction components. The exact estimation of p_{tr} for heterogeneous reactions is practically impossible and tentative values in the range 0.01 through 0.001 per minute were assigned and checked.

In most kinetic studies of radical polymerizations , it has been assumed that the values of all rate constants are scarcely influenced by the chain length of radicals. Recently,however, several workers(4 - 8) reported that the value of k_t decreases appreciably with chain length and that the rate constant for recombination between radicals with degrees of polymerization n and m can roughly be expressed as follows ;

$$k_{t,nm} = k_{t0} \cdot n^{-\alpha} \cdot m^{-\alpha} \quad (8)$$

The α value for bulk polymerization of styrene at 60°C was found to be ca. 0.1 (4) and it is considered to be influenced by solvent natures and viscosity of reaction medium. Hence, several simulations were performed to study this effect with $\alpha = 0.0$, 0.2 , 0.33 and 0.5. When $k_{t,nm}$ can be expressed by eq(8) ,

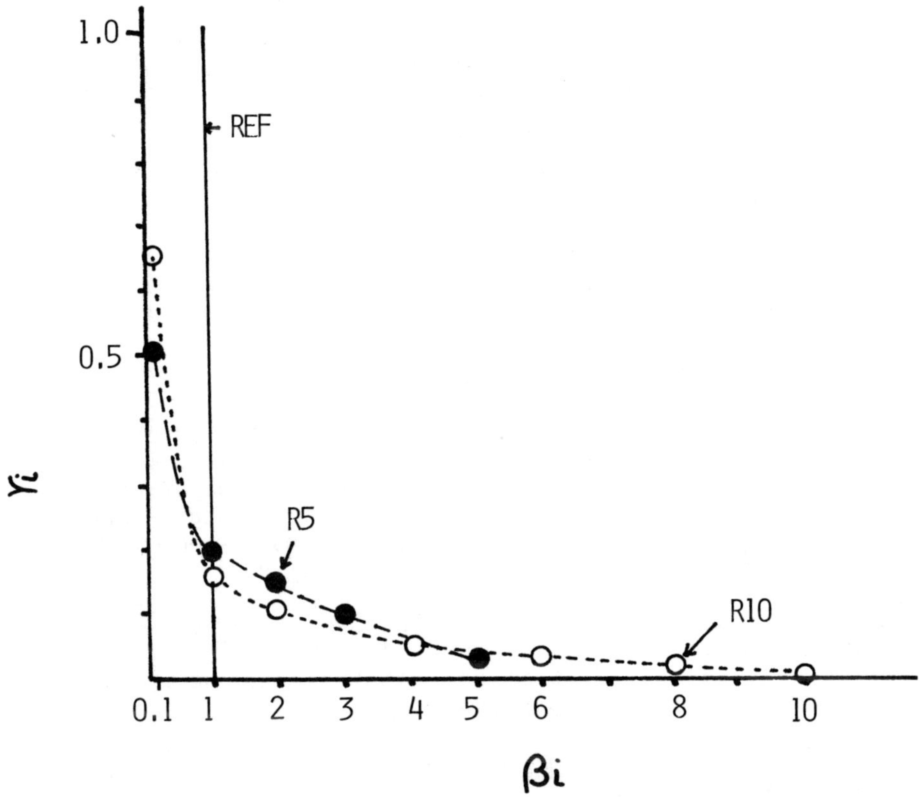

Figure 3. Initiation rate spectra employed in the present study. Key: REF, reference spectrum without divergence; R5, growth rates diverge from 0.1 GR to 5 GR; R10, growth rates diverge from 0.1 GR to 10 GR.

the term $p_{c,i}(L,t)$ in eq(7) can be rewritten as follows;

$$\sum_L p_{c,i}(L,t)\cdot n_i(L,t) = p_{co,i} \sum_L L^{-\alpha}\cdot n_i(L,t) \qquad (9)$$

where $p_{c0,i}$ is the normalization coefficient specific to each region.

The termination by recombination is a diffusion-controlled reaction and is , accordingly , sensitive to the nature of environments of active sites. Hence, it would be probable that the probability of termination by recombination is correlated with the degree of divergence of growth rate :e.g., graft radicals located on the surface of micro-crystals may have an enhanced termination rate by virtue of the relatively high degree of swelling by monomer solutions, as compared with those radicals trapped inner parts of crystals. In order to examine the effect of this correlation , a parameter δ is introduced. By employing this parameter the overall rate of termination by recombination can be expressed by

$$p_{co,i} = C_0 \cdot \beta_i^{\delta} \qquad (10)$$

overall rate of recombination =

$$C_0 \sum_i \beta_i^{\delta} \left(\sum_L L^{-\alpha} \cdot n_i(L,t) \right) \qquad (11)$$

The simulation of grafting reactions and calculation of MWD was performed as follows : starting from t = 0 with all $n_i(L,0)$= 0.0 , the time variation of all $n_i(L,t)$ was calculated at every minute of reaction by combining eqs(6 - 11) . Concurrently ,the chain length distribution of inert graft chains formed by recombination and chain transfers in espective minutes of reaction was calculated and was separately accumulated. At the final stage of reaction (60 minutes) , all graft radicals still active were terminated uni-molecularly and were added to the accumulated records of MWD data. This procedure was performed independently on each region and the resulting local MWD was added to obtain the overall MWD.

Results

In Figure 4 are shown MWD calculated for reaction time 60 minutes with three initiation rate spectra given in Figure 3 and α = 0.0 (i.e., termination rate is assumed to be independent of chain length). The MWD obtained with crude model REF where all graft radicals have the same growth rate and termination probability is much narrow relative to observed one , whereas the MWD obtained with R10 spectrum agrees fairly well. This result indicates that reactivities of graft radicals are diverged considerably.

Figure 5 shows the MWD calculated with R10 spectrum and α = 0.0 for δ = o.o , o.2 and 0.33 . With any value of δ , model

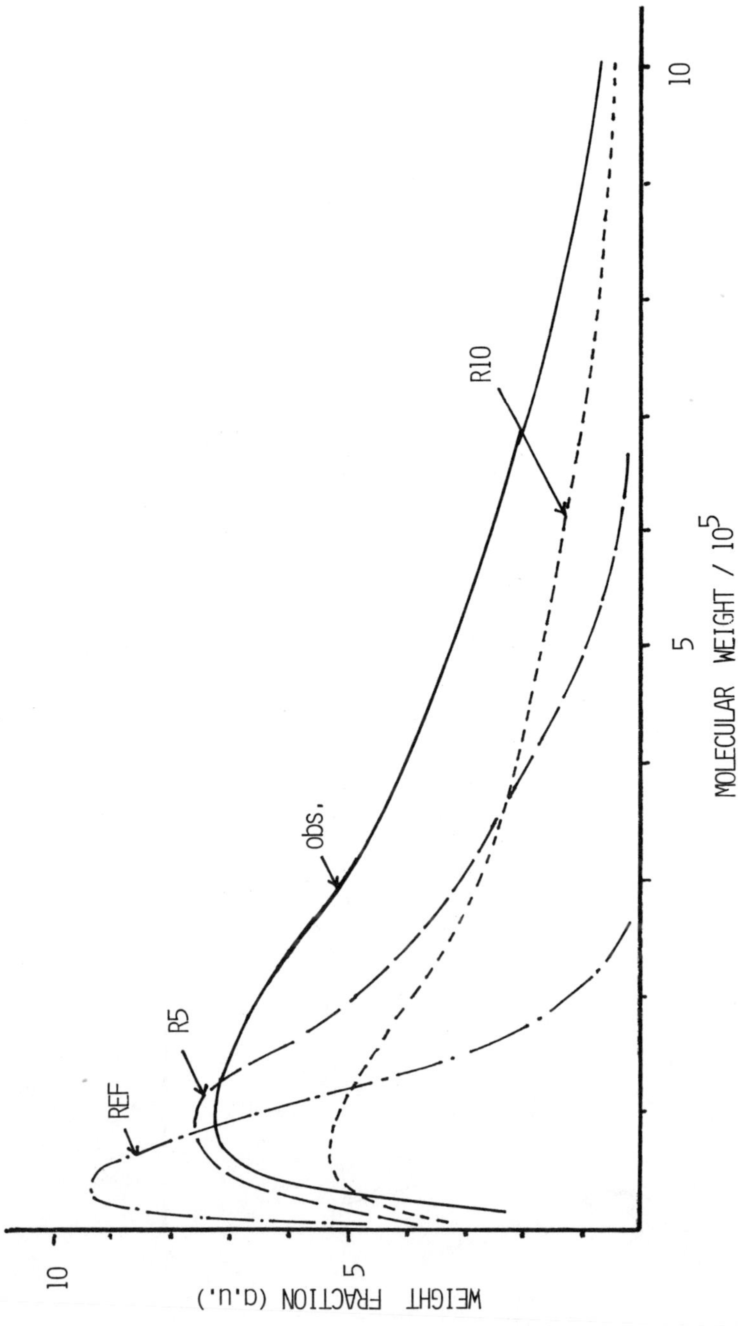

Figure 4. MWD calculated with three initiation rate spectra shown in Fig. 3 for reaction time of 60 min. Conditions: $\alpha = 0.0$, $\delta = 0.0$, $p_{tr} = 0.001$*/min.*

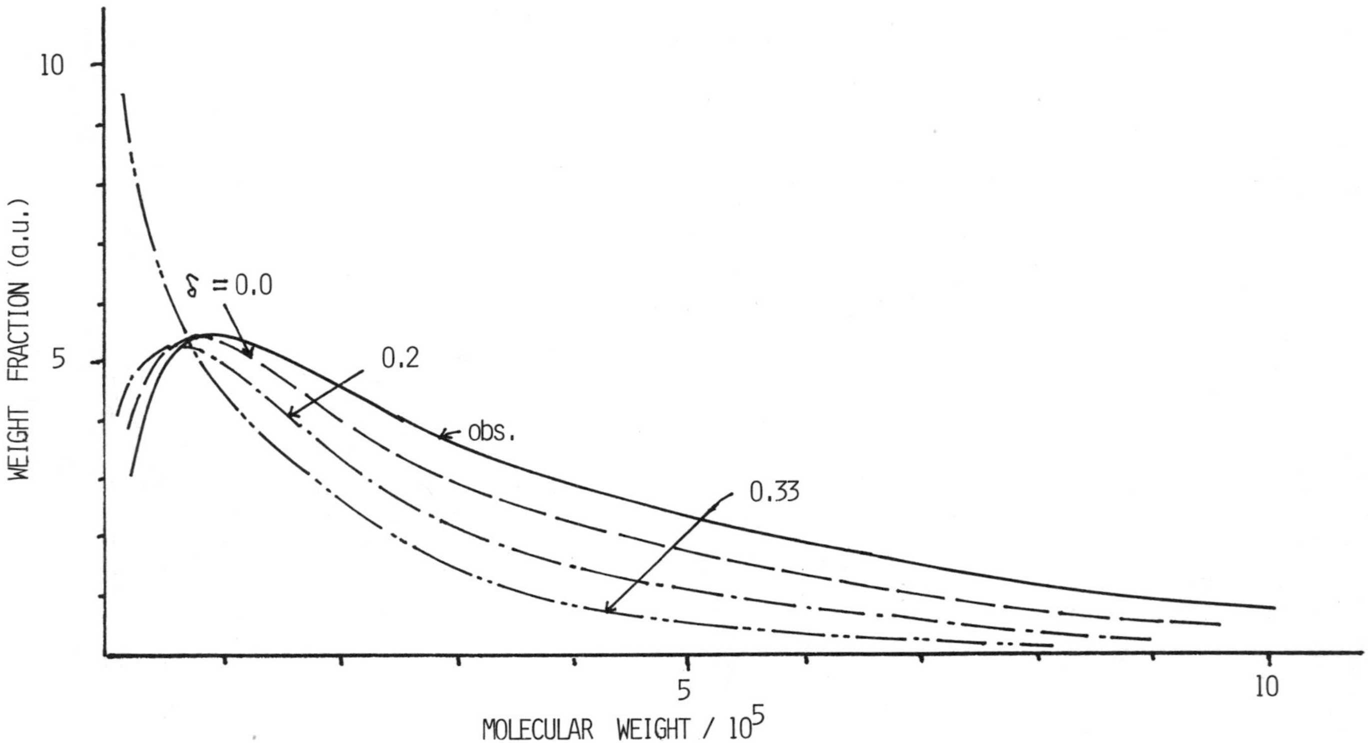

Figure 5. The effects on MWD of the correlation between local recombination rates and local growth rates. Calculated with R10 spectrum. Conditions: $\alpha = 0.0$; $p_{tr} = 0.001$; and $\delta = 0.0$, 0.2, and 0.33.

systems where the termination rate of graft radicals is assumed to be independent of chain length fail to give satisfactory agreements with the observation. In Figures 6 and 7 are shown the MWD calculated with $\alpha = 0.33$ and 0.5 for various values of δ Reasonable agreements with the observation were obtained by the combination of parameter values in ranges $\alpha = 0.33 - 0.5$ and $\delta = 0.2 - 0.33$. The same sets of parameter values give satisfactory MWD for copolymers obtained from 1/3 and 3/1 styrene/MeOH solutions,too. The effects of these parameters are closely correlated one another and,consequently, it is difficult to choose a certain set of values as the definitive one. The present results,however, indicate clearly that the growth rate of graft radicals is largely diverged and that the probability of termination by recombination is somewhat chain-length dependent. Furthermore,it is suggested that there is a weak positive correlation between local termination probabilities and local growth rates.

Figure 8 shows partial MWD in respective regions calculated with R10 spectrum , $\alpha = 0.5$ and $\delta = 0.2$. It would be noteworthy that the numbers of graft chains in regions V - VII are 3 , 2 and 0.5 % of the total chains,respectively,while fairly large weight fractions of graft chains are found in these regions by virtue of large molecular weight of chains.

Discussion

Swelling of amorphous parts of CA films by monomer solutions under the conditions of present work is rapid and is nearly completed within a few minutes and the styrene/MeOH ratios within films are practically equal to those of outer solutions(9).Hence, slow rates of initiation DA(t) shown in Figure 1 indicate that active sites for grafting are mainly those radicals trapped within and around the surface of micro-fibrils and that grafting reactions start when monomer solutions swell crystalline parts and monomer molecules arrive at active sites as was shown for pre-irradiated polyethylene systems (3). With the progression of grafting , the swelling of crystalline parts advances and CA radicals trapped inner parts of micro-crystals come to react. Mannan and Luftfar (10) observed that the micro-structure of cellulose fibre changes in parallel with the degree of grafting.

Slow growth rates of graft radicals indicate that effective concentrations of monomer around active ends of graft radicals are quite low. The rate-determining step would be the diffusion of monomer molecules within and around the surface of micro-crystals. Hence, it would be reasonable to assume that growth rate of graft radicals within crystalline parts is much slower than those located outside throughout reactions.

In the present study , the value of γ_i which designates the distribution of initiation rates in respective regions is assumed to remain constant in the course of reactions. Evidently , this assumption is crude :i.e., in the initial stages of reactions , most graft chains might be formed in the vicinities of the sur-

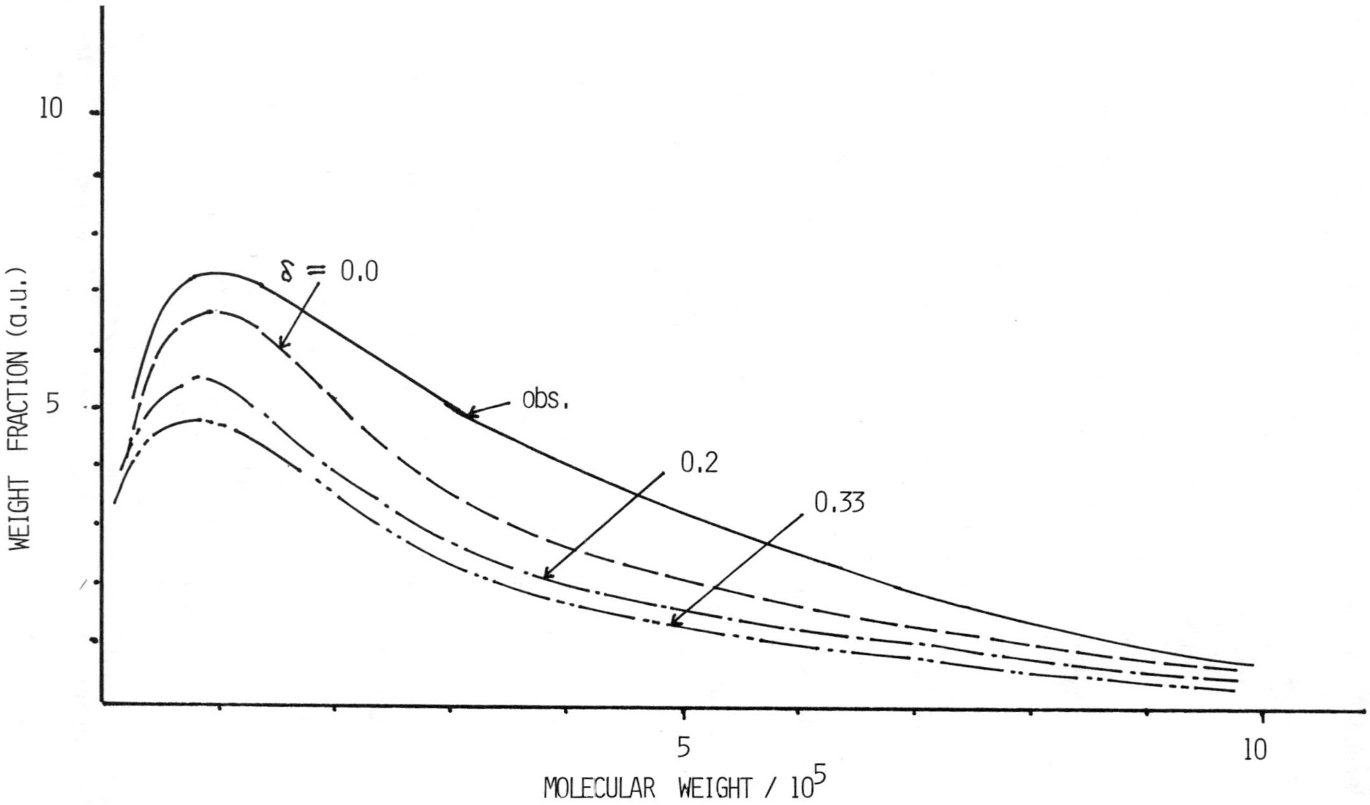

Figure 6. The effects on MWD of the correlation between local recombination rates and local growth rates. Calculated with R10 spectrum. Conditions: $\alpha = 0.330$; $p_{tr} = 0.001$; *and* $\delta = 0.0$, *0.2, and 0.33.*

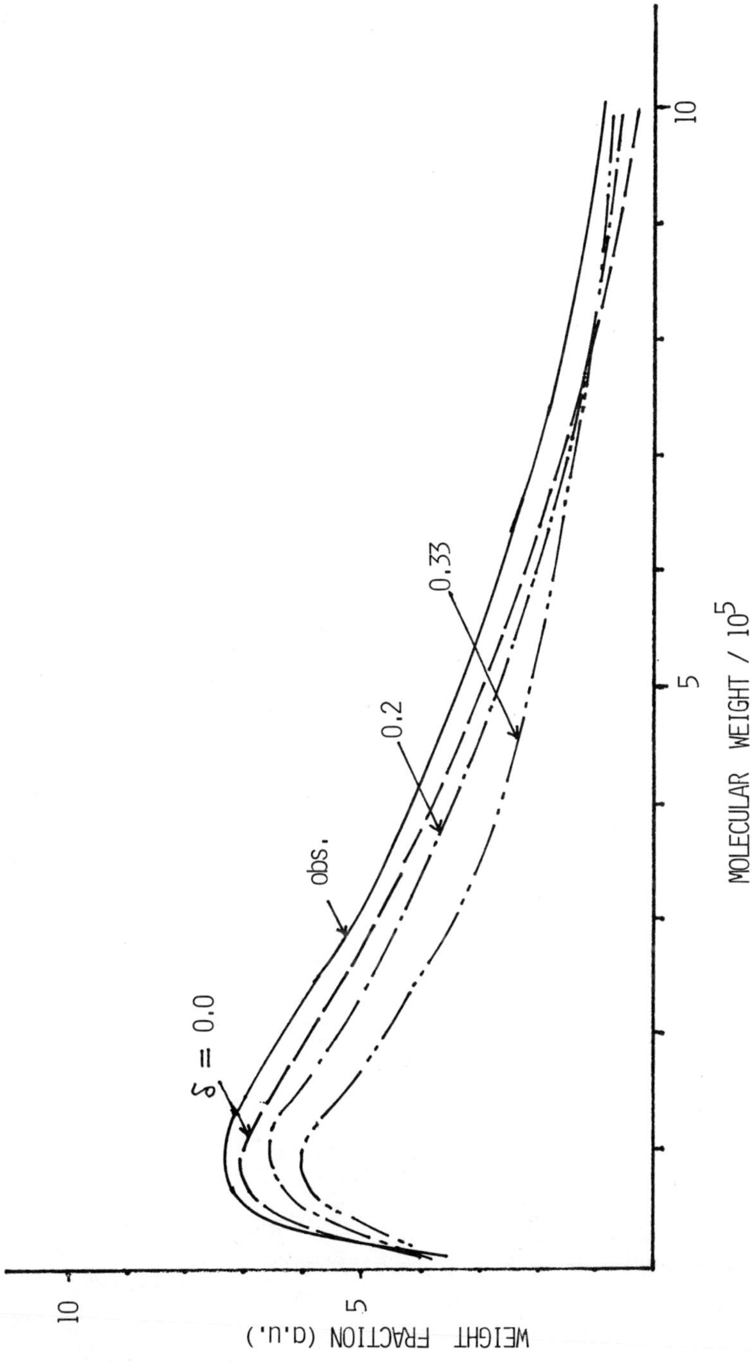

Figure 7. The effects on MWD of the correlation between local recombination rates and local growth rates. Calculated with R10 spectrum. Conditions: $\alpha = 0.5$; $p_{tr} = 0.001$; *and* $\delta = 0.0$, *0.2, and 0.33.*

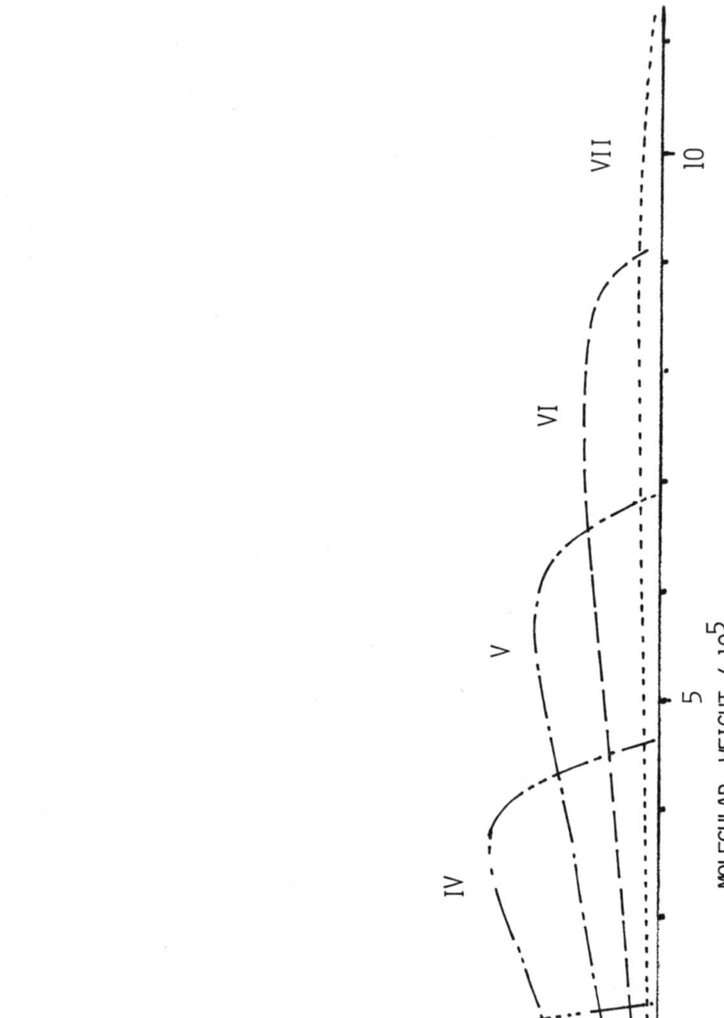

Figure 8. Partial MWD in respective regions. Calculated with R10 spectrum. Conditions: $\alpha = 0.5$, $\delta = 0.2$, *and* $p_{tr} = 0.001$.

face of micro-fibrils, whereas new graft radicals may principally be formed within crystalline parts in the latter stages. Thus, eq(5) does not hold strictly in the latter stages and the value of γ_i should be variated with time. More elaborate studies on this line are now in progress.

Figure 8 indicates that there are a small number of high molecular weight graft chains around the surface of micro-crystals, whereas many short graft chains are distributed within crystalline parts. Thus, the spacial distribution of graft components is far from uniform and due regards may well be paid to this heterogeneous distribution in the discussion of some properties of copolymers sensitive to microstructures.

Literature Cited

1. Yasukawa,T.;Murakami,K. J.Polymer Sci.Polymer Chem.Edition, 1975 , 13 , 17.
2. Yasukawa,T;Takahashi,T.;Murakami,K. Ibid. 1973,11,2547.
3. Yasukawa,T.;Takahashi,T.;Murakami,K.;Araki,K.;Sasuga,T; Ohmichi,H. Ibid. 1972 , 10 , 259.
4. Yasukawa,T;Takahashi,T.;Murakami,K. J.Chem.Phys. 1973 , 59 , 3937.
5. D'Odriscoll,K.F.;Mahabadi,H.K. J.Polymer Sci. Polymer Chem. Edition, 1976 , 14 , 869.
6. Horie,K.;Mita,I. Macromolecules , 1978 , 11 , 1175.
7. Yasukawa,T;Murakami,K. Polymer , 1980 , 21 , 1423.
8. Yasukawa,T.;Murakami,K. Macromolecules , 1981 , 14 , 227.
9. Yasukawa,T.;Sasaki,Y.;Murakami,K. Makromol.Chem. 1972 , 153 , 323.
10. Mannan,Kh.M.; Lutfar,L.B. Polymer , 1980 , 21 , 777.

RECEIVED December 24, 1981.

GRAFT COPOLYMERIZATION OF STARCH AND BAMBOO

13

Cross-linking in Saponified Starch-g-Polyacrylonitrile

GEORGE F. FANTA, ROBERT C. BURR, and WILLIAM M. DOANE

Northern Regional Research Center, Agricultural Research Service, U.S. Department of Agriculture, Peoria, IL 61604

Evidence, based on solubility data in DMSO, was obtained for crosslinking during the graft polymerization of acrylonitrile onto starch, and this crosslinking apparently occurs by way of chain combination of growing PAN macroradicals. Crosslinking also occurs between starch and PAN during alkaline saponification. Soluble (and thus uncrosslinked) starch-*g*-PAN polymers were rendered partially insoluble by alkaline saponification, as was a synthetic mixture of starch and PAN. Although PAN will crosslink with itself if saponifications are carried out in ethanol-water systems containing predominantly ethanol, PAN:PAN crosslinking does not take place during aqueous saponifications. Since the individual grafted granules of starch are crosslinked, these granules maintain their integrity after alkaline saponification and exist in the saponificate as highly swollen gel particles. When the saponificate is dried, these gel particles coalesce into either films or macroparticles which will not redisperse back into gel when placed in water. These properties are analogous to those of crosslinked rubber latexes and may be similarly explained by assuming interdiffusion of polymer chain ends on the surfaces of individual gel particles, followed by hydrogen bonding between these polymer chains. If dry saponified starch-*g*-PAN is heated, its ability to absorb water is reduced; however, this absorbency loss is not related to loss of hydrogen-bonded water from the graft copolymer.

Acrylonitrile graft polymerizes readily with either granular or pasted starch to yield starch-*g*-polyacrylonitrile (PAN), and reactions are generally carried out in water with ceric ammonium nitrate as the initiating system (1). Reaction of starch-*g*-PAN with alkali at elevated temperatures converts the nitrile substituents of PAN to a mixture of carboxamide

0097-6156/82/0187-0195$6.25/0
© 1982 American Chemical Society

and alkali metal carboxylate and yields a family of final products that are useful as thickeners (2) and as water absorbents (3, 4). Although saponified (hydrolyzed) starch-g-PAN (HSPAN) is hydrophilic and exhibits extensive swelling in aqueous systems, Taylor and Bagley (5) have shown that it exists largely as insoluble gel particles. We later showed by scanning electron microscopy (6) that these particles maintain the outward appearance of the original PAN-grafted starch granules. Since both starch and saponified PAN are soluble in aqueous alkali, the fact that a copolymer of the two components is not only insoluble but still bears some physical resemblance to the original starch-g-PAN precursor, as it swells and imbibes water, is indicative of crosslinking in the final saponified product. The purpose of this research was to gain some insight as to how this crosslinking might take place, and also to explain some of the properties of HSPAN films and particles. Although gelatinized starch gives thickening agents and water absorbents with superior properties, we used granular, unpasted starch in nearly all of this study because final products are easier to handle. Any conclusions based on results obtained with granular starch would certainly be applicable to gelatinized starch.

Results and Discussion

Crosslinking During Graft Polymerization. Perhaps the most obvious reaction path leading to crosslinking in HSPAN is the combination of two growing PAN macroradicals during the graft polymerization step. Combination of a PAN and a starch macroradical would give the same final results. Since the high degree of swelling of HSPAN in water indicates a low crosslink density, only a few such chain combinations would be sufficient to produce the observed physical properties. Ceric-initiated aqueous polymerizations of acrylonitrile in the presence of polyols have been studied, and evidence has been presented for termination by chain combination at low ceric ion concentrations (7, 8).

To gain some insight into whether crosslinking by free radical combination indeed takes place during graft polymerization reactions with starch, a number of these polymerizations were run under carefully controlled conditions, and the solubilities in dimethylsulfoxide (DMSO) of the resulting starch graft copolymers were determined at 75°C. Since starch and PAN are individually soluble under these conditions, we assumed that a totally uncrosslinked graft copolymer of the two components would be soluble also, and that the solubility of a particular polymer would thus decrease as its crosslink density increased. Polymers prepared with ceric ammonium nitrate initiation (Ce^{+4}:starch anhydroglucose unit [AGU] molar ratio of 1:100) and their respective solubilities in DMSO are shown in Table I. When corn starch was allowed to react with

TABLE I

Polymer Solubilities in Dimethylsulfoxide at 75°C. Ceric Initiation

No.	Polymer	Add-on,[a] %	$\bar{M}_v$ of grafted PAN[b]	Grafting frequency, AGU[c]/graft	Solubility, %
1	Starch	--	---		100
2	Ce^{+4}-Treated starch	--	---		36
3	Starch-g-PAN	52	83,000	473	8
4	Starch-g-PAN ($C_6H_{13}SH$)	50	17,000	100	34
5	Ce^{+4}-Treated starch ($C_6H_{13}SH$)	--	---		28

[a] Determined by weight loss on acid hydrolysis.

[b] Calculated from intrinsic viscosity in dimethylformamide.

[c] Anhydroglucose unit.

ceric ammonium nitrate in the absence of acrylonitrile (No. 2), its solubility was reduced from 100% to 36%: a result which parallels an earlier study with wheat starch (9). Although starch can thus be rendered partially insoluble by reaction with ceric ion, we will later show that by allowing ceric-treated starch to react with alkali under the conditions of saponification, we obtain a product that is highly soluble in water. Therefore, any crosslinks introduced by reaction with Ce^{+4} apparently have no large bearing on the final properties of HSPAN.

When starch was graft polymerized with acrylonitrile to give a product containing 52% grafted PAN, the solubility of the resulting starch-g-PAN in DMSO was only 8% (No. 3, Table I), a result which immediately suggests crosslinking during the polymerization reaction. If this crosslinking is indeed due to termination of growing PAN macroradicals by chain combination, graft polymerization in the presence of a chain transfer agent should not only lower the molecular weight of grafted PAN but should increase solubility in DMSO as well. Exactly this effect was observed when the graft polymerization was repeated in the presence of 1-hexanethiol (No. 4). Graft molecular weight was decreased by about a factor of 5, while solubility in DMSO increased approximately fourfold. When 1-hexanethiol was included in the reaction mixture when starch was allowed to react with Ce^{+4} in the absence of acrylonitrile, we obtained a starch product that was only 28% soluble in DMSO (No. 5).

Since the partial insolubility of the starch-ceric ammonium nitrate reaction product complicates the interpretation of solubility data, we ran a second series of graft polymerizations using cobalt-60 as an initiator in an attempt to remove this variable (Table II). If combination of PAN macroradicals is occuring during graft polymerization, it should occur during cobalt-60 initiated polymerizations as well as in those initiated by ceric ammonium nitrate. In the first four reactions of Table II, starch was irradiated as a water slurry under graft polymerization conditions, but in the absence of acrylonitrile, to determine the influence of different doses of irradiation on starch solubility. An effect of total dose on starch solubility was indeed observed, and this can perhaps be ascribed to competing radiation-induced degradation and crosslinking (10, 11). One path leading to crosslinking might be acetal and hemiacetal formation through carbonyl groups formed by irradiation (12). Since maximum solubility (95%) was observed at 0.1 Mrad, this total dose was used in subsequent graft polymerizations with acrylonitrile.

TABLE II

Polymer Solubilities in Dimethylsulfoxide at 75°C. Cobalt-60 Initiation

No.	Polymer	Dose, Mrad	AN,[a] g	Add-on,[b] %	$\bar{M}_v$ of grafted PAN[c]	Grafting frequency, AGU[d]/graft	Solubility, %
1	Starch	0	--	--	---		100
2	Starch	0.05	--	--	---		62
3	Starch	0.1	--	--	---		95
4	Starch	0.2	--	--	---		80
5	Starch-g-PAN	0.1	5	13	30,000	1,200	34
6	Starch-g-PAN	0.1	10	27	110,000	1,800	29
7	Starch-g-PAN	0.1	15	38	180,000	1,800	26
8	Starch-g-PAN ($C_6H_{13}SH$)	0.1	15	28	30,000	480	45
9	Starch-g-PAN ($HOCH_2CH_2SH$)	0.1	15	17	29,000	870	53

[a] AN = acrylonitrile.

[b] Determined by weight loss on acid hydrolysis.

[c] Calculated from intrinsic viscosity in dimethylformamide.

[d] Anhydroglucose unit.

In numbers 5-7 of Table II, starch-g-PAN copolymers containing different amounts of grafted PAN were prepared by the mutual irradiation of starch and varying amounts of acrylonitrile in water. Although values for % add-on in these three reactions ranged from 13-38%, and graft molecular weights varied from 30,000 to 180,000, solubilities in DMSO were not greatly different (26-34%). The roughly threefold reduction in solubility observed by the incorporation of as little as 13% PAN into the starch matrix is a good indication that graft polymerization had led to crosslinking. Although complete DMSO solubility was not achieved when graft polymerizations were run in the presence of 1-hexanethiol (No. 8) or 2-mercaptoethanol (No. 9), both of these chain transfer agents enhanced solubility. Numbers 5 and 9 of Table II make an especially good comparison to show the effect of a chain transfer agent on solubility, since both PAN content and graft $\bar{M}_v$ in the two polymers are similar.

Crosslinking During Alkaline Saponification. Although granules of starch-g-PAN are apparently crosslinked, the PAN moiety is not itself a network polymer, since removal of the starch component gives a material that is readily soluble (e.g., in DMF, for $\bar{M}_v$ determination). We, therefore, needed to know initially whether the saponified PAN moiety was likewise soluble or whether aqueous alkaline saponification causes the PAN portion of starch-g-PAN to crosslink with itself. A ceric-initiated starch-g-PAN containing 50% PAN ($\bar{M}_v$:64,000) was thus prepared and saponified in aqueous alkali, and the resulting HSPAN was isolated by methanol precipitation and dried. Enzymatic hydrolysis of this product removed the starch component and afforded saponified PAN with $\bar{M}_n$ = 44,000, as determined by membrane osmometry. Since saponification of the PAN moiety gives soluble polymer with a molecular weight not greatly different from its PAN precursor, PAN:PAN crosslinking is apparently not occurring to a large extent during alkaline saponification in water.

We next examined the effect of running the saponfications in aqueous ethanol instead of in water, since alcohol-containing solvent systems are also being used for some saponified starch-g-PAN preparations (13). Contrary to aqueous saponfications, we found evidence that crosslinking of the PAN moiety does take place in this solvent system, although it is highly dependent on the ethanol:water ratio. Initially, we examined the saponification of PAN homopolymer in solvent combinations ranging from 95% ethanol to a 1:1 mixture of 95% ethanol and water (Table III). Crosslinking in the different solvents was compared by determining percent soluble polymer in the undried saponification reaction mixtures after exhaustive dialysis against distilled water. Brookfield viscosities of these water dispersions also served as an

Table III. Influence of Ethanol:Water Ratio on Alkaline Saponification[a]

95% EtOH (ml)	H_2O (ml)	Saponified PAN[b] dispersion: Solubles (%)	Saponified PAN[b] dispersion: Viscosity cp[c] (% solids)	Saponified starch-g-PAN[d] dispersion: Solubles (%)	Saponified starch-g-PAN[d] dispersion: Absorbency of insoluble gel[e]
8.9	0	64	4,600 (0.25)	31	150
8.0	0.88	84	300 (0.27)	27	140
7.56	1.32	71	200 (0.26)	--	--
4.44	4.44	100	19 (0.24)	28	310

[a]2.00 g of 50% NaOH added to each solvent system. 0.50 g polyacrylonitril (PAN) or 1.00 g starch-g-PAN added, and dispersion heated under reflux for 3 h. Saponified dispersions were dialyzed against distilled water.

[b]$\bar{M}$ of PAN: 49,000.

[c]Brookfield viscosity at 30 rpm.

[d]No. 3, Table I.

[e]Expressed as grams of water-swollen gel per gram of dry polymer.

indicator for crosslinking; since the viscosity of a crosslinked, partially soluble, swollen gel network would be higher than the corresponding uncrosslinked polymer solution. Although crosslinking clearly occurs in 95% ethanol, as indicated by a water solubility of only 64% for saponified PAN and a Brookfield viscosity of 4600 cp for the aqueous reaction mixture at 0.25% solids, solubilities progressively increase and Brookfield viscosities decrease with increasing water contents of the solvent systems. When saponified in a 1:1 solution of 95% ethanol and water, PAN apparently does not crosslink to a significant extent, since the resulting saponified PAN is 100% water soluble, and the Brookfield viscosity of the solution is only 19 cp.

The saponification of a ceric-initiated starch-g-PAN was likwise examined at the different ethanol-water ratios, and these results are also shown in Table III. Once again, solubilities were determined for undried water dispersions after exhaustive dialysis. Brookfield viscosity was not used as a measure of crosslinking for these HSPAN preparations, because individual gel particles could not absorb all of the available water at 0.22% solids and thus could not exist in the closely packed state which is necessary for the attainment of meaningful viscosities (5). To more reliably estimate the extent of crosslinking, the insoluble gel was separated from excess solution by filtration and freeze dried. The weight of freeze-dried solid was determined, and water absorbency was then expressed as grams of swollen gel per gram of dry polymer. Low crosslink densities would thus give highly water-swollen gels and high absorbency numbers. Although the percent solubles did not vary greatly with the ethanol content of the saponification mixture, 1:1 ethanol (95%)-water produced an insoluble HSPAN gel fraction with about twice the absorbency of those prepared at higher ethanol contents. We can therefore assume that some PAN:PAN crosslinking likewise has taken place at high ethanol concentrations, as was observed for PAN homopolymer.

Although the formation of PAN:PAN crosslinks during alkaline saponification of starch-*g*-PAN takes place to a significant extent only when alcohol is the major component of the saponification solvent system, some crosslinking between starch and PAN occurs even when saponifications are carried out in water or in 1:1 ethanol (95%)-water (Table IV). Evidence for this starch:PAN crosslinking was obtained by saponifying two totally DMSO-soluble (and thus not crosslinked) starch-*g*-PAN polymers and then showing that: (1) The resulting saponified polymers had only a partial solubility in water, and (2) they gave viscous aqueous dispersions of gel at low concentrations (about 0.25% solids). The first soluble starch-*g*-PAN (No. 2, Table IV) was prepared by dispersing starch graft copolymer No. 1 in hot DMSO and then subjecting the dispersion to treatment with ultrasound to mechanically degrade the polymer. Residual gel was removed from the DMSO solution by filtration, and the soluble polymer was then isolated by alcohol precipitation. The second soluble starch-*g*-PAN (No. 3) was prepared by subjecting another starch graft copolymer to a mild acid hydrolysis to partially depolymerize the starch moiety. The ethanol-water solvent system apparently promotes a greater number of starch:PAN crosslinks than water, as evidenced by the lower solubilities (45 vs. 71% and 42 vs. 65%) and the higher viscosities for water dispersions (2400 vs. 920 cp and 690 vs. 220 cp).

The next logical step to prove crosslinking between starch and PAN was to saponify a 1:1 physical mixture of starch and PAN and to show that the resulting saponificate had solubility and viscosity properties different than those of the saponificate of the individual components. Numbers 4, 5, and 6 (Table IV) show the results of saponifying starch, ceric-treated starch, and PAN homopolymer. Saponifications in water and in ethanol-water gave complete or nearly complete solubilities as well as low viscosities for the dialyzed solutions. Although saponification of the physical mixture (No. 7) in water gave about the same results as the individual components,

Table IV. Alkaline Saponifications. Properties of Saponified Dispersions

No.	Polymer	Saponified in water[a]			Saponified in EtOH-H_2O[d]		
		Solubles, (%)	Viscosity, cp[b] (% solids)	Absorbency of insoluble gel[c]	Solubles, (%)	Viscosity, cp[c] (% solids)	Absorbency of insoluble gel[c]
1	Starch-g-PAN (Ce^{+4})[e]	26	---	270	28	---	310
2	Sonified starch-g-PAN (Ce^{+4})[f]	71	920 (0.25)	---	45	2400 (0.25)	---
3	Acid-treated starch-g-PAN (Ce^{+4})[g]	65	220 (0.23)	---	42	690 (0.25)	340
4	Starch	99	2.4 (0.25)	---	100	3.0 (0.28)	---
5	Ce^{+4}-treated starch	100	3.0 (0.25)	---	94	2.8 (0.25)	---
6	PAN ($\overline{M}_v$: 49,000)	100	16 (0.25)	---	100	19 (0.24)	---
7	Starch-PAN phys. mixture	100	22 (0.31)	---	85	1000 (0.25)	---
8	Starch-g-PAN (^{60}Co)[h]	---	---	---	28	3900 (0.23)	480
9	Sonified starch-g-PAN (^{60}Co)[i]	---	---	---	51	2900 (0.25)	---
10	Sonified starch-g-poly-acrylamide (^{60}Co)[j]	---	---	---	100	11 (0.25)	---

[a]Either 1.0 g starch graft copolymer or 0.5 g homopolymer (starch or PAN) saponified in 9.0 ml 0.7 N NaOH for 3 h at 95–100°C.

[b]Brookfield viscosity at 30 rpm.

[c]Expressed as g water-swollen gel/g dry polymer.

[d]Polymers (same quantities as footnote a) saponified for 3 h at reflux in a solution of 2.0 g 50% NaOH, 4.44 ml water, and 4.44 ml EtOH.

[e]Prepared under the same conditions as No. 3, Table I, PAN content: 51%; $\overline{M}_v$ of PAN: 64,000.

[f]PAN content: 53%; $\overline{M}_v$ of PAN: 36,000.

[g]PAN content: 46%; $\overline{M}_v$ of PAN: 35,000.

[h]No. 7, Table II.

[i]PAN content: 37%; $\overline{M}_v$ of PAN: 100,000.

[j]Polyacrylamide content: 35%.

saponification in ethanol-water gave a product that was 85% soluble and that had a Brookfield viscosity of 1000 cp at 0.25% solids in water.

Although the number of starch:PAN crosslinks in HSPAN is too low to permit their identification by spectroscopy, we can propose a reasonable crosslinking mechanism based on the known reactions of PAN and starch with alkali. When PAN is heated with alkali, a deep red-brown color is produced due to nucleophilic-initiated polymerization through the nitrile groups to yield a polyimine type of conjugated ring structure (14, 15). This conjugated ladder-type polymer is then opened in a second reaction to give a polyacrylamide, which is then further hydrolyzed to the observed poly(acrylic acid salt-*co*-acrylamide). Under the conditions of Table IV, a crosslinked final product is not observed in the absence of starch, since $^{\ominus}OH$ (and to a lesser extent, $^{\ominus}OC_2H_5$) is the nucleophile that initiates the nitrile polymerization and then reacts to open the conjugated ring structure. In the presence of starch, however, starch alkoxide can also serve as the nucleophile, thus leading to the observed crosslinking. The close proximity of starch and PAN, when the two polymers are chemically bonded through graft polymerization, would favor these reactions.

The role of the conjugated polyimine intermediate in the observed crosslinking can be seen by comparing the saponification of a starch-*g*-PAN (No. 9, Table IV) with that of a starch-*g*-polyacrylamide prepared under the same conditions (No. 10). Although polyacrylamide is an intermediate in the conversion of PAN to poly(acrylic acid salt-*co*-acrylamide), the polyimine structure is of course not formed during its saponification. Both starch-*g*-PAN and starch-*g*-polyacrylamide were prepared by cobalt-60 initiation, since acrylamide does not graft efficiently in the presence of ceric ammonium nitrate (16). Both graft copolymers were irradiated with ultrasound to give totally soluble products; the ultrasonic treatment of starch-*g*-PAN was carried out in DMSO, while that of starch-*g*-polyacrylamide was run in water.

Saponifications of these graft copolymers were carried out in 1:1 ethanol (95%)-water (Table IV). Reaction No. 8 shows the saponification results for starch-*g*-PAN before treatment with ultrasound; and although the insoluble gel fraction was more highly swollen, the % solubles was the same as for starch-*g*-PAN prepared with ceric initiation (No. 1). After solubilization by treatment with ultrasound, starch-*g*-PAN gave a saponification product that was only 51% water-soluble and that showed a Brookfield viscosity of 2900 cp at 0.25% solids in water. The corresponding starch-*g*-polyacrylamide, however, gave a totally water-soluble saponification product with low viscosity, indicating that crosslinking did not occur.

Film Formation. From our first studies of HSPAN, it was apparent that continuous films can be formed by simply allowing water dispersions of the polymer to evaporate to dryness (17). These films are unique in that they do not revert back to the original gel particles when placed in water. Rather, the entire film swells as a continuous entity as it imbibes water; and within a few minutes, a highly swollen sheet of gel is formed that has enough tensile strength to allow it to be carefully manipulated without breaking. If these films are redispersed in water with rapid stirring (e.g., Waring Blendor for short periods of time), the resulting dispersion will again form a film with similar properties if it is allowed to evaporate to dryness. Swollen films derived from gelatinized starch graft copolymers possess more strength and continuity than similar films prepared from granular starch.

Bonding of individual gel particles with each other to give larger conglomerates of highly absorbent polymer is also observed when other isolation techniques are used, e.g., alcohol precipitation or drum drying (4). However, if alcohol precipitation is carried out by slow addition of alcohol to the water dispersion, HSPAN tends to precipitate as individual gel particles rather than as larger chunks, and the resulting dried polymer therefore gives very smooth water dispersions when hydrated (18).

To verify that these film- and conglomerate-forming properties of HSPAN were not related to the sticking together of individual gel particles by the minor amount of soluble polymer present in saponification mixtures, films were cast from water dispersions of HSPAN polymers, prepared from either granular or gelatinized starch, from which the solubles had been extracted with water. These films were indeed continuous, and they absorbed water to form continuous gel sheets. We also observed that addition of 10% of either saponified PAN or soluble starch to HSPAN prepared from granular starch did not outwardly alter film properties.

To explain the fact that HSPAN swells in water to form gel sheets or macroparticles rather than disintegrating into a gel dispersion, we initially felt that chemical bonding must take place between individual particles of water-swollen gel as water evaporates. Although we cannot totally eliminate this possibility, the proposal of primary chemical bonding is not necessary to explain the behavior of these films and conglomerates. For example, Voyutskii (19) has reviewed the formation of films from vulcanized rubber latexes and concludes that film formation in these systems is observed because of interdiffusion of ends of individual macromolecules in adjacent latex particles. This diffusion can take place even though individual latex particles are crosslinked, 3-dimensional networks; and the continuity of the resulting films, even when partially swollen in a hydrocarbon solvent, can be ascribed to Van der Waals forces between interwoven polymer ends. Bradford and Vanderhoff (20) have also prepared films from crosslinked latex particles. These authors studied a 65:35 styrene-butadiene copolymer crosslinked with varying amounts of divinylbenzene and found that although the incorporation of divinylbenzene retarded the coalescence of latex particles, these particles did indeed coalesce, presumably due to a similar interdiffusion of polymer chain ends.

The analogy between crosslinked latex particles and the particles of water-swollen HSPAN is obvious. In addition to Van der Waals forces, hydrogen bonding between the hydrophilic polymer chains can also take place, and some of these hydrogen bonds could be quite strong, as evidenced by the well-known phenomenon of amylose retrogradation (21) and the resistance of retrograded amylose to water solution. An analogy to the vulcanized rubber latexes can also be seen in the facile conversion of HSPAN film to gel particles by a short period of high-speed stirring in water, followed by the reformation of a film similar to the original by simply drying the water dispersion. Voyutskii (22) found that a film prepared from vulcanized latex gave a continuous rubber sheet on cold milling, presumably because the absence of primary chemical bonding between individual particles allows them to easily slide with respect to each other; the same type of secondary (Van der Waals) bonds can then readily reform. In contrast, a rubber film that was vulcanized after it had been prepared, and was thus crosslinked throughout the entire polymer matrix, was converted to a rubber crumb on milling.

If the formation of continuous films from HSPAN gel is indeed largely due to hydrogen bonding and Van der Waals forces between interdiffused polymer chains, rather than to the formation of primary chemical bonds, this property should not be unique to the starch graft copolymers but should be exhibited, at least to some extent, by other water-swollen, hydrophilic gels as well. This probability was verified experimentally for a commercial thickening agent (Carbopol 941) and a commercial water absorbent (Permasorb 10). Water dispersions of both of these polymers (neither of which contain starch) formed films when dried; and although the water-swollen films were more easily disintegrated than those from HSPAN, they did remain intact if they were not agitated. Carbopol is a particularly good polymer for comparison, since Taylor and Bagley (5, 23) have studied its rheology and gel content and in these respects have found it to be quite similar to HSPAN.

Heat-Induced Crosslinking of Dry Polymer. When a dry sample of HSPAN is heated, the polymer gradually loses its capacity to absorb water (24), and this absorbency loss increases with both temperature and the duration of the heat treatment. Physical mixtures of starch and saponified PAN also crosslink when heated (25). An increase in hydrogen bonding between polymer chains (e.g., through starch hydroxyls) due to removal of bound water molecules at the higher temperatures could explain these findings; however, if this mechanism were operating, loss of absorbency should parallel the loss of water when HSPAN is heated. This is not what we observed experimentally (Table V). When samples of polymer,

Table V. Influence of Water Loss on Absorbency of Saponified Starch-g-PAN[a]

Heating conditions			
Temp., °C	Time, h	Weight Loss, %	Absorbency[b], g/g
25 (control)	=	---	730
60	1	2.4	680
80	1	6.8	700
100	1	10.4	680
120	1	12.8	680
140	1	14.8	480
100	3	12.4	700
100 (vacuum)	3	14.1	680

[a]A commercial sample prepared from gelatininzed starch (SGP 502S, Henkel Corp., screened to 100-200 mesh, water content: 14.2%.

[b]Distilled water.

containing 14.2% water, were heated in an oven for 1 hr at temperatures ranging from 60 to 120°C, absorbencies of the heated polymers were not greatly affected, even though nearly all of the water had been driven off at 100-120°C, as evidenced by weight loss of the samples. The fact that, even at these temperatures, a small amount of water was still retained by the polymer was apparently not the reason that absorbency did not decrease. When two separate polymer samples were heated in the oven for 3 hr at 100°C, the first at atmospheric pressure and the second under a vacuum of about 3 mm Hg, both polymers showed about the same absorbency, even though the vacuum-heated polymer had virtually all of its water removed.

When a sample of HSPAN was heated for 1 hr at 140°C (Table V), its weight loss was actually slightly greater than the original water content of 14.2%, and there was also a significant decrease in water absorbency. A condensation type of crosslinking reaction with loss of volatiles might account for these observations, and either hydroxyl, carboxamide, or carboxyl substituents could participate in such reactions at this high temperature. A choice among the several mechanistic pathways must await further research.

Conclusions

There are a number of factors which simultaneously contribute to give HSPAN its unique properties. First of all, crosslinking occurs during the graft polymerization reaction of acrylonitrile with starch, probably by combination of PAN macroradicals. Further crosslinking within the granule structure takes place during alkaline saponification. Crosslinking of PAN with itself can occur when saponifications are carried out in systems composed predominantly of alcohol; however, starch:PAN crosslinking is the major reaction in solvents which contain mainly water. As a result of these crosslinking reactions, individual granules of starch-*g*-PAN do not disintegrate when saponified with hot alkali, but retain their integrity and exist in water as highly swollen gel particles. Aqueous dispersions of these gels thus have high viscosity, when all the free water has been absorbed and the gel particles are in a closely packed state.

When viscous aqueous dispersions of HSPAN are spread onto a Teflon-coated tray and allowed to dry, individual gel particles knit together to form continuous films. Gel particles also coalesce to form larger conglomerates when the polymer is isolated from water by other techniques, such as alcohol precipitation or drum drying. Dry HSPAN does not redisperse back into gel particles when placed in water but swells as either a continuous film or as macroparticles as it imbibes water. The formation of primary chemical bonds between individual gel particles need not be proposed to account for these properties. The behavior of HSPAN is similar in many respects to the published behavior of crosslinked rubber latexes, and HSPAN properties may be similarly explained by assuming interdiffusion of polymer chain ends on the surfaces of gel particles, followed by hydrogen bonding between these polymer chains.

When dry HSPAN is heated, its ability to absorb water is diminished. This loss of absorbency is not related to loss of bound water and therefore cannot be due to additional hydrogen bonding. A condensation type of crosslinking reaction is probably occurring between hydroxyl, carboxamide, or carboxyl substituents at high temperatures.

Experimental

Materials. Granular Globe Pearl corn starch (Globe 3005; water content about 10%) was from CPC International. Acrylonitrile (Eastman practical grade) was fractionated at atmospheric pressure through a 15-in. Vigreux column, and a center cut was collected. Ceric ammonium nitrate was Fisher certified ACS grade. PAN homopolymer was Polymeric Acrylonitrile Type A from DuPont.

Graft Polymerization. Ceric Initiation. A stirred slurry of 10 g of starch in 200 ml of water was sparged with a slow stream of nitrogen for 1 hr at room temperature; 15 g of acrylonitrile was then added, followed after 5 min by a solution of 0.338 g of ceric ammonium nitrate in 3 ml of 1N nitric acid. In the reaction run with 1-hexanethiol, ceric ammonium nitrate solution was added first, followed after 5 min by a solution of 1.0 g of mercaptan in 15 g of acrylonitrile. The resulting mixture was allowed to stir for 2 hr at 25-27°C (temperature maintained with ice-water) and then was diluted with 200 ml of ethanol. The pH was adjusted to 6-7 with sodium hydroxide solution, and the polymer was isolated by filtration, washed with water and ethanol, and vacuum dried at 60°C. Ungrafted PAN was removed by repeated extraction of the polymer with dimethylformamide (DMF) at room temperature.

Graft Polymerization. Cobalt-60 Initiation. The cobalt-60 source was a Gammacel 200 unit from Atomic Energy of Canada, Ltd. Dose rate at the center of the chamber was 0.44-0.42 Mrad/hr, as calculated from the original dosimetry data provided by the manufacturer and the decay rate of cobalt-60.

A stirred slurry of 10 g of starch in 100 ml of water was sparged with a slow stream of nitrogen for 30 min at room temperature, and acrylonitrile (5, 10, or 15 g) was added. In reactions run with mercaptans, either 1.0 g of 1-hexanethiol or 0.1 g of 2-mercaptoethanol was dissolved in 15 g of acrylonitrile. The reaction mixture was stirred magnetically and irradiated with cobalt-60 to a total dose of 0.1 Mrad. The reaction mass thickened with the onset of polymerization, and most reaction mixtures were too thick to stir by the end of the irradiation period. Reaction temperature reached 40-50°C due to the exothermic polymerizations. The reaction mass was allowed to stand for 2 hr at ambient temperature, and the polymer was isolated by filtration, washed with water and with ethanol, and vacuum dried at 60°C. Ungrafted PAN was removed from the polymer by repeated extraction with DMF at room temperature.

A similar procedure was used to prepare starch-*g*-polyacrylamide, and 15 g of acrylamide (97%, Polysciences) was used in the polymerization. After irradiation, the reaction mass was extracted several times with water to remove ungrafted polyacrylamide, and the graft copolymer was washed with ethanol and vacuum dried at 60°C. Weight percent polyacrylamide in the graft copolymer (% add-on) was 35%, based on the weight gain of starch.

Characterization of Graft Copolymers. Starch was removed from starch-*g*-PAN by heating 2 g of graft copolymer in 150 ml of 0.5*N* hydrochloric acid for 1.5 hr under reflux. The residual PAN was separated by filtration, washed with water and with ethanol, and dried. Infrared analysis showed the amount of carbohydrate remaining to be near the limit of detectability of about 5%. Percent add-on was calculated by weight loss on acid hydrolysis, and the molecular weight of PAN ($\bar{M}_v$) was determined from its intrinsic viscosity in DMF solution (26). Grafting frequency, expressed as the average number of anhydroglucose units (AGU) per grafted PAN chain, was calculated from % add-on and $\bar{M}_v$ of PAN.

Solubility Determination in Dimethylsulfoxide (DMSO). About 2-2.5 g of polymer, which had been previously oven dried, was allowed to stand overnight at room temperature over water in a closed container. Water contents of these moistened polymers ranged from 13-22%, depending on PAN content. About 1.2 g (accurately weighed) of moistened polymer was dispersed in 200.0 g of DMSO, and the resulting mixture was stirred and heated to 75°C. The flask was stoppered, and the dispersion was heated with occasional stirring in a 75°C oven for 22 hr and then was cooled to room temperature. The mixture was centrifuged, and the supernatant was gravity filtered through fluted Whatman 54 paper. A weighed portion of filtrate was evaporated to near dryness, and the polymer was precipitated with 95% ethanol, washed with 95% ethanol, and dried. Percent solubility was calculated from the weight of dry soluble polymer and the initial weight of polymer used, after correcting for % moisture.

Preparation of DMSO-Soluble Starch-g-PAN Samples and Water-Soluble Starch-g-Polyacrylamide. Five grams of starch-*g*-PAN (prepared by either ceric- or cobalt-60 initiation) was dispersed in 333 g of DMSO, and the dispersion was heated for 2 hr at 115°C. The cooled dispersion was then treated with ultrasound for 10 min to mechanically rupture gel particles and to render the polymer soluble (Branson Sonifier, Model S 125;

set for maximum output) the treated dispersion then was allowed to filter through fluted Whatman 54 paper. The filtrate was concentrated, and the polymer was precipitated with 95% ethanol. The polymer was removed by filtration, washed with 95% ethanol, and dried under vacuum at 60°C. A portion of sonified starch-g-PAN was subjected to acid hydrolysis, and % add-on by weight loss and $\bar{M}_v$ of grafted PAN were determined. Total solubility in DMSO at 75°C was confirmed for the ceric-initiated starch-g-PAN preparation.

Starch-g-PAN with 40% add-on ($\bar{M}_v$ of PAN:35,000) was prepared by ceric initiation as described above, with the exception that only 10 g of acrylonitrile was used. Five grams of this graft copolymer was dispersed in 200 ml of water, and concentrated hydrochloric acid was added to give a pH of 2.0. The mixture was heated under reflux for 15 min, and the polymer was isolated by filtration. The polymer was suspended in water, the pH was adjusted to 6.5 with sodium hydroxide solution, and the polymer was again separated by filtration. The polymer was washed with water and with ethanol and dried under vacuum at 60°C. The yield was 4.357 g of starch-g-PAN, which was 99% soluble in DMSO at 75°C. The % add-on was 46%, as calculated from the original 40% add-on and the loss of 0.643 g of starch by acid hydrolysis.

Five grams of starch-g-polyacrylamide was dispersed in 333 ml of water and the mixture was heated for 4.5 hr at 95-100°C. The cooled dispersion was then treated with ultrasound, as described for starch-g-PAN in DMSO. The resulting solution was gravity-filtered through fluted Whatman 54 paper, and the filtrate was freeze-dried to give 4.4 g of polymer. To give a denser, more compact product, which might more closely resemble soluble starch-g-PAN, the freeze-dried polymer was dispersed in 20 ml of water, and the polymer was precipitated from the thick paste by addition of ethanol. The polymer was separated by filtration, washed with ethanol, and vacuum dried at 60°C.

Preparation of Starch-PAN Synthetic Mixture. Solutions of both starch and PAN homopolymer were prepared in DMSO at 1% solids. Equal weights of the two solutions were combined and then concentrated under vacuum. The polymer mixture was precipitated with ethanol, and the precipitated solid was washed with ethanol and dried under vacuum at 60°C. PAN content of the mixture was 45%, as determined by weight loss on acid hydrolysis.

Alkaline Saponifications in Water. A suspension of 1.0 g of starch-g-PAN or starch-PAN physical mixture in 9 ml of 0.7N sodium hydroxide (in a 25 ml Erlenmeyer flask) was heated on a steam bath for 5-10 min until the mixture thickened sufficiently to preclude settling. Reactions with either starch or PAN homopolymer were run with 0.5 g of polymer. The flask was loosely stoppered (to permit escape of ammonia) and heated for 3 hr in a 95-100°C oven. The reaction mass was then diluted with water (75 or 150 ml, depending on the weight of polymer saponified) and dialyzed against distilled water. A known weight of the dialyzed dispersion (pH about 7) was freeze dried to determine % solids. The dispersion was diluted with water to 0.25% solids, and the Brookfield viscosity was obtained at 30 rpm. With some dispersions, a well-drained insoluble gel fraction could be isolated by gravity filtration through fluted Whatman 54 paper. A weighed portion of this insoluble gel was freeze-dried, and the absorbency (expressed as grams of water-swollen gel per gram of dry polymer) was calculated from the weight of freeze-dried solid. To reduce viscosity and polymer swelling, the dispersion was acidified with hydrochloric acid to pH 4.5, and a portion of this dispersion was allowed to gravity filter through fluted Whatman 54 paper. The percent soluble polymer was calculated from the % solids in both the filtrate and the unfiltered dispersion, as determined by freeze-drying.

Alkaline Saponifications in Aqueous Ethanol. A solution of 2 g of 50% sodium hydroxide in 4.44 ml of water and 4.44 ml of 95% ethanol was prepared, and 1 g of either starch-g-PAN or starch-PAN physical mixture was added. Reactions with starch or PAN homopolymer were run with 0.5 g of polymer. The mixture was heated under reflux for 3 hr, and the resulting reaction mass was dispersed in water, dialyzed, and analyzed as described for saponifications in water. Saponifications run at higher ethanol:water ratios were carried out in a similar manner.

Enzymatic Hydrolysis of HSPAN. HSPAN for this particular experiment was prepared as follows. Five grams of starch-g-PAN (prepared from granular starch, 50% add-on, $\bar{M}_v$ of PAN:64,000) in 45 ml of 0.7N sodium hydroxide solution was heated for 3 hr at 95-100°C. The reaction mass was added to 500 ml of methanol, and the mixture was subjected to high-shear stirring (Waring Blendor) for 1 min. HSPAN was separated by filtration, washed with methanol, and vacuum dried at 60°C.

Two grams of HSPAN was added to 500 ml of water, the pH was adjusted from 9 to 6.5 with dilute hydrochloric acid, and 0.1 ml of Thermamyl 60L enzyme solution (Novo Enzyme Corp.) was added. The resulting mixture was heated at 95°C for 21 hr, and the clear yellow solution was exhaustively dialyzed against distilled water. Freeze drying yielded 1.235 g of polymer, which contained only about 5% residual carbohydrate (by infrared analysis). The number average molecular weight of this polymer was 44,000, as determined in 0.15N sodium chloride solution on a Melabs Model CSM-2 membrane osmometer equipped with a B-19 membrane (Schleicher and Schuell Co.).

Film Formation. Starch-*g*-PAN samples used for this study were prepared as described above, except that the small amounts of PAN homopolymer were not removed by DMF extraction. For the graft polymerization with gelatinized starch, the starch-water slurry was heated for 30 min at 85°C before reaction at room temperature with acrylonitrile and ceric ammonium nitrate. Saponifications in these experiments were carried out by stirring 55 g of graft copolymer with 450 ml of 0.7*N* sodium hydroxide solution in a sigma mixer for 2 hr at 90–100°C. The reaction mass was stirred with 3 l. of methanol, and HSPAN was isolated by filtration. The polymer was washed with 90:10 methanol-water, dried under vacuum at 40°C, and ground to pass 40 mesh (yield: 63–64 g).

Either 1.0 g of HSPAN from granular starch or 0.5 g of the polymer derived from gelatinized starch was added to 800 ml of water, and the dispersion was slowly stirred overnight at room temperature. The mixture was centrifuged (1500 X *g*) and the supernatant decanted. The gel fraction was subjected to three more water extractions and was then spread onto a Teflon-coated tray and dried in a forced air oven at about 40°C.

HSPAN films containing 10% by weight of either saponified PAN or soluble starch were prepared by separately saponifying starch-*g*-PAN (prepared from granular starch by ceric-initiation) and PAN homopolymer and then dialyzing the saponificates. Starch solution was prepared by dissolving an acid-modified starch (Clinton 290B) in water. Solutions/dispersions were then combined in the desired ratios (after determining concentrations by freeze drying) and were dried on Teflon-coated trays.

A film from Carbopol 941 (B. F. Goodrich) was prepared by dispersing 0.5 g of polymer in 200 ml of water. Sodium hydroxide solution was added to adjust the pH to 7.5, and the dispersion was then dried on a Teflon-coated tray.

A film from Permasorb 10 (National Starch & Chemical Corp.) was prepared by dispersing 0.50 g of polymer in 200 ml of water and drying the dispersion on a Teflon-coated tray.

Influence of Heating on Absorbency and Moisture Loss. The HSPAN used in this study was a commercial sample prepared from gelatinized starch (SGP 502 S, Henkel Corp.), which we screened to give a fraction that passed 100 mesh but was retained by 200 mesh. The water content was 14.2%, as determined by drying in an Abderhalden drying apparatus for 4 hr under vacuum (1 mm Hg) at 100°C over P_2O_5.

Duplicate 0.5 g samples of HSPAN were accurately weighed into small glass-stoppered vials, and the open vials were kept for 1 hr in an oven heated to the desired temperature. The vials were stoppered, allowed to cool in a desiccator, and then weighed to determine the percentage of the original sample weight lost through volatilization. Percent weight loss of two duplicate samples generally did not differ by more than 0.2.

To determine water absorbency, an accurately weighed 5 mg sample of HSPAN was allowed to soak for 30 min in 50 ml of distilled water. The swollen polymer was then separated from unabsorbed water by screening through a tared 280 mesh sieve, which was 4.8 cm in diameter. The polymer on the sieve was allowed to drain for 20 min, and the sieve was then weighed to determine the weight of water-swollen gel. Absorbency was calculated as grams of water per gram of polymer. No correction for residual moisture in the polymer was applied. Absorbencies were run in duplicate, and results agreed to within 10%.

Literature Cited

1. Fanta, G. F.; Bagley, E. B. Encyclopedia of Polymer Science and Technology, Supplement, Vol. 2, N. M. Bikales, Ed., Wiley, New York, 1977, p. 669.
2. Gugliemelli, L. A.; Weaver, M. O.; Russell, C. R.; Rist, C. E. J. Appl. Polym. Sci. 1969, 13, 2007.
3. Fanta, G. F.; Weaver, M. O.; Doane, W. M. Chem. Technol. 1974, 4, 675.
4. Weaver, M. O.; Montgomery, R. R., Miller, L. D.; Sohns, V. E.; Fanta, G. F.; Doane, W. M. Staerke, 1977, 29, 413.
5. Taylor, N. W.; Bagley, E. B. J. Appl. Polym. Sci. 1974, 18, 2747.
6. Fanta, G. F.; Baker, F. L.; Burr, R. C.; Doane, W. M.; Russell, C. R. Staerke 1977, 29, 386.
7. Rout, A.; Rout, S. P.; Singh, B. C.; Santappa, M. Makromol. Chem. 1977, 178, 639.
8. Mohanty, N.; Pradhan, B.; Mahanta, M. C. Eur. Polym. J. 1980, 16, 451.
9. Fanta, G. F.; Burr, R. C.; Russell, C. R.; Rist, C. E. Cereal Chem. 1970, 47, 85.
10. Hofreiter, B. T. J. Polym. Sci., Polym. Chem. Ed. 1974, 12, 2755.
11. Whistler, R. L.; Ingle, T. R. Starch: Chemistry and Technology, Vol. 1, R. L. Whistler and E. F. Paschall, Eds., Academic Press, New York, 1965, p. 414.
12. Hofreiter, B. T.; Russell, C. R. Staerke 1974, 26, 18.
13. Smith, T. U.S. Patent 3,661,815, May 9, 1972.
14. Peebles, L. H., Jr. Encyclopedia of Polymer Science and Technology, Supplement, Vol. 1, H. F. Mark and N. M. Bikales, Eds., Wiley, New York, 1977, p. 2.
15. Takata, T.; Hiroi, I.; Taniyama, M. J. Polym. Sci. A, 1964, 2, 1567.

16. Fanta, G. F.; Burr, R. C.; Doane, W. M.; Russell, C. R. J. Appl. Polym. Sci. 1971, 15, 2651.
17. Weaver, M. O.; Bagley, E. B.; Fanta, G. F.; Doane, W. M. Appl. Polym. Symp. 1974, No. 25, 97.
18. Antholz, P.; Swarthout, E. J. U.S. Patent 4,221,684, Sept. 9, 1980.
19. Voyutskii, S. S. Autohesion and Adhesion of High Polymers, Polymer Reviews, Vol. 4, H. F. Mark and E. H. Immergut, Eds., Wiley, New York, 1963, p. 80.
20. Bradford, E. B.; Vanderhoff, J. W. J. Macromol. Chem. 1966, 1, 335.
21. Foster, J. F., Starch, Chemistry and Technology, Vol. 1, R. L. Whistler and E. F. Paschall, Eds., Academic Press, New York, 1965, p. 350.
22. Voyutskii, S. S. Autohesion and Adhesion of High Polymers, Polymer Reviews, Vol. 4, H. F. Mark and E. H. Immergut, Eds., Wiley, New York, 1963, p. 87.
23. Taylor, N. W.; Bagley, E. B. J. Polym. Sci., Polym. Phys. Ed. 1975, 13, 1133.
24. Bagley, E. B.; Taylor, N. W. Ind. Eng. Chem., Prod. Res. Dev. 1975, 14, 105.
25. Fanta, G. F.; Doane, W. M. U.S. Patent 4,116,899, Sept. 26, 1978.
26. Onyon, P. F. J. Polym. Sci. 1959, 37, 315.

RECEIVED December 17, 1981.

14

Grafting Reaction of Dextran onto Polymer Surface

M. TANIGUCHI, R. K. SAMAL, M. SUZUKI, H. IWATA, and Y. IKADA[1]

Institute for Chemical Research, Kyoto University, Uji, Kyoto 611, Japan

In an attempt to make a non-biological polymeric material biocompatible, dextran was covalently coupled onto the surface. The synthetic substrate material selected is an ethylene-vinyl alcohol copolymer which is crystalline and not soluble in water. Hydroxyl groups at and near the surface were first urethanated with hexamethylene diisocyanate to introduce the reactive isocyanate groups onto the surface. Dextran or aminodextran was then grafted onto the surface via urethane or urea linkage formation by using the introduced isocyanate groups. The extent of grafting reaction could be followed, though qualitatively, by measuring the contact angle of the grafted surface against water, since the surface became increasingly hydrophilic as the grafting proceeded. Observed streaming potentials also supported the grafting reaction of dextran onto the surface. However, the surface properties of grafted materials were found to depend markedly on the reaction conditions under which the isocyanate groups were introduced to the surface. Evaluation of the blood compatibility using whole blood showed that the polymeric surface became more blood-compatible when grafted with a sufficient amount of dextran.

Polymer complexes or composites from synthetic and naturally occurring macromolecules sometimes provide materials possessing novel, attractive properties, particularly when we are able to successfully combine the biological activity of naturally

[1] To whom correspondence should be addressed.
Current address: Research Center for Medical Polymers and Biomaterials, Kyoto University, Uji, Kyoto-Fu 611, Japan.

0097-6156/82/0187-0217$6.00/0
© 1982 American Chemical Society

occurring polymers with the excellent stability, physical and chemical, of synthetic polymers (1). One of the typical examples is the enzyme-immobilized complex, which has been used in biochemical basic researches and industries.

Recently, considerable attention has been directed on biomaterials that are used in contact with living tissue and biological fluids for medical applications. One of the primary requisites for the biomedical materials is not to adversely affect the biological components and the entire living organism. This is called biocompatibility. It has been reported that a synthetic material becomes blood-compatible when it is attached or complexed with urokinase, a plasminogen activator (2,3) and with heparin, which inhibits formation of fibrin (4,5). Shimizu and his coworkers have demonstrated that coating of polymeric surfaces with collagen drastically enhances their histocompatibility (6,7). We have recently revealed that blood compatibility may be attained when the material has a diffuse surface which differs from the rigid clear surface of conventional materials (8)(See the model inserted in Figure 1). It seems likely that the diffuse material, once hydrated, will interact with the living tissue only to an insignificant extent, in other words, be inert to the living tissue with which it is in contact, because there exists a dilute aqueous polymer layer at the interface, behaving like a cushion.

Such a diffuse surface may be most easily generated by graft copolymerization of a water-soluble monomer or by coupling reaction of a water-soluble macromolecule onto a material with a clear surface. The graft chains generated should not exert any positive interaction with the living components and hence preferably be nonionic, nontoxic, and stable against the attack of hydrolytic enzymes in biological fluids. Among the water-soluble macromolecules which may satisfy these demands is dextran. This nonionic and water-soluble poly-α-1,6-D-glucose has been clinically used as a blood plasma substitute (9).

In the present work we have used, therefore, dextran as the water-soluble macromolecule to be coupled and a film of an ethylene-vinyl alcohol copolymer (EVAL) as the substrate material. This copolymer has hydroxyl groups but is insoluble in water and can be injection-molded to yield a material with good mechanical properties. The coupling reaction of dextran onto the film surface will be achieved by using di-isocyanate, since dextran, as well as EVAL, has hydroxyl groups which readily react with isocyanate under formation of urethane linkage.

Experimental

Materials. The film of EVAL which contains the ethylene

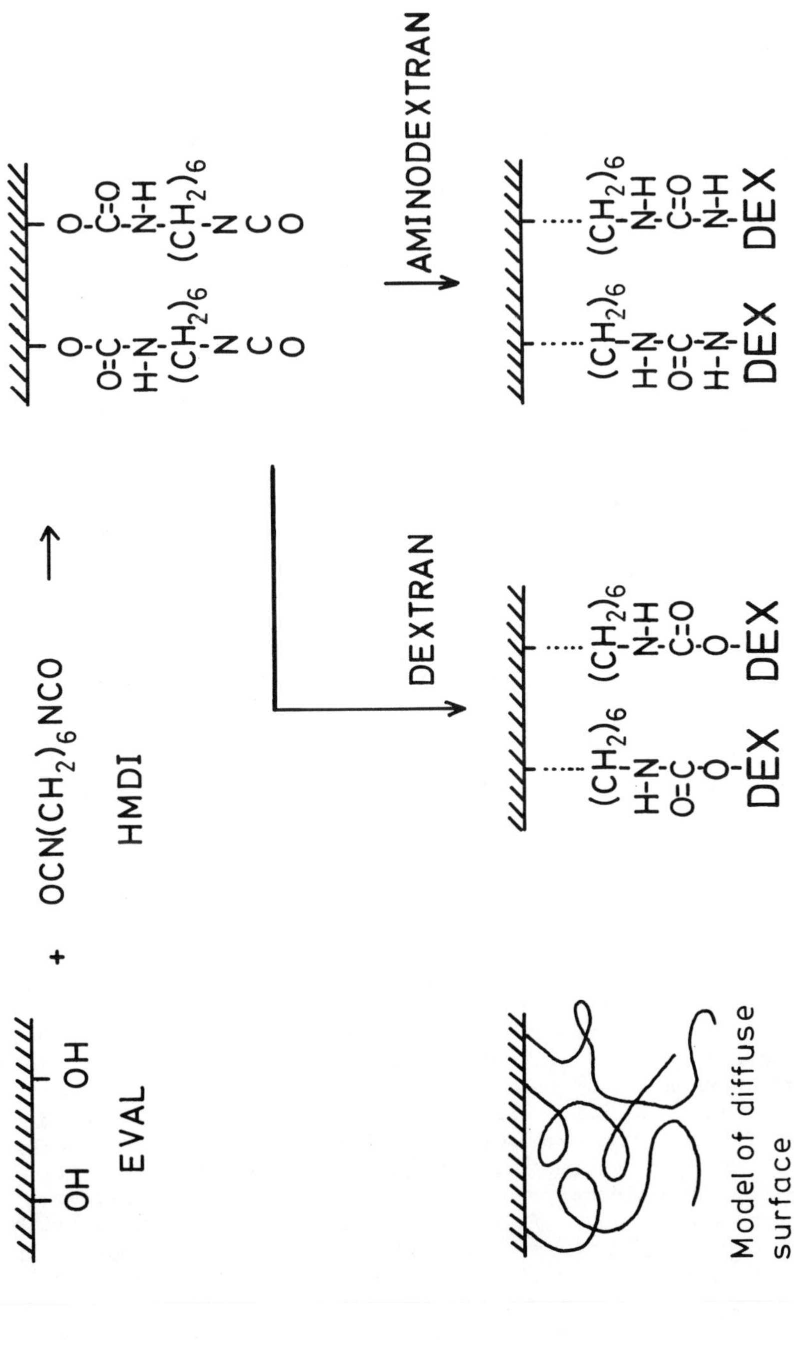

Figure 1. Reaction scheme for coupling of dextran or aminodextran to produce a diffuse surface. Key: EVAL, ethylene–vinyl alcohol copolymer and HMDT, hexamethylene diisocyanate.

unit by 33 mol% and has a thickness of 20 μm, was donated by Kuraray Co. Ltd., Osaka, Japan. It was first cut into 1 x 2 cm, size. The film strips were immersed in 1 N NaOH solution at room temperature for two days, then washed successively with distilled water followed by methanol, and finally subjected to Soxhlet extraction with methanol for two days. The purified films were dried under vacuum at room temperature and stored in a desiccator before used. Dextran powder of molecular weight 200,000 - 300,000, donated by Meito Sangyo Co. Ltd., Nagoya, Japan, was first dissolved in water and reprecipitated with ethanol. After filtration the residue was dried at room temperature and stored until used. Aminodextrans were synthesized by Williamson's reaction (10) which is briefly described as follows. To 30 ml of 5N NaOH containing 7 g of dextran, 7 g of 2-chloroethylamine was added and the reaction mixture was kept at 80°C in a nitrogen atmosphere for given durations. After neutralization with HCl, the solution was subjected to dialysis with deionized water. The degree of amino substitution was determined with the conductivity measurement. Hexamethylene diisocyanate was distilled under reduced pressure and stored in ampoules. Solvents such as toluene, acetone, n-hexane were distilled by normal procedures. Ferric chloride and di-n-butyltin dilaurate were used without further purification.

Coating of glass capillaries and tubes with EVAL. The glass capillaries with diameter of 0.5 mm and length of 15 cm, and glass tubes with length of 5 cm were washed with dichromate in concentrated sulfuric acid for 1 hr at 100°C, followed by washing with distilled water, and preserved by immersing in water or dimethyl sulfoxide (DMSO). The cleaned capillaries and tubes were dipped in 5 wt% EVAL solution in DMSO and then dried at 100°C.

Reaction. Urethanation of EVAL films and EVAL-coated glass capillaries and tubes was carried out at 30 °C with either 100% hexamethylene diisocyanate (HMDI) or 10 vol% HMDI solution in toluene, both in the presence of di-n-butyltin dilaurate (0.07 vol%) as catalyst. The urethanated surface was thoroughly washed with pure acetone to wash away the unreacted HMDI and the catalyst, and then immersed in 3.8 wt% aqueous solution of aminodextran without catalyst or in 10 wt% aqueous solution of the unmodified dextran containing 2 wt% ferric chloride which is an urethanation catalyst. The ungrafted dextran was removed from the surface by washing thoroughly the graft material with water. The surfaces grafted with dextran or aminodextran were subjected to the measurement of contact angle (with films) and streaming potential (with glass capillaries) and to evaluatioin of blood compatibility by *ex vivo* test using whole blood from dogs (with glass tubes).

The method of the surface characterizations and the animal experiment are the same as those described in a previous work(8).

Results and Discussion

Grafting reaction. Figure 1 shows the scheme of the coupling reaction of dextran, with or without primary amino group, onto the EVAL surface. The coupling reaction is to take place via urethane linkage when the unmodified dextran is used and via urea linkage when dextran possesses amino groups, because urea is formed with a much faster rate than urethane if catalyst is not present. We employed both the unmodified dextran and the amino dextran, which was obtained by substitution of a small amount of hydroxyl groups with primary amine obtained by Williamson's method. Dependence of the degree of substitution on the reaction time is given in Figure 2. No attempt was made to increase the degree of substitution to a higher extent than that reached in this experiment, since too high substitution must lead to a significant alteration of the original nature of dextran.

Introduction of the isocyanate group onto the EVAL was carried out by reacting the EVAL either with pure HMDI or with HMDI diluted with toluene to a mixing ratio of HMDI/toluene= 1/9 by volume. As described below, the addition of toluene to HMDI gave a large effect on the surface properties of dextran-coupled EVAL. Therefore, it is convenient to present the experimental results by dividing them into two, i.e., urethanation with 100% HMDI and urethanation with 10% HMDI. In both cases we did not determine the exact extent of urethanation. If necessary, it is possible to determine the amount of urethanated hydroxyl groups by fluorescence spectroscopy, as described in a previous work (11).

Urethanation with 100% HMDI: Change in contact angles against water for the films urethanated with 100% HMDI, followed by coupling with aminodextran having 5.38 mol% of amino group, is given in Figures 3 and 4. The advancing and receding contact angles of grafted films were measured by the inverted bubble method throughout this work in order to avoid the film surface from drying (12). Figure 3 represents the contact angles for the films urethanated with 100% HMDI for different durations but coupled with the aminodextran for 1 hr. On the other hand, Figure 4 includes the contact angle plots for the films urethanated with 100% HMDI for 3 hr, but coupled with the aminodextran for different periods. Therefore, the contact angles at the zero reaction time in Figure 3 are identical to those for the starting virgin film of EVAL, while the contact angles at the zero reaction time in Figure 4 correspond to those for the film urethanated with HMDI. Strictly speaking, the latter film has no isocyanate any more but amino groups because the measurement of contact angles was made after hydro-

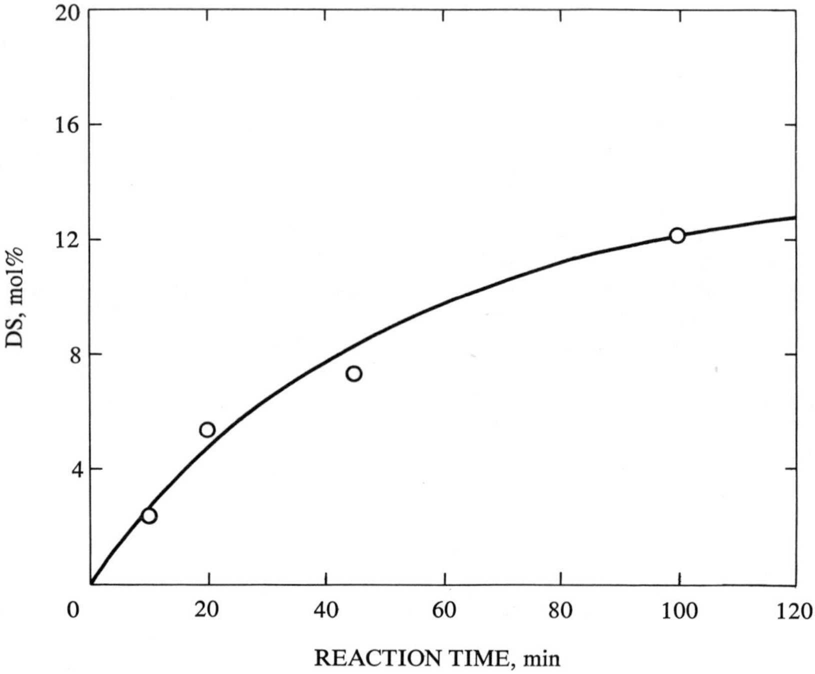

Figure 2. Degree of substitution (DS) of hydroxyl groups as a function of reaction time for the reaction of dextran with 2-chloroethylamine (80°C).

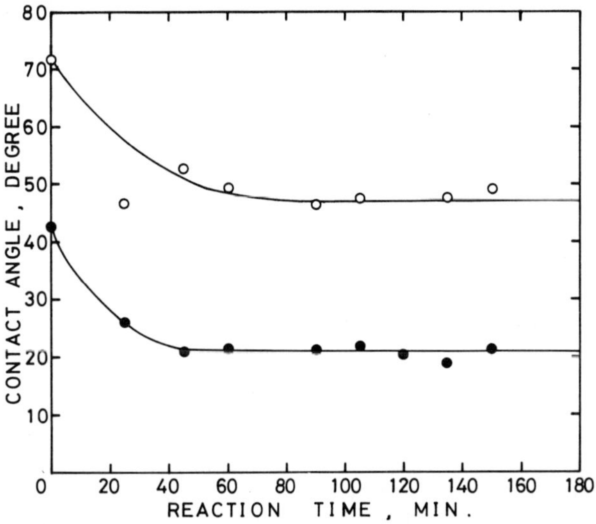

Figure 3. Effect of duration of urethanation with 100% HMDI on the contact angles of EVAL films grafted with aminodextran for 1 h at a polymer concentration of 3.8 wt; DS = 5.38 mol%. Key: ○, advancing; ●, receding.

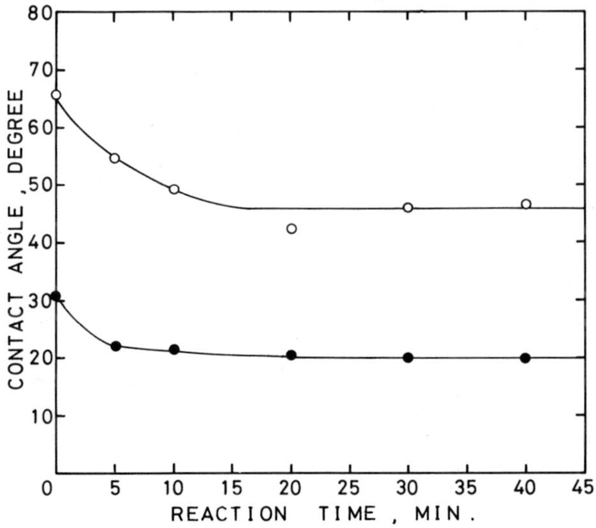

Figure 4. Effect of duration of coupling reaction with aminodextran (DS = 5.38 mol%) on the contact angles of EVAL films grafted. Urethanation was carried out for 3 h with 100% HMDI. Key: ○, advancing; ●, receding.

lysis of the isocyanate to the amino groups, as follows:

$$EVAL\text{-}OCONH(CH_2)_6NCO + H_2O \longrightarrow EVAL\text{-}OCONH(CH_2)_6NH_2 + CO_2$$

Comparison of the contact angles at the zero reaction time reveals that conversion of the hydroxyl groups on the EVAL surface to amino hydrocarbon enhances the film wettability to a small extent. From Figures 3 and 4 it is concluded that urethanation of the film comes to a maximum level within 60 min and the coupling reaction within 20 min. It should be noted that the decrease in contact angles is actually due to the covalent bonding of dextran on the EVAL, but not due to a physical adsorption of dextran, because the contact angles after keeping the urethanated film in contact with aminodextran, are dependent on the duration of urethanation.

Although the extent of reaction cannot be quantitatively estimated from the contact angle data, it provides a number of information on the physico-chemical state of the surface. For instance, the remarkable difference in the advancing and receding contact angle observed in Figures 3 and 4 implies that the surface structure may be considerably different from that of conventional polymers such as polyethylene and cellulose which exhibit no large difference in the advancing and receding contact angle. The phenomenon that the advancing contact angle is significantly different from the receding contact angle is referred to as the hysteresis of contact angle and often reported to be observed, for instance, for water-swollen gels (13). There is, however, no approved explanation for the reason of the contact angle hysteresis, but it seems likely that a distinct hysteresis of contact angle will appear if the surface has such unstable structure as hydrophilic segments or chains at the surface may migrate into the interior of the material upon drying. The EVAL films urethanated with 100% HMDI and then coupled with aminodextran gave a slight increase in contact angle when drying treatment was applied to the grafted films, which, otherwise, were in any case kept immersed in water throughout the present work.

The large hysteresis of contact angle, together with the noticeable influence of drying on the contact angle, suggests that the density of graft dextran chains at the EVAL surface is rather low. Thus, to roughly learn the extent of urethanation, we measured the infrared spectrum of the urethanated film. The result is given in the curve b of Figure 5. Comparison of this spectrum (curve b, Figure 5) with the one for the starting EVAL (curve a, Figure 5), reveals that no peak characteristic of urethane linkage can be detected in the infrared spectrum of the urethanated film. This result denotes that urethanation took place to a very small extent, probably restricted to the region at the film surface.

In addition to the coupling reaction with aminodextran, we attempted to couple the unmodified dextran to the urethanated EVAL film using ferric chloride, a water-soluble catalyst, for urethanation. The coupling reaction of dextran through urethanation was found to take place almost similarly to that of aminodextran (the leveling-off contact angles were 45° for advancing and 20° for receding).

Urethanation with 10% HMDI. The infrared spectrum of the EVAL film urethanated with 10% HMDI in toluene is given in the curve c of Figure 5. As is obvious, the spectrum has the peak characteristic of urethane linkage around 1700 cm^{-1}. This is in marked contrast with that of film urethanated with 100% HMDI. Presence of urethane bonds in such a conventional infrared spectrum strongly indicates that urethanation has taken place not only at the surface but also in the interior of the film. This is highly reasonable because toluene is diffusable into the films without difficulty, enabling the HMDI molecules to react with the inner hydroxyl groups.

Figure 6 shows the contact angles of films urethanated with 10% HMDI for various durations but coupled with dextran in every case for 4 min, while the contact angles of films urethanated with 10% HMDI for 140 min but coupled with dextran for various durations are represented in Figure 7. In both cases the unmodified dextran was used instead of the aminodextran and ferric chloride was added to the reaction mixture by 2 wt% as catalyst. It is apparent from Figures 6 and 7 that urethanation of the film with 10% HMDI reaches the highest level within 100 min and grafting with dextran within 1 min. The reaction rate should, of course, depend on the catalyst concentration.

Some interesting features are recognized in the contact angle plots. By comparing the contact angles at the zero reaction time in Figure 7 with those in Figure 4 it is obvious that the film urethanated with 10% HMDI followed by hydrolysis has lower contact angles than that through urethanation with 100% HMDI. Moreover, coupling of dextran onto the urethanated film leads to a remarkable decrease in contact angles. It is noteworthy that the hysteresis of contact angle substantially disappears as the coupling reaction of dextran approaches to saturation The leveling-off contact angles (15° and 13°) are close to the advancing and receding contact angle of cellulose (12.0° and 11.8°)(11). These values seem to be the lowest contact angles that nonionic polymeric surfaces may have.

The above findings strongly suggest that a high density of dextran chains at the film surface will be attained when dextran is coupled to the EVAL film which has been urethanated with HMDI in the presence of toluene. It appears that the dextran grafting reaction is not limited merely to the film surface, but at variance with the simple model depicted in Figure 1, has taken place also in the region near the surface. This is in contrast with the coupling reaction onto the film urethanated with 100% HMDI. Apparently, the films grafted with dextran through urethanation with 10% HMDI were much more slippery in a hydrated state than those with 100% HMDI.

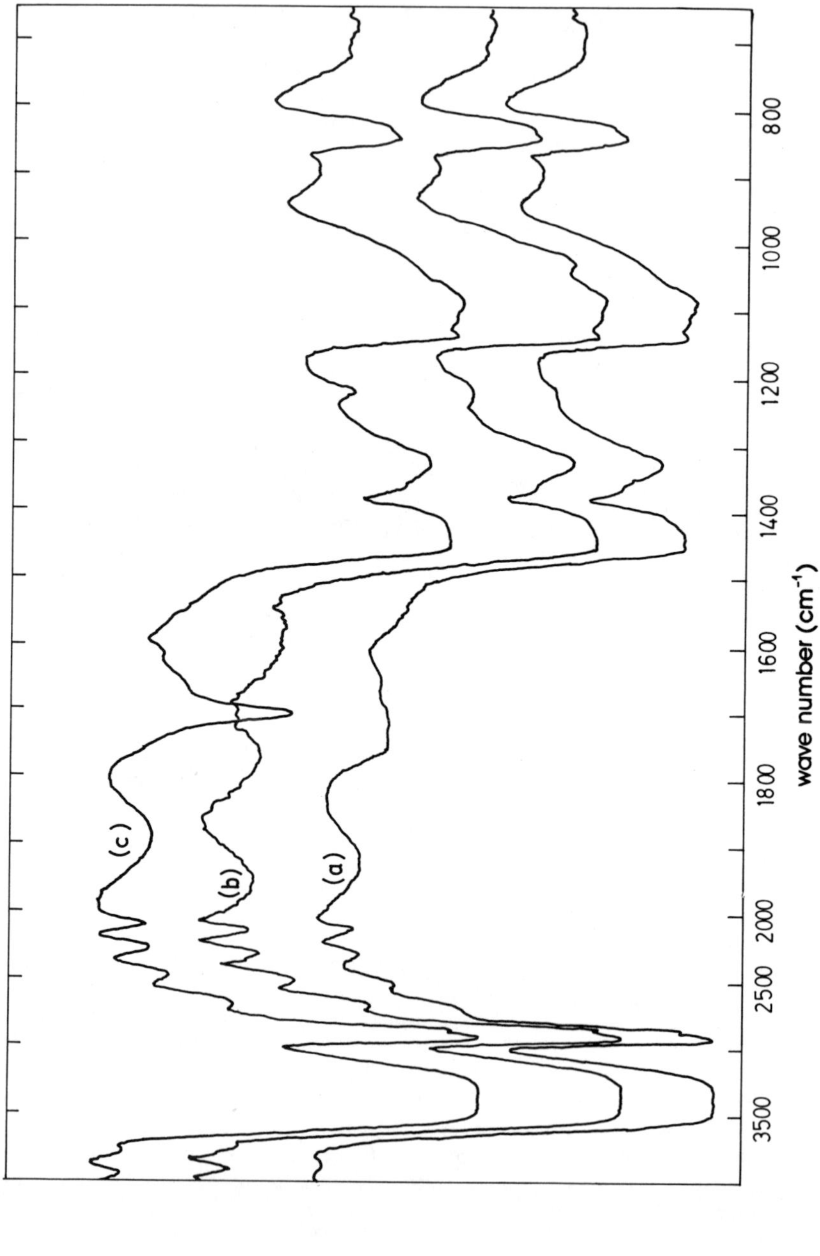

Figure 5. IR spectra of *a*, starting EVAL; *b*, EVAL urethanated with 100% HMDI; and *c*, EVAL urethanated with 10% HMDI in toluene.

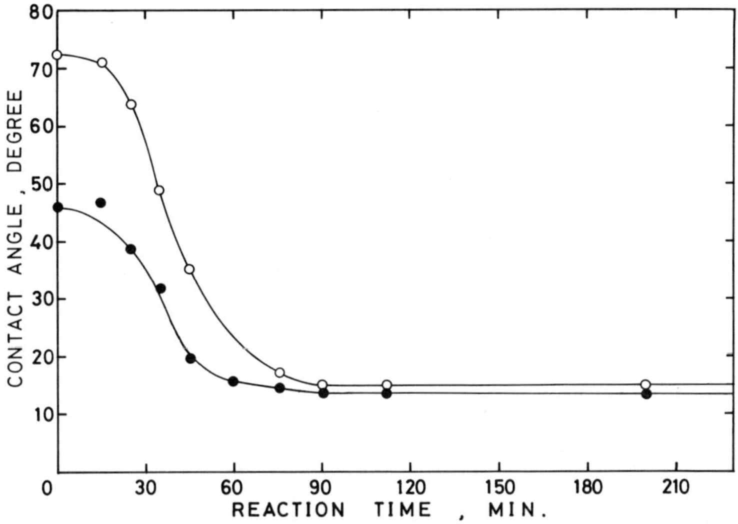

Figure 6. Effect of duration of urethanation with 10% HMDI in toluene on the contact angles of EVAL films grafted with dextran for 4 min at a polymer concentration of 10 wt%. Key: ○, advancing; ●, receding.

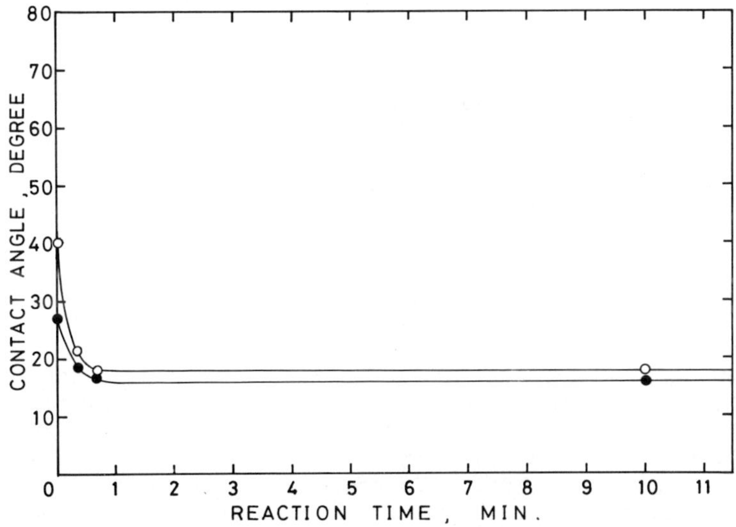

Figure 7. Effect of duration of coupling reaction with dextran on the contact angles of EVAL films grafted. Urethanation was carried out for 140 min with 10% HMDI in toluene. Key: ○, advancing; ●, receding.

Surface properties. As described above, the contact angles of films against water are decreased upon grafting with dextran, indicating that the hydrophobic surface of EVAL film has transformed to hydrophilic. The change in surface properties can be also studied by determining the zeta potential. As we have previously demonstrated, it provides a valuable information regarding the interfacial structure when the material is in contact with an aqueous environment (8). The zeta potential is defined as the potential at the slipping plane or the plane of shear located at the zone from which ions are displaced by the flowing electrolyte solution.

Figure 8 shows the zeta potentials against pH of the flowing electrolyte solution for the virgin EVAL and grafted with pure dextran after urethanation with 100% HMDI. The pH of flowing solution was adjusted by mixing HCl with NaOH, the ionic strength being kept to 10^{-3}. As is seen in Figure 8, the zeta potentials are negative over the whole range of pH studied. This may be due to selective adsorption of negative ions on the surface as is the case for most of the nonionic polymers such as polyethylene. It is expected that contribution of the adsorbed ions to the zeta potential will be decreased with the increasing extent of dextran covering on the material surface because of shifting of the slipping plane. The rather small variation in zeta potential upon grafting with dextran implies that the surface is not densely grafted by the dextran chains.

On the other hand, the zeta potentials of the EVAL urethanated with 10% HMDI, followed by coupling with the unmodified dextran, have much smaller negative values than those of the EVAL urethanated with 100% HMDI and then coupled with dextran. The result is given in Figure 9, where the zeta potentials of the EVAL urethanated, followed by hydrolysis, are also given together with those of the EVAL coupled with the aminodextran whose degree of substitution is 5.38 mol%. It is seen that the surface grafted with unmodified dextran through urethanation with 10% HMDI gives the zeta potentials which are much close to zero in comparison with those urethanated with 100% HMDI and then coupled with dextran. This suggests a higher surface density of dextran in the modification with 10% HMDI, in accordance with the result of the contact angle. Expectedly, zeta potentials of positive value are observed over a wide range of pH for the surface having the grafted aminodextran chains. The difference in zeta potential between the urethanated and subsequently hydrolyzed surface and the virgin EVAL is rather small to suggest that the surface, urethanated and then hydrolyzed, carries a relatively small amount of negative fixed charge.

Finally, to examine the biocompatibility of the modified

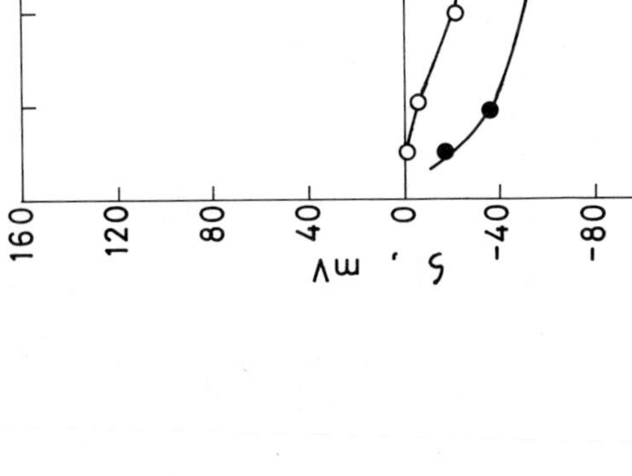

Figure 8. pH dependence of zeta potentials of urethanation with 100% HMDI. Key: ●, *virgin EVAL;* ○, *EVAL urethanated for 3 h with 100% HMDI grafted with dextran for 1 h.*

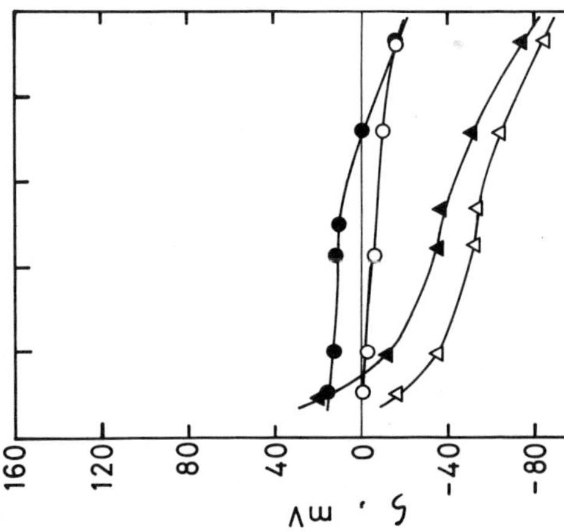

Figure 9. pH dependence of zeta potentials of urethanation with 10% HMDI. Key: △, *virgin EVAL;* ▲, *EVAL urethanated and then hydrolyzed;* ○, *EVAL urethanated for 140 min with 10% HMDI in toluene grafted with dextran for 4 min;* ●, *EVAL urethanated for 140 min with 10% HMDI in toluene grafted with aminodextran for 4 min.*

surfaces, we followed thrombus formation which occurred when they were allowed to come into contact with the whole blood of dog for different durations. Figures 10 and 11 represent the result for the dextran-coupled surfaces, together with that of glass which was used as reference. Thrombus formation was not studied for the aminodextran-coupled ones, because the ionic charge could affect the blood compatibility. The ordinate of Figures 10 and 11 is the optical density at 540 nm of the aqueous solutions obtained by filtrating the mixtures made from pure water and the blood which had been brought into contact with the various surfaces. The more the thrombus is formed, the lower is the optical density, because the formed thrombi occlude the red blood cells which would, otherwise, release hemoglobin molecules through hemolysis. As can be seen from Figures 10 and 11, the grafting reaction of dextran through urethanation with 100% HMDI results in formation of more amounts of thrombi than before grafting. On the other hand, it is apparent that the EVAL grafted with dextran through urethanation with 10% HMDI in toluene is good in blood compatibility in comparison with the virgin EVAL. Further evaluation of blood compatibility with a different method is currently being carried out for the modified surfaces, since the method applied in this work is not sufficiently sensitive for the evaluation of blood compatibility.

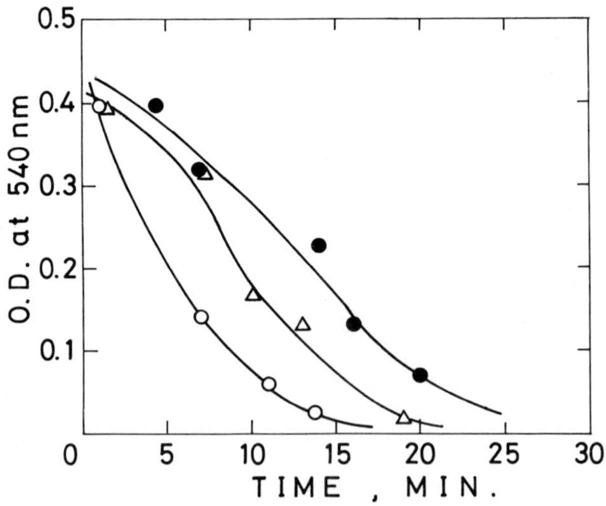

Figure 10. Ex vivo evaluation of blood compatibility of urethanation with 100% HMDI. Key: ○, glass; ●, virgin EVAL; △, EVAL urethanated for 3 h with 100% HMDI grafted with dextran for 1 h.

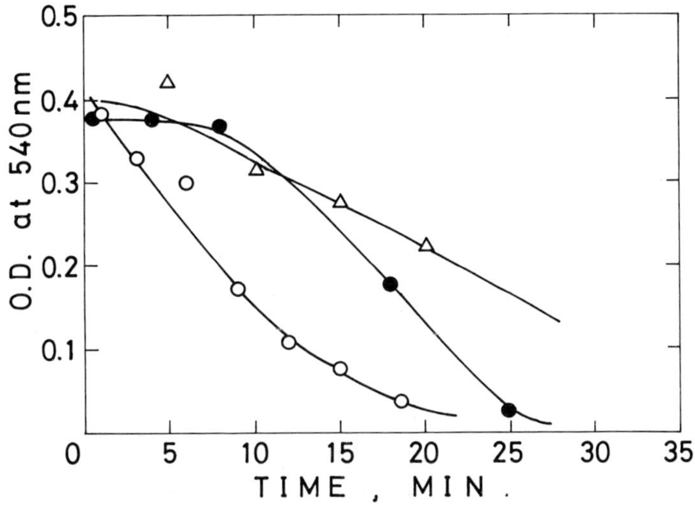

Figure 11. Ex vivo evaluation of blood compatibility of urethanation with 10% HMDI in toluene. Key: ○, glass; ●, virgin EVAL; △, EVAL urethanated for 140 min with 10% HMDI in toluene grafted with dextran for 4 min.

Conclusions

Aminodextran and dextran can be coupled rapidly to the urethanated EVAL film through urea or urethane formation, as evidenced by contact angle and zeta potential. IR assay shows that the coupling reaction has taken place not only at the film surface but also in the interior.

Dextran-coupled surfaces give rise to thrombus formation to a lesser extent than the virgin hydrophobic surface. Although antithrombogenicity of the modified surfaces should be further estimated with more effective methods *in vitro* as well as *in vivo* we may conclude that the grafting reaction of dextran onto the EVAL surface is promising for preparing the blood-compatible material, especially when we recall that the crystalline EVAL has good chemical and mechanical properties as implantable material. If dextran has short lifetime in contact with body fluids, we can graft water-soluble polymers much more stable against hydrolysis, such as poly(vinyl alcohol).

Literature Cited

1. Hixson, H. F., J_R.; Goldberg, E. P., Eds.; "Polymer Grafts in Biochemistry"; Marcel Dekker, Inc, New York and Basel, 1976.
2. Iwata, H.; Ikada, Y. Abstr. 6th Intern. Congress Rad. Res., Tokyo, 1979,; P.60.
3. Sugitachi, A.; Tanaka, M.; Kawahara, T.; Takagi, K. Trans. Am. Soc. Artif. Intern. Organs, 1980, 26, 274.
4. Chang, T.M.S. in "Biomedical Applications of Polymers", Gregor, H. P., Ed., Plenum Press, New York and London, 1975; p.147.
5. Mori, Y.; Nagaoka, S.; Masubuchi, Y; Itoga, M.; Tanzawa, H.; Kikuchi, T.; Yamada, Y.; Yonaha, T.; Watanabe, H.; Idezuki, Y. Trans. Am. Soc. Artif. Intern. Organs, 1978, 24, 736.
6. Shimizu, Y.; Abe, R.; Teramatsu, T.; Okamura, S.; Hino, T. Biomat., Med. Dev., Art. Org. 1977, 5, 49.
7. Shimizu, Y.; Miyamoto, Y.; Teramatsu, T.; Okamura, S.; Hino, T. Biomat., Med. Dev., Art. Org. 1978, 6, 375.
8. Ikada, Y.; Iwata, H.; Horii, F.; Matsunaga, T.; Taniguchi, M.; Suzuki, M.; Taki, W.; Yamagata, S.; Yonekawa, Y.; Handa, H. J. Biomed. Mater. Res. 1981, 15, 697.
9. Gronwall, A. "Dextran and its Use in Colloidal Infusion Solutions", Academic Press, Inc., New York, 1957; p. 156.
10. Jakubovic, A.O.; Brook, B.N. Polymer. 1961, 2, 18.
11. Matsunaga, T.; Ikada, Y. in "Modification of Polymers", Carraher, C.E.; Tsuda, M. Eds., ACS Symposium Series, 1980. 121, 391.
12. Hamilton, W.C., J. Colloid Interface Sci. 1972, 40, 219.
13. Holly, F.J.; Refojo, M.F. J. Biomed. Mater. Res. 1975, 9, 315.

RECEIVED December 24, 1981.

15

Graft Copolymerization of Acrylonitrile onto Bamboo Using IR Spectroscopic Technique as a Probe

K. C. LIN, Y. S. WU, T. H. TAN, and T. L. TAI

National Taiwan University, Department of Chemistry, Taipci, Taiwan, ROC

Graft copolymerization of acrylonitrile onto bamboo using ceric salt as an initiator was investigated. Evidence for the formation of graft copolymer was obtained by means of the infrared method. Both spectroscopic and gravimetric determinations were performed in evaluating the grafting percentages. Results indicate that, due to the possible formation of ceric-adsorbed species, grafting percentages evaluated by the latter method were found to be less reliable. In the presence of relatively high concentration of ceric ion, severe deviation from the true value of grafting percentage might even be the result of oven-drying the initial products of graft copolymerization before removing homopolymers from the grafting system. The grafting reaction was found to depend strongly on the concentration of ceric ion and the maximum grafting occurred in a range around 0.15 mol/l at 40°C. In addition, the polymer branches of graft copolymers were characterized by gel permeation chromatography. The effect of lignin contained in bamboo on graft copolymerization reaction was also examined.

Bamboo, a lignocellulosic material, is an abundant natural resource in some parts of the world. Although the use of ceric salts for initiating graft copolymerization of vinyl monomers on cellulosic material has been extensively investigated (1-16), information concerning grafted bamboo is rather scarce in the literature. The graft copolymerization of methyl methacrylate onto bamboo using ceric salts as an initiator has previously been studied by us using infrared spectroscopy (17-19). The formation of graft copolymers could be detected by the appearance of some characteristic absorptions attributed to the polymer branches of the graft copolymers. The grafting was found to be not only dependent strongly on the concentration of ceric ion, but also

0097-6156/82/0187-0233$6.00/0
© 1982 American Chemical Society

influenced by some other factors such as the concentration of monomers, the amount of swelling agent, the acidity of the aqueous solution, the reaction temperature, the reaction time and even the grafting procedures (18). The grafting of methyl methacrylate onto bamboo was shown to occur at the cellulose component.

In the present investigation, the grafting of acrylonitrile onto bamboo was conducted in a similar manner. By optimizing all other factors which could affect the grafting, the effect of the concentration of ceric ion on the grafting was examined. The grafting percentages were evaluated with both gravimetric and spectroscopic methods, and the reliability of the data obtained was checked by hydrolyzing grafted bamboo samples and isolating the polyacrylonitrile branches of graft copolymers. Efforts were also made to obtain information concerning the effect of lignin contained in bamboo on the graft copolymerization reaction. In addition, the polymer branches of the graft copolymers were characterized by gel permeation chromatography.

Experimental

Samples and Reagents. The bamboo samples used were three-year old Muli (Melocanna baccifera), obtained from the Taiwan Forest Research Institute. The samples were prepared by stripping off the outermost waxy surface and then grinding them into powder. The bamboo powder was then screened through a 60 mesh sieve and dried in an oven at 110°C to constant weight.

Acrylonitrile monomer (for synthesis) obtained from E. Merck was extracted with 6-8% aqueous sodium hydroxide solution three times in a separation funnel for removing inhibitor. The liquid monomer was then washed with distilled water several times and dried over anhydrous calcium chloride.

Ceric ammonium nitrate (GR grade) was used without purification. Acetone and N,N-dimethylformamide (DMF) were freshly distilled under reduced pressure before use.

Holocellulose was isolated from bamboo powders by removing lignin and resinous compounds using alternate treatments of glacial acetic acid and sodium chlorite at 70-80°C for three hours according to the method of Ku and Chiou (20).

Graft Copolymerization. Dried bamboo sample (0.25 g) was first swollen in 40 ml distilled water for at least 15 minutes at room temperature in a 50 ml, stoppered, Erlenmeyer flask. The flask was then placed in a thermostated water bath maintained at 40 ± 1°C. By adding 0.8 ml of dilute nitric acid (0.005 M) and 1.25 ml of acrylonitrile monomer into the flask with stirring, followed by the adding of a certain amount of initiator into the reaction mixture, copolymerization reaction was initiated. The grafting reaction was discontinued one hour later by cooling the reaction mixture down to room temperature and the reaction mixture was then allowed to stand overnight. The reaction products were filtered and washed well with distilled water several times. They were then soaked and extracted repeatedly with 200 ml N,N-dimethylformamide several times during a period of 12-16 hours until the grafted sample showed a negligible loss on further extraction. The grafted sample was finally washed repeatedly with acetone and then either vacuum-dried or dried in an oven at 100°C to constant weight.

For the purpose of evaluating the weight of homopolymer occluded in the reaction product, the initial products of graft copolymerization were ovendried at 100°C prior to the soaking-extraction treatment. However, it was found that even though thoroughly extracted with N,N-dimethylformamide, the sample obtained could gain considerable weight owing to the formation of ceric-adsorbed species.

Evaluation of Grafting Percentage. In the gravimetric method, all the results are based on the increase in weight of the substrate after grafting, followed by thorough extraction and washing. The grafting percentage is computed as the percent increase in the weight of grafted sample over the original weight of bamboo sample as follows:

$$\text{Grafting Percentage} = \frac{W_g - W_o}{W_o} \times 100\ \%$$

where W_g is the weight of grafted bamboo and W_o is the original weight of bamboo sample.

In the spectroscopic method, the grafting percentages were evaluated with the aid of a calibration curve. The calibration was obtained by plotting the grafting percentages calculated on the basis of the percent in weight of polyacrylonitrile over that of bamboo versus the band intensity ratio between the nitrile absorption of polyacrylonitrile and the hydroxyl absorption of bamboo. For each calibration point, a certain given amount of polyacrylonitrile and bamboo were used and the KBr pellets were prepared quantitatively for infrared measurements. The infrared spectra were recorded on a Perkin-Elmer 580 infrared spectrophotometer in absorbance mode. Similar procedures were carried out to obtain the calibration curve for holocellulose-polyacrylonitrile copolymers.

Hydrolysis of Bamboo-Polyacrylonitrile Copolymer. In order to isolate the polymer branches of the graft copolymers and to check the accuracy of the grafting percentages evaluated in the gravimetric and the spectroscopic methods, the hydrolysis of graft copolymers was carried out. A mixture of 0.25 g of grafted bamboo sample and 125 ml of 75% aq. $ZnCl_2$ solution was stirred overnight with a magnetic stirrer. An equal volume of 6 N HCl was added, and the mixture was stirred for 1-2 hours. The contents of the flask were continuously heated for another 4 hours with constant stirring at 100°C. After the completion of the reaction, polyacrylonitrile was precipitated in methanol-H_2O (3:1) mixed solvent, filtered, washed with distilled water, and dissolved in N,N-dimethylformamide. Polyacrylonitrile could again precipitate out by pouring the solution into water with stirring. After repeating washing and dissolving procedures three times, the precipitate of polyacrylonitrile was finally dried in oven at 60°C to constant weight, and the grafting percentage could thus be calculated.

Determination of Adsorbed Ceric Ion. In determining the amount of ceric ion adsorbed on the graft copolymers, the volumetric titration was conducted. Here, a given excess amount of ferrous sulfate solution (0.01 N) was added into the aqueous solution of hydrolysis and then the solution was back-titrated with 0.01 N ceric ammonium sulfate in a 1 N sulfuric acid solution using o-phenanthroline as an indicator. In so doing, the amount of ceric ion adsorbed on the grafted bamboo sample could thus be determined.

Molecular Weight Determination. The molecular weight distributions of the isolated polyacrylonitrile branches were obtained with a Waters Associates Model ALC/GPC 244 gel permeation chromatography unit equiped with a Model 730 data module. Either two μ-bondagel E-linear columns or four μ-styragel columns, of pore sizes 10^4, 10^3, 500 and 100 Å, were used to obtain the gel permeation chromatograms, which were calibrated with seven standard polystyrene samples in the range from 2.2×10^3 to 2.0×10^5. The chromatrograms of the grafted polyacrylonitrile samples were obtained by injecting 300 μl of a 0.5-1.0 % solution in N,N-dimethylformamide. N,N-dimethylformamide was chosen as the eluting solvent, and the flow rate was maintained at 1.0 ml/min.

Results and Discussion

Infrared Spectra of Grafted Bamboo. It has been demonstrated that the occurrence of graft copolymerization of methyl methacrylate onto bamboo can be accertained by the presence of characteristic absorptions of polymer branches in the infrared spectrum (17-19), in addition to the weight increased in bamboo samples. In this study, similar procedures were conducted for the grafting of acrylonitrile. The formation of graft copolymer could easily be detected by the presence of the distinctive and characteristic nitrile absorption at around 2260 cm^{-1} or the CH_2 deformation vibration at around 1460 cm^{-1}.

A typical infrared spectrum of bamboo which was graft-copolymerized with acrylonitrile is shown in Figure 1. For the purpose of comparison, the background spectrum of bamboo and that of polyacrylonitrile are accompanied with it. As being observed in the case of methyl methacrylate, the vibrational absorptions of the polyacrylonitrile branches of grafted bamboo occurred at about the same frequency positions as those of polyacrylonitrile (21-26). Bands belonging to polymer branches could easily be identified and assigned. As lignin can react quite easily with oxidizing agent such as ceric ion, bands at 1604, 1510, 1460, 1440, 1340, 1250 and 843 cm^{-1} which are assigned to the absorptions of lignin would decrease in their intensities considerably as expected. It is of interest to note that an intense band which was observed at 1670 cm^{-1} may be attributed to $-CONH_2$ (27) and probably not to a C=N stretching absorption resulting from the partial nitrile polymerization.

It is worthwhile to mention here that occasionally a peculiar sharp and strong band might appear at 1386 cm^{-1} in the spectrum of grafted bamboo. Since this band was not observed in the spectrum of grafted holocellulose, we thought that either benzene-alcohol extract or lignin might be responsible for the appearance of this band. As shown in Figure 2, the comparative examination of the infrared spectra observed for the extract obtained by Soxhlet-extraction of bamboo samples with benzene-alcohol mixed solvent (67/33 v/v %) for 72 hours, and that for the reaction product between the extract and ceric ion could evidently demonstrate the origin of the band.

Graft Copolymerization of Acrylonitrile onto Bamboo. The effect of ceric ion concentration on the graft copolymerization of acrylonitrile onto bamboo is shown in Figure 3. As can be seen, the grafting percentage increases to a maximum and then decreases with further increase in the concentration of the initiator. The optimum concentration of the initiator was found at around 0.15 M.

The grafting percentages on both curve (A) and curve (B) were calculated gravimetrically, while those on curve (C) were evaluated spectroscopically. The data points on curve (A) were obtained in the case of the reaction products of the grafting reaction being ovendried at 100°C before extracting homopolymers from the grafting system. Those on curve (B) and curve (C) were obtained without drying treatment. In fact, the band intensities of the characteristic nitrile absorption for graft copolymers on curve (A) do not increase properly and correspondingly with the increase in weight of graft copolymers even though the homopolymers were carefully

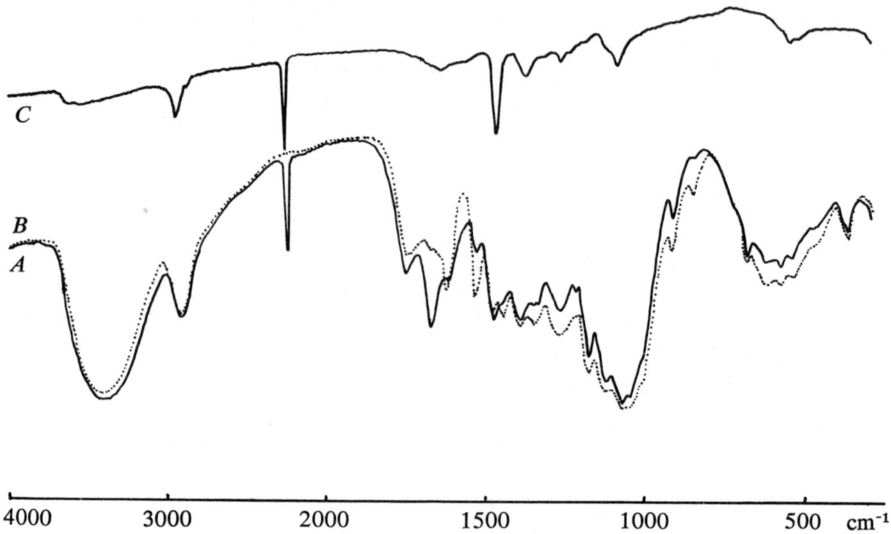

Figure 1. IR spectra. Key: A, ——, bamboo grafted with acrylonitrile; B, – – –, background spectrum of bamboo; C, ——, background spectrum of polyacrylonitrile.

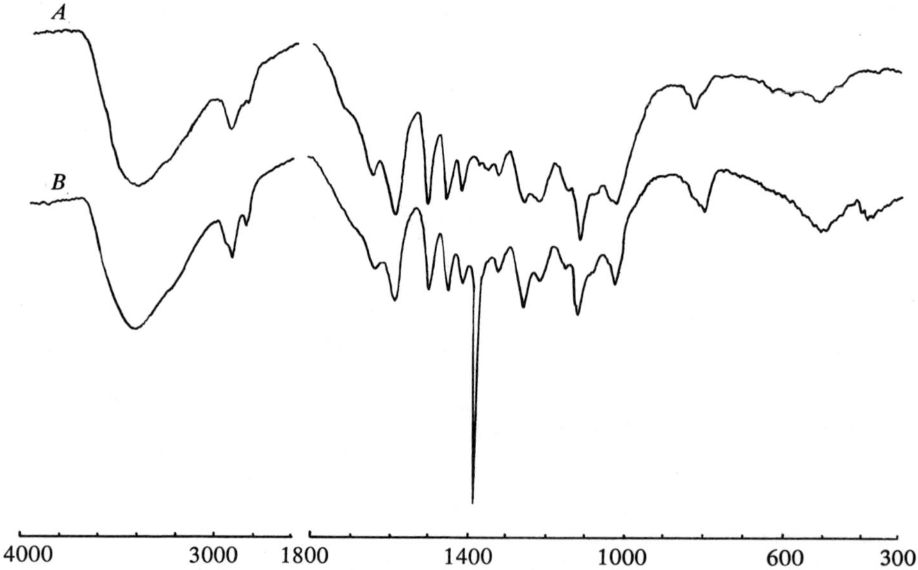

Figure 2. IR spectra. Key: A, benzene–alcohol extract obtained from bamboo; B, benzene–alcohol extract after reacting with ceric ion.

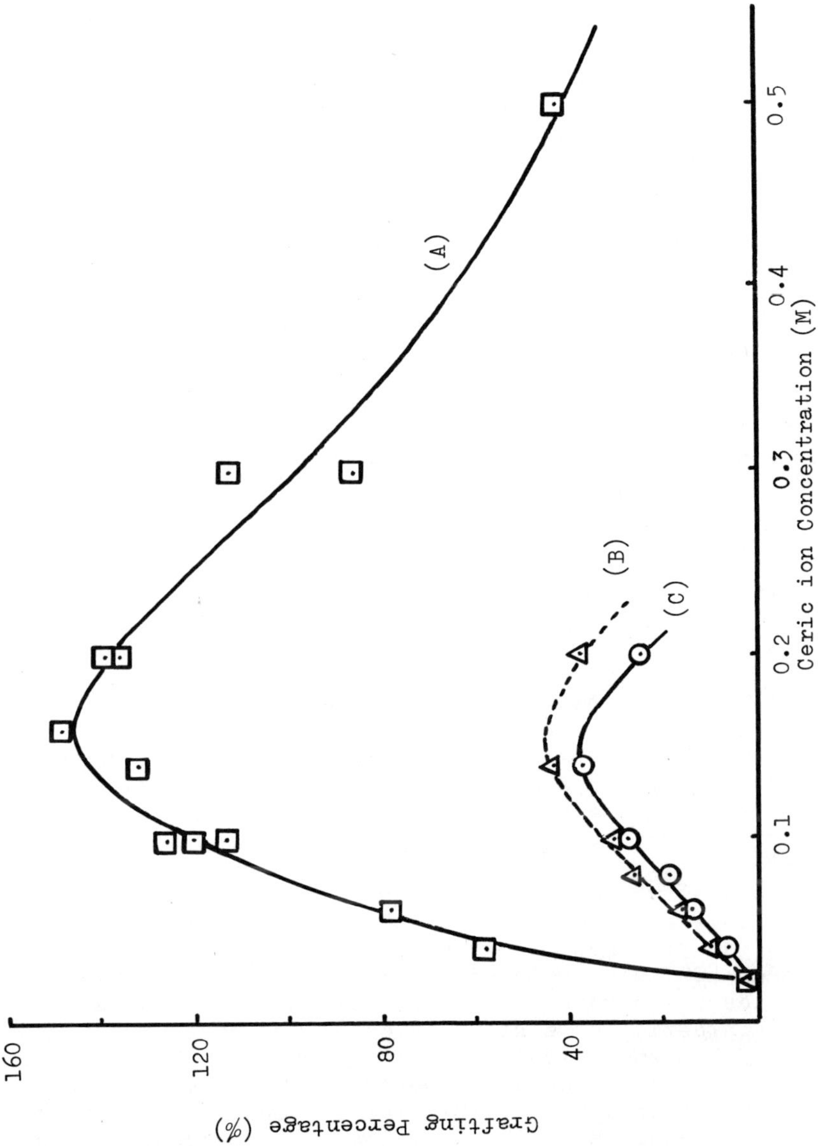

Figure 3. Effect of ceric ion concentration on the graft copolymerization of acrylonitrile onto bamboo. Key: A, using gravimetric method with drying treatment; B, using gravimetric method without drying treatment; C, using spectroscopic method.

and thoroughly extracted. This may indicate that the reliability of grafting percentages on curve (A) is questionable.

By isolating the polymer branches of graft copolymers, the true grafting percentages for grafted bamboo samples on curve (A) could be evaluated. As bamboo samples could be hydrolyzed in an acid solution containing acetic anhydride, the grafted bamboo samples were hydrolyzed and the precipitates of polyacrylonitrile could be obtained. In fact, the grafting percentages computed from the precipitates of polyacrylonitrile give a good approximation to the true values. The grafting percentages so evaluated decrease remarkably but are agreeable with the results shown on curve (B).

To find the cause of this discrepancy, the determination of the amount of ceric ion adsorbed on the bamboo samples was attempted. The weight of adsorbed ceric ions in each graft copolymer was determined and converted into the grafting percentages based on the weight of original bamboo samples. Results of the amount of ceric-adsorbed species and polyacrylonitrile precipitate and the sum of these two values are given in Table I. Obviously, the high grafting percentages shown on curve (A) are not true and can be accounted for satisfactorily as the sum of ceric-adsorbed species and grafted polyacrylonitrile. In consequence, severe deviation from the true value of grafting percentage could be a result of drying treatment of the reaction products before removing homopolymers from the grafting system, especially in the presence of relatively high concentration of ceric ion.

The grafting percentages of grafted bamboo samples evaluated without giving any heat treatment are given in Table II. Values obtained from the gravimetric method are somewhat higher than those obtained from the spectroscopic method. This is probably due to the difficulty in removing both homopolymer and adsorbed ceric ion thoroughly.

As shown in Figure 3, the grafting percentage increases with increasing concentration of the initiator up to 0.16 M. Thereafter, further increase in ceric ion concentration brings about lower grafting yields. This phenomenum was ascribed to fast termination of the growing polymer chains owing to the presence of excess ceric ion (15). Varma and Narasimham (28) have pointed out that in the copolymerization of cellulose, the rate of polymerization should increase with increase in ceric ion concentration at low concentration of ceric ion, but the situation would be reversed at higher concentrations of ceric ion. Therefore, a maximum grafting percentage is obtained at a particular optimum concentration of ceric ion, beyond which the grafting percentage decreases progressively. The present results are in agreement with our previous finding for methyl methacrylate-grafted bamboo (18), and are also consistent with the results obtained for grafted cellulose by other investigators (15,28,29).

Table I. Grafting percentage of grafted bamboo evaluated by gravimetric method.

Ceric Ion Concentration (M)	Grafting* Percentage (%)	Polyacrylonitrile Precipitate (%)	Ceric Ion Adsorbed (%)	Sum of columns 3 & 4 (%)
0.02	2.1	2.0	1.0	3.0
0.06	80.1	16.1	60.4	76.5
0.10	121.4	29.4	86.9	116.3
0.14	132.0	38.2	96.4	124.6
0.16	144.3	40.1	107.1	147.2
0.20	136.7	27.8	105.8	133.6

Weight of bamboo sample: 0.25 g

* Data on curve (A)

Table II. Grafting percentages of grafted bamboo.

Ceric Ion Concentration (M)	Grafting Percentage (%)	
	Spectrophotometric Method	Gravimetric Method
0.02	0.7	0.7
0.04	6.2	8.5
0.06	14.2	14.8
0.08	19.2	26.4
0.10	28.1	28.8
0.12	34.7	39.2
0.14	36.8	43.2
0.16	39.4	46.8
0.18	35.9	35.7
0.20	25.5	37.2

Weight of bamboo sample: 0.25 g

The maximum grafting percentage for acrylonitrile-grafted bamboo was found to fall in the range 40-50%. The maximum values are quite comparable with the data reported by Hebeish et al. (6) for grafted cotton.

For the purpose of the characterization of the graft copolymers, polyacrylonitrile branches were isolated from the grafted bamboo through hydrolysis. Gel permeation chromatography was used to determine the molecular weight distributions. Figure 4 shows some of the typical gel permeation chromatograms obtained. As can be seen, the molecular weights of polyacrylonitrile branches of grafted bamboo are rather low and the molecular weight distributions are quite broad, comparing with those of polymer branches of grafted cellulose.

It is of interest to note that, as demonstrated in Figure 4, a drastic change in the pattern of the molecular weight distribution was observed as ceric ion concentration changed from 0.04 M to 0.08 M, while the pattern of the molecular weight distribution remained about the same thereafter, except in the lowering of the molecular weight distribution. In fact, the molecular weight distribution of isolated polyacrylonitrile branches exhibits only a diffused broad band centered at around 1.2×10^5 at low ceric ion concentration, but exhibits two separated bands at higher ceric ion concentration. A weak broad band and an intense broad band centered at around 1.4×10^5 and 5.4×10^3, respectively, were observed in the gel permeation chromatograms with ceric ion concentration at 0.14 M. Apparently, the polymer chains of grafted bamboo are severely shortened as a result of fast termination of copolymerization which seems to occur at about 0.08 M of ceric ion concentration. The number and the broadness of the band observed for molecular weight distribution of grafted bamboo may not only indicate the complexity of the grafting reaction but may also reveal the possible existence of multiple grafting sites or the formation of two different types of polymeric species.

Graft Copolymerization of Acrylonitrile onto Holocellulose. In order to investigate the influence of lignin upon the grafting, the graft copolymerization of acrylonitrile onto holocellulose was undertaken. Figure 5 shows the infrared spectrum obtained for grafted holocellulose. As can be seen, the absorption bands due to lignin disappear and the characteristic features of the polymer branches are exhibited in the spectrum. In fact, without the presence of lignin, the grafting reaction could favorably proceed between holocellulose and acrylonitrile and the higher grafting percentages could thus be obtained. Results of graft copolymerization of acrylonitrile onto holocellulose were shown in Figure 6 and the experimental data was given in Table III. The maximum grafting percentage could reach around 60-70%

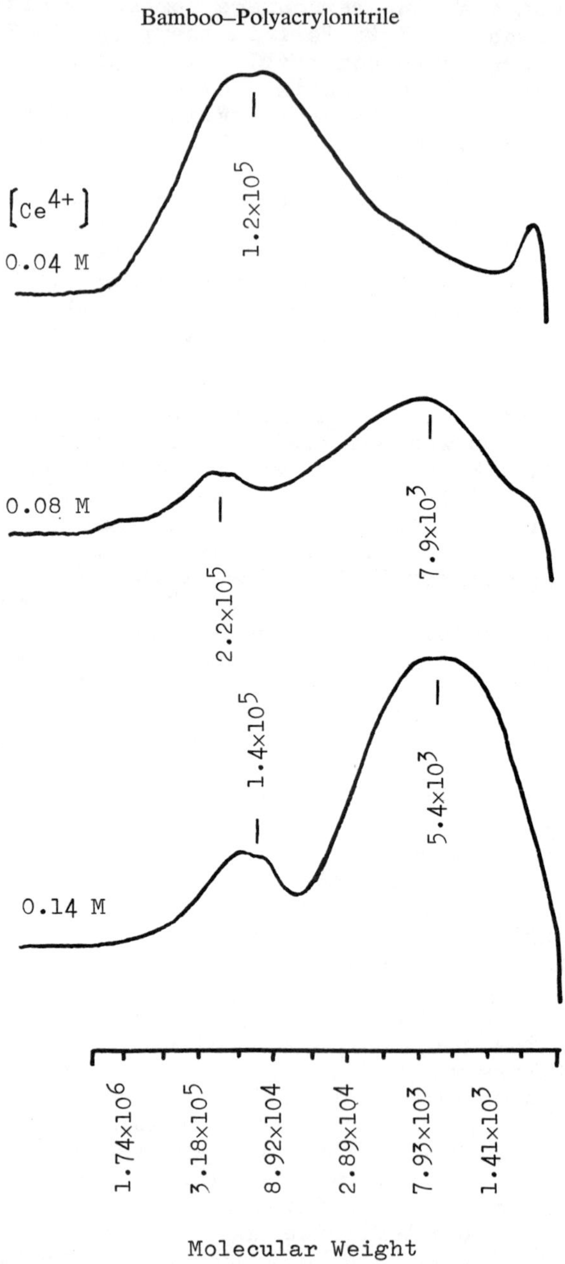

Figure 4. Molecular weight distribution for polyacrylonitrile branches of grafted bamboo.

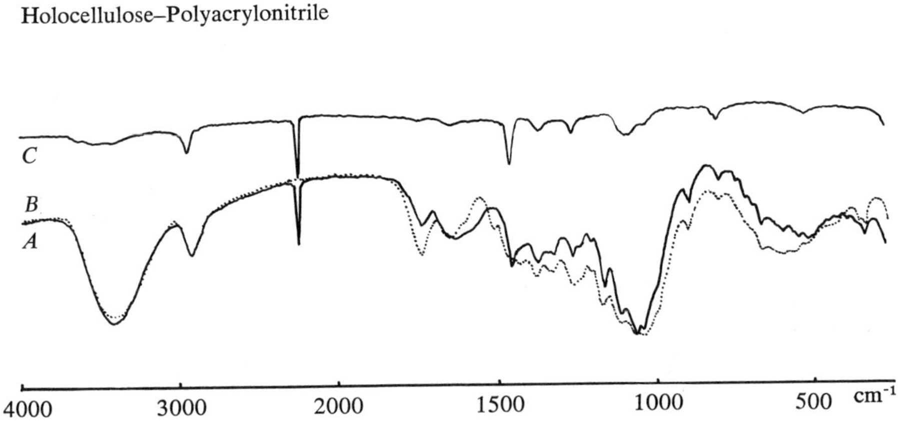

Figure 5. IR spectra. Key: A, ——, holocellulose grafted with acrylonitrile; B, – – –, background spectrum of holocellulose; C, ——, background spectrum of polyacrylonitrile.

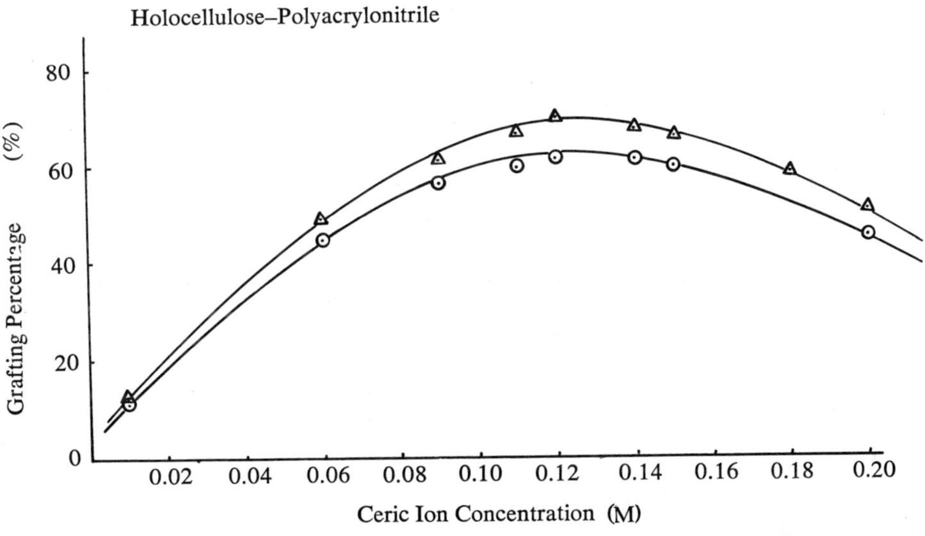

Figure 6. Effect of ceric ion concentration on the graft copolymerization of acrylonitrile onto holocellulose. Key: △, using gravimetric method; ⊙, using spectroscopic method.

Table III. Grafting percentages of grafted holocellulose.

Ceric Ion Concentration (M)	Grafting Percentage (%)	
	Spectrophotometric Method	Gravimetric Method
0.01	11.9	12.1
0.06	44.9	49.1
0.09	56.1	60.8
0.11	59.1	66.4
0.12	61.3	69.6
0.14	61.1	67.2
0.15	59.1	65.6
0.18	58.1	58.9
0.20	44.6	50.2

Weight of holocellulose sample: 0.25 g

which is about a half fold increase in comparison with the maximum grafting percentage of grafted bamboo. As Muli bamboo contains 21% lignin and 4.8% benzene-alcohol extract, the increase in grafting percentages for grafted holocellulose can not be accounted for purely based on the difference in chemical compositions between bamboo and holocellulose. Evidently, the graft copolymerization of bamboo with acrylonitrile would be retarded considerably and that the polymer chains of grafted bamboo would be shortened greatly in the presence of lignin.

For characterizing the polyacrylonitrile branches of grafted holocellulose, gel permeation chromatographic measurements were also taken. Also, the effect of lignin on the graft copolymerization between bamboo and acrylonitrile was examined by comparing molecular weight distribution for grafted bamboo and grafted holocellulose. Some typical gel permeation chromatograms for grafted holocellulose are shown in Figure 7. Similar phenomenum was observed for grafted holocellulose as for grafted bamboo. As can be seen in Figur 7, molecular weight distributions for grafted holocellulose are much higher than those for grafted bamboo. In fact, a single band centered at 4.5×10^5 was observed for molecular weight distribution of grafted holocellulose at low ceric ion concentration (0.01 M). However, as the concentration of ceric ion increased, the maximum position of the band shifted toward lower molecular weight, and the band became broader on the side of lower molecular weight. Actually, for grafted holocellulose with ceric ion concentration at 0.12 M, the first band shifted down to 2.2×10^5 and a second band centered at around $9.0 \times 10^4 - 1.0 \times 10^5$ appeared, which is comparable with the value obtained for grafted cellulse (8). Accordingly, the difference in molecular weight distributions for grafted bamboo and grafted holocellulose is quite remarkable and may reflect the influence of lignin on the grafting copolymerization of bamboo.

Conclusions

The results of the present study on the graft copolymerization of acrylonitrile onto bamboo can be summarized as follows:

(1) The appearance of the nitrile absorption at 2260 cm^{-1} can be used as a criterium for the formation of graft copolymers.

(2) The grafting reaction depends strongly on the concentration of ceric ion and the maximum grafting occurs in a range around 0.15 M for grafted bamboo and around 0.12 M for grafted holocellulose at 40°C.

(3) Grafting percentages evaluated with the spectro-

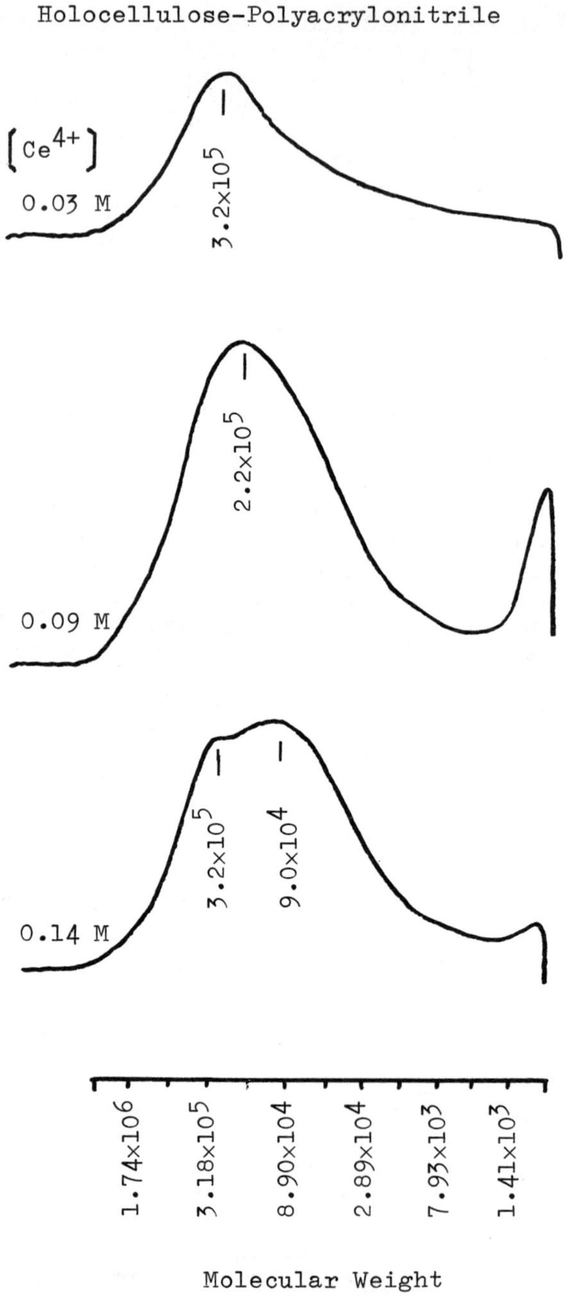

Figure 7. Molecular weight distribution for polyacrylonitrile branches of grafted holocellulose.

scopic method are probably more reliable than those evaluated with the gravimetric method due to the possible contamination of ceric adsorbed species.

(4) As the consequence of the use of large amount of ceric salt in copolymerization reaction, severe deviation from the true value of grafting percentage may result if the initial products of graft copolymerization are oven-dried before removing homopolymers from the grafting system.

(5) The maximum grafting percentage for acrylonitrile grafted bamboo was found to fall in the range of 40-50%, while the maximum for grafted holocellulose was around 60-70% for the experimental condition used.

(6) By comparing the difference in molecular weight distribution and the difference in grafting percentages for grafted bamboo and grafted holocellulose, evidence was obtained that the graft copolymerization of acrylonitrile onto bamboo would be retarded considerably and the polymer chains of grafted bamboo would be shortened greatly in the presence of lignin.

Acknowledgements

Financial support from National Science Council of the Republic of China is gratefully acknowledged.

Literature Cited

1. Mino, G.; Kaizerman, S. J. Polymer Sci. 1958, 31, 242.
2. Iwakura, Y.; Kurosaki, T.; Imai, Y. J. Polymer Sci. 1965, A. 3, 1185.
3. Kulkarni, A. Y.; Mehta, P. C. J. Appl. Polymer Sci. 1965, 9, 2633.
4. Arthur, J. C.; Baugh, P. J.; Hinozosa, O. J. Appl. Polymer Sci. 1966, 10, 1591.
5. Kulkarni, A. Y.; Mehta, P. C. J. Appl. Polymer Sci. 1968, 12, 1321.
6. Hebeish, A.; Mehta, P. C. J. Appl. Polymer Sci. 1968, 12, 1625.
7. Ogiwara, Y.; Ogiwara, Y.; Kubota, H. J. Polymer Sci. 1968, A-1, 6, 1489.
8. Morris, N. J.; Blouin, F. A.; Arthur, J. C. J. Appl. Polymer Sci. 1968, 12, 373.
9. Huang, R. Y. M.; Chandramouli, P. J. Appl. Polymer Sci. 1968, 12, 2549.
10. Huang, R. Y. M.; Chandramouli, P. J. Polymer Sci. 1969, A-1, 7, 1393.
11. Rao, S. R.; Kapur, S. L. J. Appl. Polymer Sci. 1969, 13, 2619.

12. Kubota, H.; Ogiwara, Y. J. Appl. Polymer Sci. 1969, 13, 1569.
13. Ogiwara, Y.; Kubota, H.; Arai, K. J. Appl. Polymer Sci. 1970, 14, 1049.
14. Kantouch, A.; Hebeish, A.; El-Rafie, M. H. J. Appl. Polymer Sci. 1971, 15, 1007.
15. Hebeish, A.; Kantouch, A.; El-Rafie, M. H. J. Appl. Polymer Sci. 1971, 15, 11.
16. Mansour, O. Y.; Nagaty, A. J. Polymer Sci., Polymer Chem. Ed., 1974, 12, 1887.
17. Lin, K. C.; Lin, Y. S.; Hus, T. C. J. Chinese Chem. Soc. 1980, 27, 83.
18. Lin, K. C.; Lin, Y. S.; Hu, C. T. Proc. Natl. Sci. Counc. B, ROC 1981, 5, 26.
19. Hu, C. T.; Wu, Y. S.; Lin, K. C. To be published.
20. Ku, Y. C.; Chiou, C. H. "Tests on Fiber Morphology and Chemical Composition of Important Bamboos in Taiwan"; Taiwan Foresty Institute; Taipei, Taiwan, 1966; p 1-7.
21. Liang, C. Y.; Krimm, S. J. Polymer Sci. 1958, 31, 531.
22. Matsubara, I. Bull. Chem. Soc., Japan, 1961, 34, 1710.
23. Yamadera, R. J. Polymer Sci. 1961, 50, S4.
24. Liang, C. Y.; Pearson, F. G.; Marchessault, R. H. Spectrochim Acta 1961, 17, 568.
25. Tadokoro, H.; Murahashi, S.; Yamadera, R.; Kamei, T. J. Polymer Sci. 1963, A-1, 1, 3029.
26. Yamadera, K.; Tadokora, H.; Murahashi, S. J. Chem. Phys. 1964, 41, 1233.
27. Polcin, J.; Rapson, W. H. Pulp and Paper Canada, 1971, 72, 69.
28. Varma, D. S.; Narasimham, Veena. J. Appl. Polymer Sci. 1972, 16, 3325.
29. Ogiwara, Y.; Ogiwara, Y.; Kubota, H. J. Polymer Sci. 1967, A-1, 5, 2791.

RECEIVED January 22, 1982.

GRAFT COPOLYMERIZATION OF LIGNOCELLULULOSIC FIBERS

16

Grafting of Some Vinyl Monomers onto Lignocellulose and Cellulose in the Presence of Lignin

OLFAT Y. MANSOUR and AHMED NAGATY
Cellulose and Paper Laboratory, National Research Centre,
Sh. El Tahrir, Dokki, Cairo, Egypt

AL-FOUNS BESHAY
Monoufya University, Faculty of Science, A.R. of Egypt

Sodium bisulfite-china clay proved to be an efficient initiator for homopolymerization and graft polymerization of methyl methacrylate onto cellulose. Grafting reactions using ceric ammonium sulfate, sodium bisulfite-soda lime glass or -china clay are inhibited or retarded on adding soda lignin to the grafting medium. This may relate to the inhibiting effect of the quinonoid groups present in lignin; also to the consumption of the ceric initiator in oxidizing of the lignin. The extent of inhibition and retardation depends on the amount of added lignin, the initiator used and the grafting temperature. Lignin in-situ of the lignocellulosic substrate effects inhibition or retardation only on using ceric ammonium sulfate as initiator. This may relate to the presence of the in-situ lignin in the form of polyhydric alcohols which on oxidation are transformed into quinones. Ceric consumed in oxidation and the quinone groups subsequently formed may both contribute for the inhibition or retardation of the grafting process. In analogy to soda lime glass and china clay, lignin of the lignocellulosic substrate acts with sodium bisulfite as an initiator system for grafting of polymethyl methacrylate onto lignocellulose. In this case, on subjecting the lignocellulosic substrates to the same chemical treatment, the lignin content of the treated samples being the governing factor for the rate and yield of grafting, while the structure of lignin is the governing factor on subjecting to different chemical treatments.

Tirzina (1) found that alkali lignin retarded the polymerization of styrene in dimethyl sulfoxide and strongly inhibited polymerization in the presence of oxygen and benzoyl peroxide. Straforelli (2) investigated the effect of lignin on the graft copolymerization of methyl methacrylate initiated by azo-

0097-6156/82/0187-0253$6.00/0
© 1982 American Chemical Society

bisisobutyronitrile. Kubota and Ogiwara (3) examined the effect of lignin in a ceric ion-initiated graft copolymerization of methyl methacrylate on high-yield pulp delignified to different lignin levels. Erdelyi (4) prepared a series of pulps by the sulfate cooking process and subsequent hypochlorite bleaching, and determined their graftability in a system using ceric ammonium sulfate as initiator and acrylonitrile as monomer. From the latter three studies, it was arrived at that a higher concentration of catalyst is required if copolymerization is to be carried out in the presence of lignin. Also, the presence of lignin generally reduces polymerization rate and gives rise to prolonged inhibition periods at higher lignin levels (5).

Given the fact that lignin is easily oxidized by most conventional oxidants which are in turn often used to initiate copolymerization, one expects that its presence will retard, if not inhibit, the reaction as the initiator will be used up preferentially in the reaction with lignin. Moreover, the reaction of lignin with oxidants gives rise to the formation of quinonoid structures which have pronounced retarding and/or inhibiting properties (5).

However, there are some indications that higher lignin content may not always be to the detriment of pulp graftability (6). Hornof et al. (7) showed that in the case of sulfite pulp copolymerization with acrylonitrile by the xanthate method, the grafting efficiency increased with rising lignin content while the total conversion remained almost constant. Kraft pulp, on the other hand, gave a higher total conversion to polymer with rising lignin content, with a simultaneous increase in the amount of homopolymer.

The same authors (5) showed also in another work that the presence of larger amounts of lignin in bisulfite pulps may have a favourable effect on grafting polyacrylonitrile using the cellulose xanthate-hydrogen peroxide redox system to initiate the copolymerization reaction. The plots of the total conversion as well as of polymer loading show a minimum centered around approximately 15% of lignin.

In the present work, the effect of adding lignin and the presence of in-situ lignin on the graftability onto cellulose has been investigated.

Experimental

Cotton linter and bagasse ground to 40 mesh and extracted with methanol-benzene 1:1, also alkaline treated bagasse and bagasse semichemical pulp were grafted with acrylonitrile, ethyl acrylate or methyl methacrylate. Grafting was carried out under purified nitrogen with occasional light-hand shaking. Ceric ammonium sulfate was used as initiator for grafting the two former monomers (8), while sodium bisulfite-soda lime glass (9) or sodium bisulfite-china clay for grafting the latter. Ceric ammonium sulfate concentration was 0.08 or 0.1 g% in 1% sulfuric acid (pH about 1.0) and sodium bisulfite concentration was 0.1 g%. Liquor, monomer and soda lime glass or china clay to cellulose ratios were 30:1, 3:1 and 1:1, respectively.

Homopolymerization reactions were also carried out with the use of methyl methacrylate monomer.

After the reaction period, the reaction mixture was filtered through a dried and weighed sintered glass crucible G_3, washed thoroughly until free from residual monomer and other reagents, dried, and then weighed. The crude grafted celluloses were extracted with a suitable solvent for the removal of the homopolymer; dimethylformamide (DMF) for polyacrylonitrile, tetrahydrofuran (THF) for poly(ethyl acrylate), and acetone for poly(methyl methacrylate). Extraction was carried out until no further homopolymer was removed. An extraction period of 48 hrs was sufficient to bring about this. Thorough washing with distilled water followed. Finally, the samples were dried and weighed.

The percentage conversion of monomer to polymer for homopolymerization reactions and the grafting yields for the grafting ones were calculated according to the following equations:

$$\text{Conversion } \% = (B-S)/M$$

$$\text{Crude grafting yield C } \% = [(F-S)-Z]/Z$$

$$\text{True grafting yield A } \% = [(G-S)-Z]/Z$$

where B is the weight of polymer; S is the weight of glass or china clay; M is the weight of monomer; Z is the weight of cellulose; F is the weight of crude grafted cellulose (before extraction); and G is the weight of grafted cellulose (after extraction).

The ceric consumption was determined according to the method of Mino and Kaizerman (10) as nanomole ceric ammonium sulfate per 100 g cellulose.

Preparation of Materials. Soda lignin was prepared from bagasse by subjecting to soda cooking conditions (11), namely, 20% sodium hydroxide based on dry raw material, liquor to cellulose ratio 5:1, at 150°C, for one hour. Lignin was precipitated and separated through centrifugation from the cooking liquor, washed with hydrochloric acid (1%) and then with water to neutrality; after washing, the lignin was obtained by centrifuging. The washed lignin preparation was found to contain less than 1% ash.

Soda lime glass (SLG) was made by El-Nasr Company, Egypt, from the reaction of Maadi sand (900 parts), sodium carbonate (300 parts), dolmite ($CaCO_3$, $MgCO_3$) (190 parts), limestone (63 parts) and sodium sulfite (12 parts). The final composition of the soda lime glass is: SiO_2, 71.65%; R_2O_3, 1.82%; CaO, 8.63%; MgO, 3.27%; Na_2O, 14.63%.

Cotton linter was purified by Kier boiling and then bleached with the 4-stage bleaching method, namely, chlorination, hot alkali-refining, hypochlorite and chlorine dioxide steps.

Bagasse ground to 40 mesh, hot extracted with methanol-benzene mixture (1:1) in a soxhlet apparatus and then delignified with acidified sodium chlorite at pH 4.5.

Bagasse of 40 mesh was also subjected to alkaline treatments with sodium hydroxide of 2 and 4% concentration and ammonium hydroxide of concentration 12.5 and 25% at a liquor ratio 20:1, at room temperature, for 24 hrs. The samples were washed thoroughly with distilled water and then treated with 10% acetic acid, rewashed with distilled water till acid free and finally left to dry at room temperature.

Semichemical pulp of bagasse prepared by the acid sulfite cooking was obtained from Edfuo Company of Egypt.

Results and Discussion

Dudzik et al. (12) found that it is possible to introduce and trap different amounts of sulfur-containing free radicals inside the framework of the synthetic zeolites.

Soda lime glass of the composition: SiO_2, 71.65%; R_2O_3, 1.82%; CaO, 8.63%; MgO, 3.27%; Na_2O, 14.63%, when used with sodium bisulfite proved to be a good initiator system for polymerization and graft polymerization of methyl methacrylate onto cellulose (9). A scheme dealing with the mechanism of initiation has been proposed assuming trapping of the bisulfite radical inside the glass framework to form a so-called sulfur-impregnated solid. Such a solid has paramagnetic properties and acts on the methyl methacrylate monomer and cellulose as any free-radical-producing source, i.e., initiator, thus leading to polymerization and graft polymerization onto cellulose.

China clay having very nearly the empirical composition: $Al_2O_3.2SiO_2.2H_2O$ (13) was used as soda lime glass with sodium bisulfite as an initiator system for grafting polymethyl methacrylate onto cellulose.

Figure 1 shows a slight increase in the rate of grafting when using sodium bisulfite-china clay initiator system compared with sodium bisulfite-soda lime glass, while the increase in the rate of homopolymerization and the conversion % was distinct (Figure 2).

When used with china clay as an initiator, sodium bisulfite prepared one day before gave higher rate of polymerization and conversion %, compared with the one freshly prepared (Figure 2). A similar result was achieved on using sodium bisulfite-soda lime glass as an initiator system (9).

Grafting in Presence of Added Lignin. Ceric ammonium sulfate is an effective initiator for grafting polyethyl acrylate and polyacrylonitrile onto pure celluloses (8), while sodium

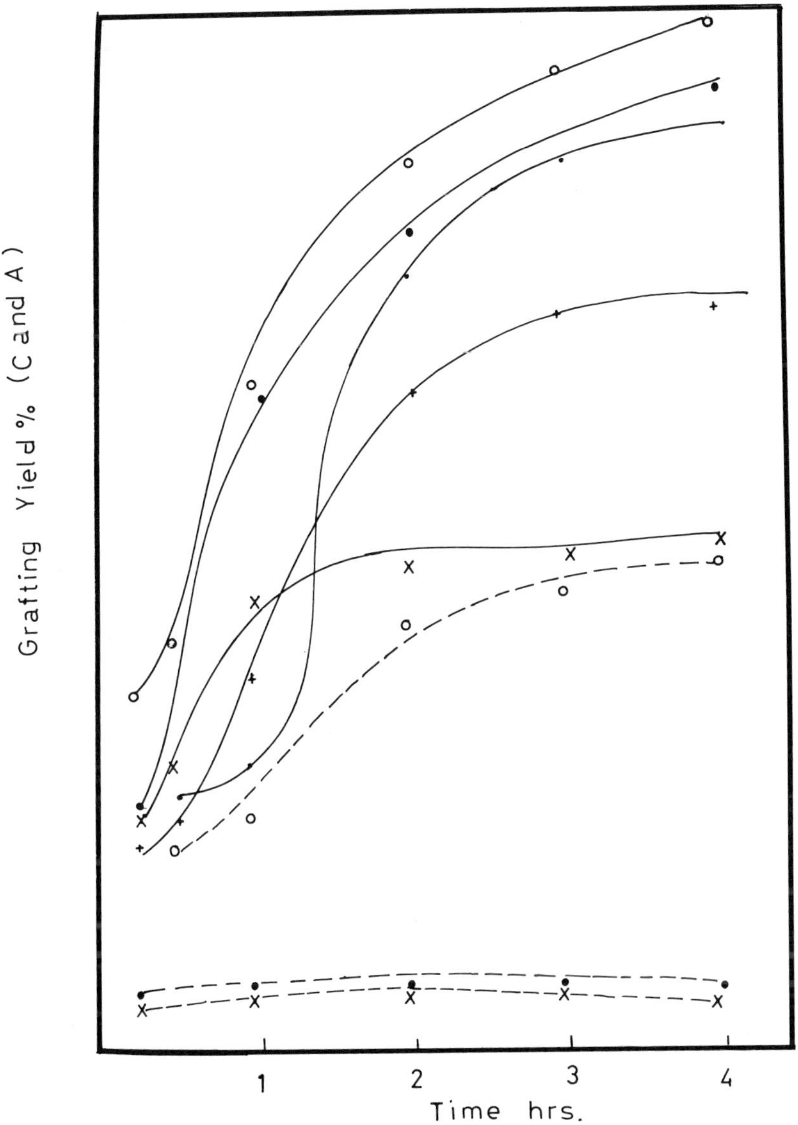

Figure 1. Grafting rate using two sodium bisulfate initiator systems. Lignin (g%) using china clay for crude grafting yield (C): ○—○, 0.0; •—•, 0.0005; ● - - - ●, 1.33; and for true grafting yield (A): ×—×, 0.0; ○ - - - ○, 0.0005. Lignin (g%) using glass for crude grafting yield (C): ●—●, 0.0; +—+, 0.0005; × - - - ×, 1.33.

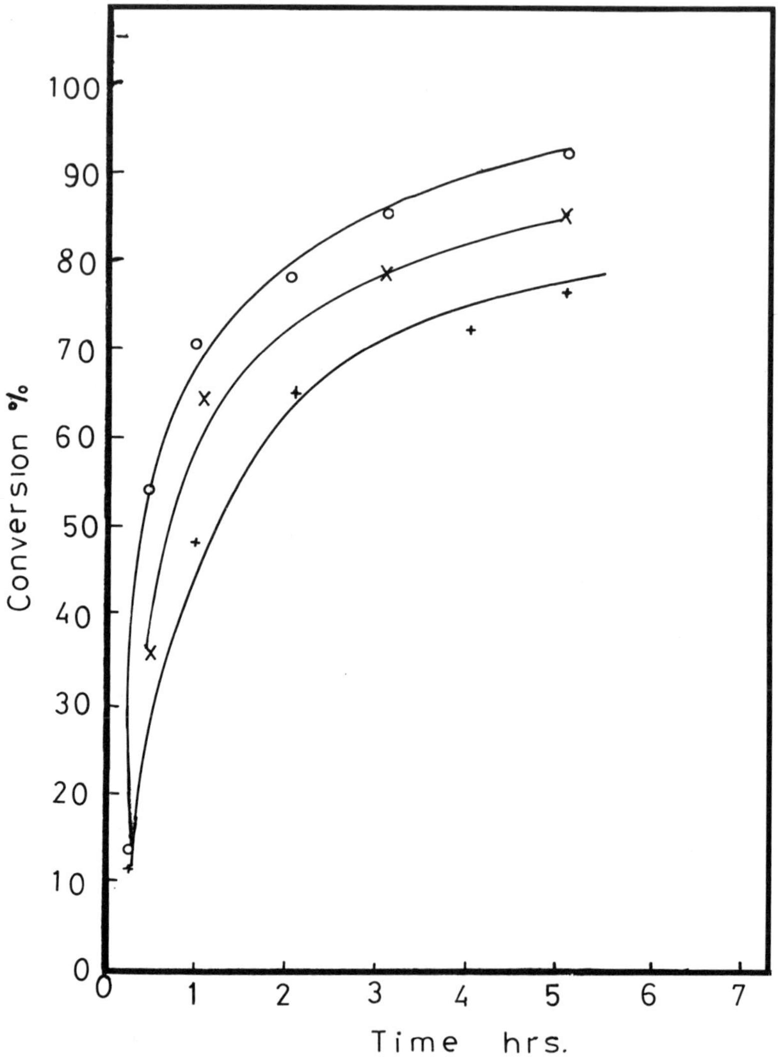

Figure 2. Homopolymerization of methyl methacrylate in presence of china clay or glass. Key (china clay): ×—×, sodium bisulfite freshly prepared; ○—○, sodium bisulfite prepared one day before. Key (glass): +—+, sodium bisulfite prepared one day before.

bisulfite-soda lime glass (9) and sodium bisulfite-china clay are effective for grafting polymethyl methacrylate. High grafting yields are thus achieved.

Addition of 1.33% of soda lignin to the grafting systems inhibited the grafting reactions completely (zero grafting yields) (Figures 1, 3, 4 and 5).

Minimizing the percent of added lignin to 0.27 resulted in grafting of the monomers onto cellulose, however, the rate and yield of graftings were sharply decreased compared with those achieved in absence of lignin (Figures 3, 4 and 5).

It was observed that there are certain solubilities of lignin in monomers. Experiments were conducted using only monomer dissolved lignin. The percent of lignin dissolved in each monomer was determined and was found 0.002% in both acrylonitrile and ethyl acrylate and 0.005% for methyl methacrylate. In the grafting systems, the percentages reach 0.0002 and 0.0005, respectively. Figures 1, 3, 4 and 5 show that the presence of such traces of lignin in the grafting media still result in sharp decrease in the rate and yield of grafting compared with those due to absence of lignin. The decrease may be as high as that due to the presence of 0.27% lignin (Figures 3 and 4).

Again, from Figures 1, 3, 4 and 5 it is clear that the presence of traces of lignin in the graft polymerization system partially retards the grafting reaction when sodium bisulfite-soda lime glass or -china clay are used as initiators. Here, retardation is attributed to the quinonoid groups present in soda lignin. Strong retardation is achieved when the initiator is ceric ammonium sulfate. The latter result relates to the preferential consumption of the initiator in oxidizing lignin (5).

It was also observed that at the same percent of the lignin added, the decrease in the rate and yield of grafting is higher on lowering the grafting temperature (Figures 3, 4 and 5). This may relate to a higher efficiency for the retarding or inhibiting properties of lignin at the lower temperature.

Grafting in the Presence of In-situ Lignin. Extracted ground and semichemical pulp of bagasse of the lignin contents 19.6 and 1.89%, respectively, were subjected to grafting with ethyl acrylate and acrylonitrile using 0.08 g ceric ammonium sulfate/100 ml as initiator. The results as illustrated in Figure 6, show absence of grafting as indicated by the zero grafting yields achieved. Insufficient rate and yield of graftings were only observed on grafting the semichemical pulp with acrylonitrile at 30°C. For the extracted bagasse with high lignin content, inhibition (zero grafting yield) took place also even on increasing the concentration of ceric ammonium sulfate to 0.1 g/100 ml. Inhibited grafting was also affected on using alkaline treated ground bagasse. Inhibition of the grafting process has been related to the preferential consumption of the initiator in oxidizing lignin. This is revealed by comparing the ceric consumed by

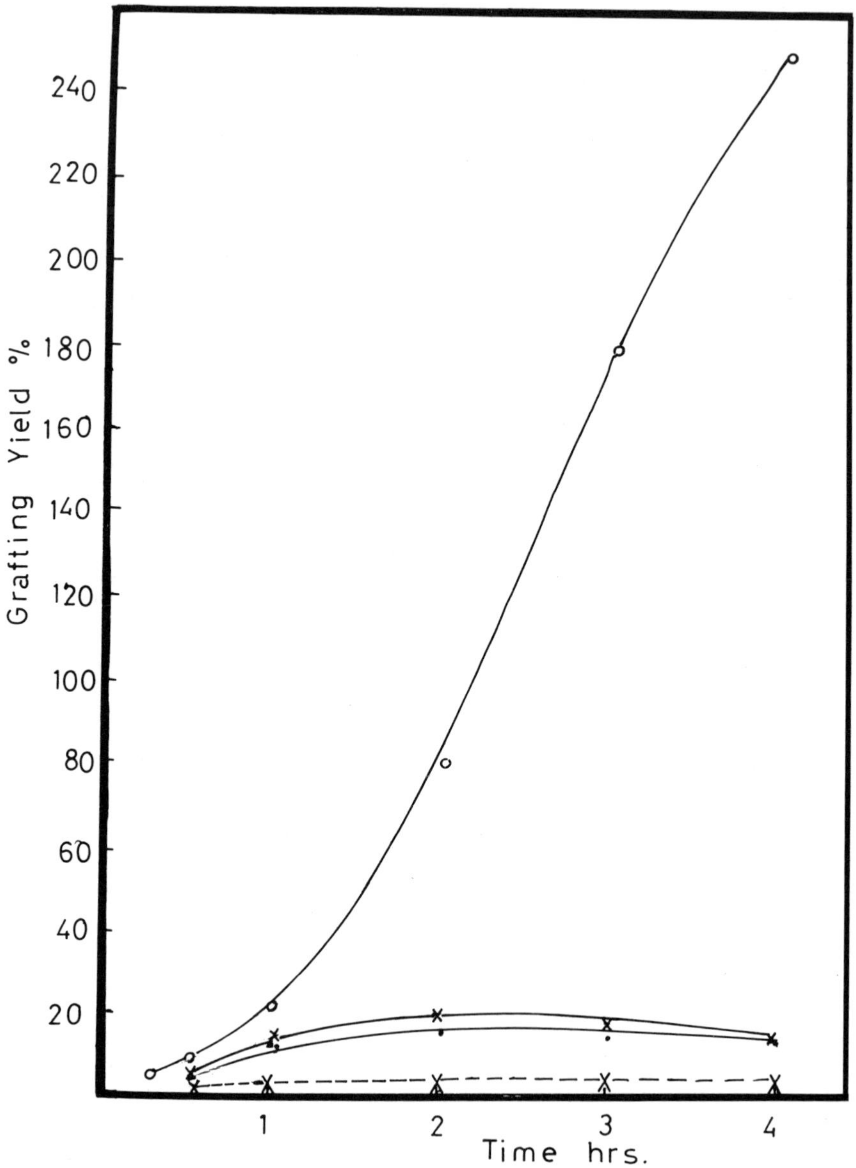

Figure 3. Rate of grafting. Conditions: initiator, ceric ammonium sulfate (0.08 g/100 mL, 1% H_2SO_4); monomer, ethyl acrylate; 30°C. Lignin (g%): ○—○, 0.0; ×–×, 0.0002; •—•, 0.27; ╳—╳, 1.33.

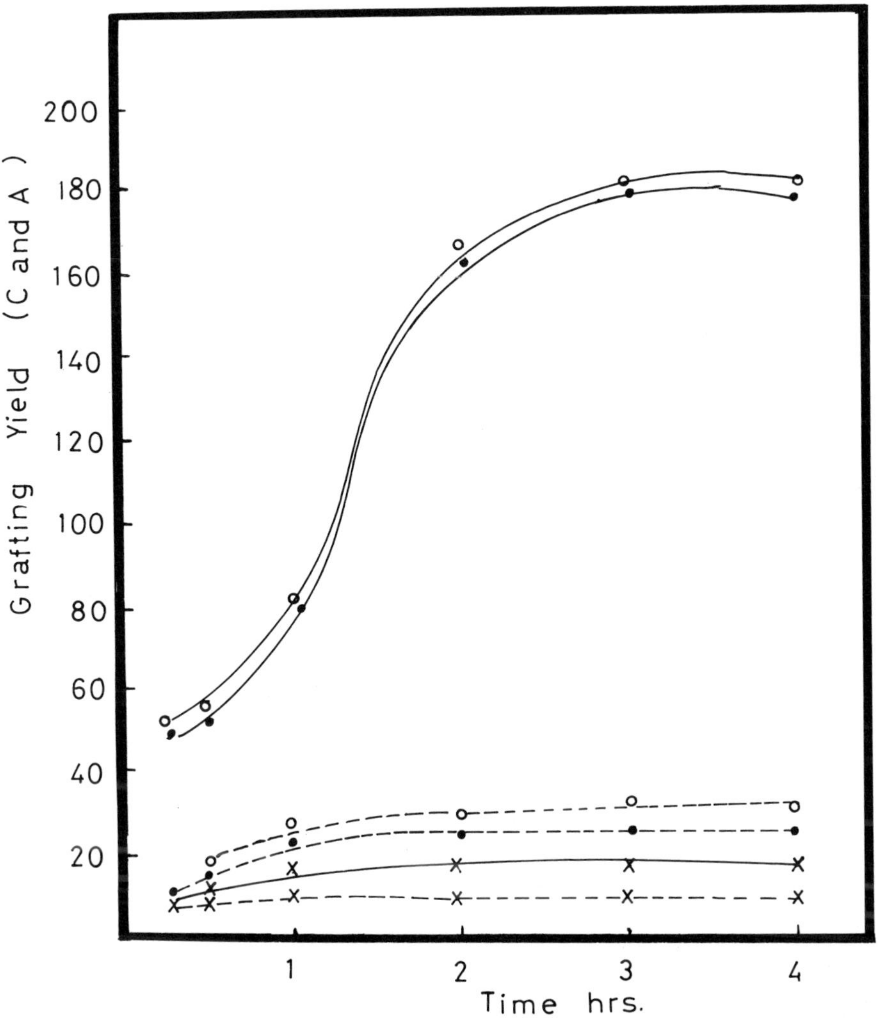

Figure 4. Rate of grafting. Conditions: initiator, ceric ammonium sulfate (0.08 g/100 mL, 1% H_2SO_4); monomer, acrylonitrile; 30°C. Lignin (g%) for crude grafting yield (C): ○—○, 0.0; ○ - - - ○, 0.0002; ×—×, 0.27; × - - - ×, 1.33. Lignin (g%) for true grafting yield (A): •—•, 0.0; ● - - - ●, 0.0002.

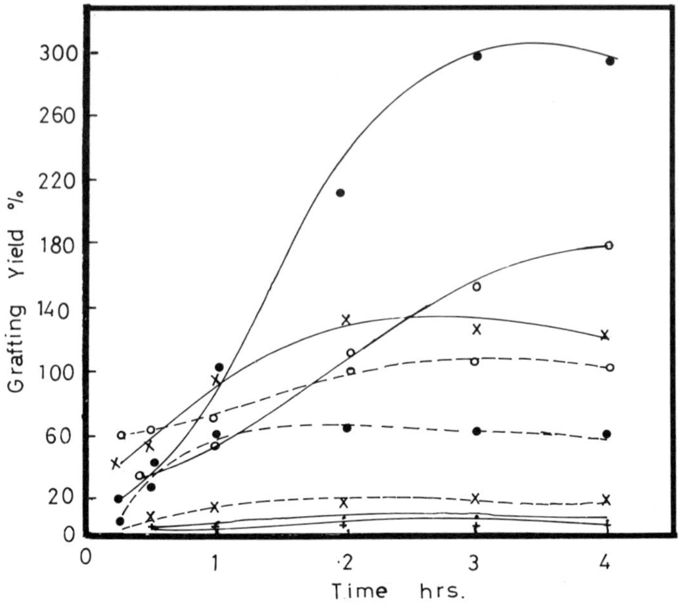

Figure 5. Grafting rate of two monomers. Conditions: initiator, ceric ammonium sulfate (0.08 g/100 mL, 1% H_2SO_4). Lignin (g%) using ethyl acrylate (40°C): ○—○, 0.0; ● - - - ●, 0.0002; +—+, 1.33. Lignin (g%) using acrylonitrile (40°C): ○ - - - ○, 0.0; × - - - ×, 0.0002; •—•, 1.33. Lignin (g%) using ethyl acrylate (60°C): ●—●, 0.0; ×—×, 0.27.

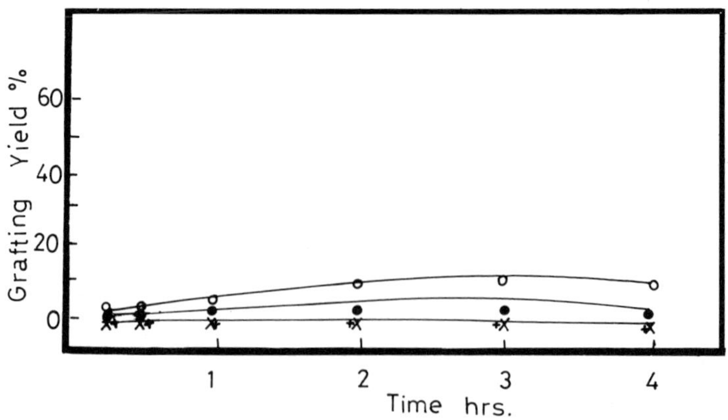

Figure 6. Grafting rate of semichemical pulp. Conditions: initiator, ceric ammonium sulfate (0.08 g/100 mL, 1% H_2SO_4; 30°C. Key (monomer): ○—○, acrylonitrile; ●—●, ethyl acrylate. Grafting rate of extracted bagasse. Conditions: monomer, ethyl acrylate; 60°C. Key (initiator, ceric ammonium sulfate): ×—×, 0.08 g/100 mL, 1% H_2SO_4; +—+, 0.10 g/100 mL, 1% H_2SO_4.

the lignocellulosic substrate bagasse semichemical pulp and the pure cellulose cotton linter, where the former substrate showed a sharp increase in the ceric consumed at the early stages (Figure 7).

On using sodium bisulfite-soda lime glass as initiator for grafting polymethyl methacrylate onto ground bagasse of the lignin content of 19.6%, low rate and yield of grafting were achieved. The same was also obtained when grafting was carried out in presence of sodium bisulfite alone, i.e., in the absence of soda lime glass. The rate and yield of graftings were as low as in the presence of soda lime glass (Figure 8).

Reducing the lignin content to 4%, then to 2.62% through delignification with acidified sodium chlorite, increased the rate and yield of grafting progressively in the absence of soda lime glass.

Semichemical pulp of lower lignin content (1.89%) showed higher rate and yield of grafting (Figure 8).

The results achieved above reveal the following:
(1) For grafting polymethyl methacrylate onto lignocellulosic substrates, sodium bisulfite may be used alone, i.e., in the absence of soda lime glass or china clay. It is worthy to mention that grafting onto pure celluloses failed to take place in absence of soda lime glass (9); (2) lignin already present in the cellulosic substrate, i.e., in-situ, does not possess retarding or inhibiting properties on the process of grafting of polymethyl methacrylate in the presence of sodium bisulfite. This relates to that lignin is present in plant as polyhydric alcohols and not as quinones. It acquires the quinonoid structure responsible for the inhibiting properties only on isolation (14).

Grafting polymethyl methacrylate successfully onto partially delignified bagasse or semichemical pulp on using sodium bisulfite alone, i.e., in the absence of soda lime glass or china clay may not be contributed to the inorganic matter, particularly silica, present in bagasse, since:

1. Silica is less efficient than soda lime glass in initiating the grafting of polymethyl methacrylate in presence of sodium bisulfite (9). The results achieved with semichemical pulp show very high rate and yield of graftings (Figure 8).
2. Crude grafting yields higher than 100% could be only achieved at soda lime glass to cellulose ratio 0.5:1 and further increased at 1:1. A maximum yield higher than 200% was achieved at 2:1 ratio (15). In the present work, the amount of inorganic matter determined as ash content is 4.029% for bagasse and 1.115% for semichemical pulp. In other words, the ratio of inorganic matter (in which silica forms one of its constituents) to cellulose is 0.04:1 for the former and 0.01:1 for the latter and hence these ratios are insufficient to produce the high grafting yields achieved with the partially delignified bagasse and the semichemical pulp.

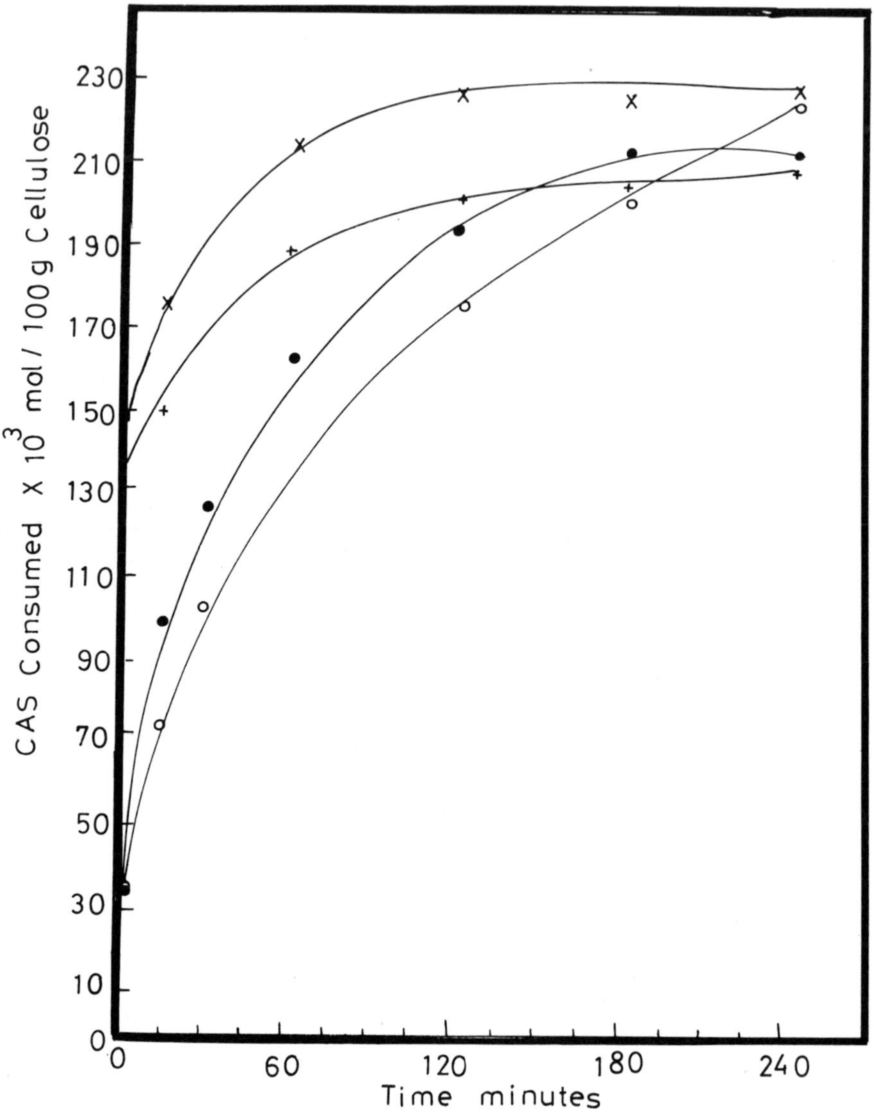

Figure 7. Ceric ammonium sulfate consumption for cotton linter: ○—○, 30°C; ●—●, 60°C; and for semichemical pulp: +—+, 30°C; ×—×, 60°C.

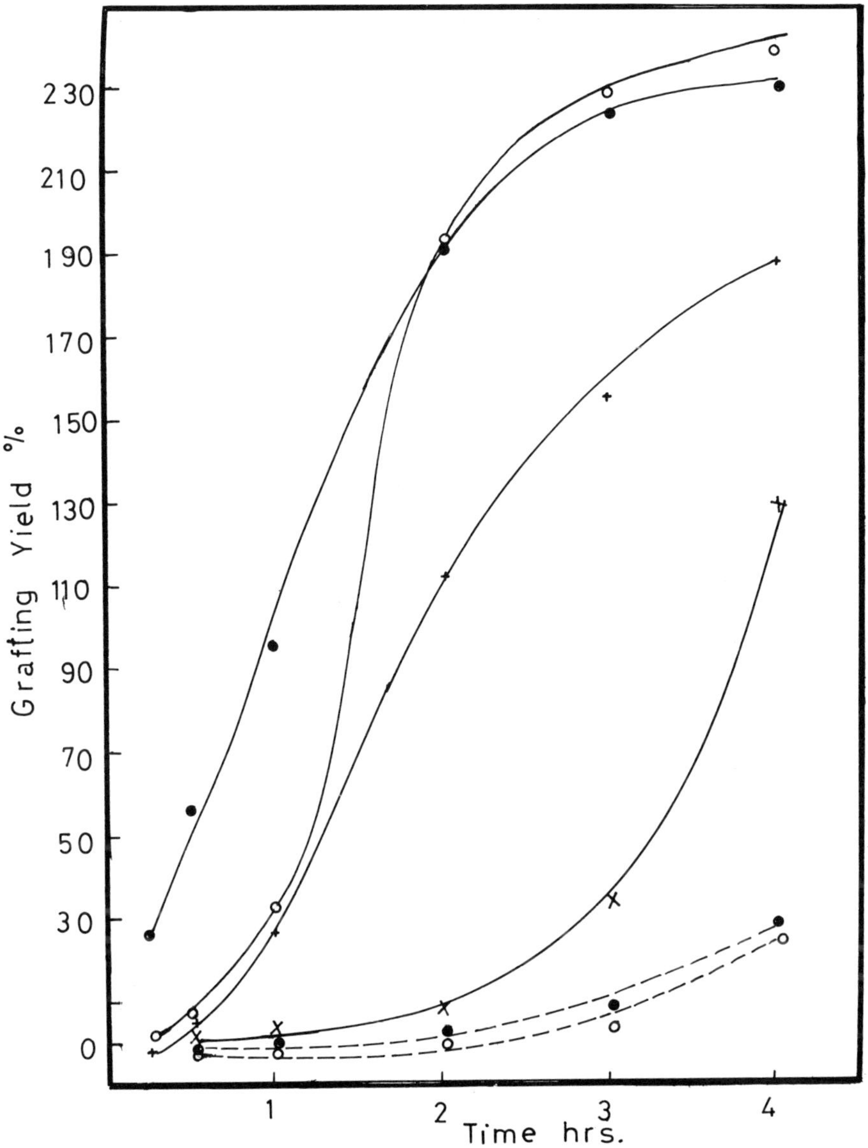

Figure 8. Graft polymerization of methyl methacrylate. Onto bagasse: ● – – – ●, in absence of glass; ○ – – – ○, in presence of glass. Onto semichemical pulp: ○—○, in absence of glass; ●—●, in presence of glass. Onto delignified bagasse: ×—×, 4.00% lignin; +—+, 2.60% lignin.

From the above, it may be concluded that in analogy to the mechanism of grafting initiated by sodium bisulfite-soda lime glass system (9), lignin with its network structure configuration may play the role of soda lime glass in trapping the bisulfite radical inside its framework forming a sulfur-impregnated solid of paramagnetic properties which act on the methyl methacrylate monomer and cellulose as any free-radical-producing source, thus leading to polymerization and graft polymerization onto cellulose.

Again, it is noteworthy that the yield of grafting was practically the same at equilibrium on carrying out the grafting process on semichemical pulp in presence and absence of soda lime glass (Figure 8).

Subjecting the ground bagasse to treating with 2 and 4% sodium hydroxide decreased the lignin content to 11.19 and 7.36%, respectively; while treating with 25% ammonium hydroxide at liquor to cellulose ratios 10:1 and 20:1, decreased it to 18.85 and 11.61%, respectively.

Figure 9 shows that for the same alkaline treatment (sodium or ammonium hydroxide), the sample of the lower lignin content showed again higher rate and yield of graftings. For different alkaline treatments, samples treated with ammonium hydroxide were of higher grafting rates and yields than those treated with sodium hydroxide, in spite of the higher lignin content of those of the former treatment. Also, the rate and yield of graftings were higher for the bagasse subjected to alkaline treatments than that treated with acid chlorite (Figures 8 and 9), in spite again of the higher lignin content of the former samples. This may relate to differences in the structure of the residual lignin due to preferential removal of different lignin fractions and combined hemicelluloses, accompanying different chemical treatments.

In conclusion, the lignin content is the governing factor for the rate and yield of grafting of polymethyl methacrylate in the presence of sodium bisulfite onto lignocellulosic substrate when the latter is subjected to the same chemical treatment. For different chemical treatments, the structure of the residual lignin is the governing factor. In other words, the extent to which lignin interferes with the grafting reaction varies from one type of cellulosic substrate to another, depending largely on its chemical history. A similar achievement was arrived at by Hornof et al. (5), although their studies were different from those of ours.

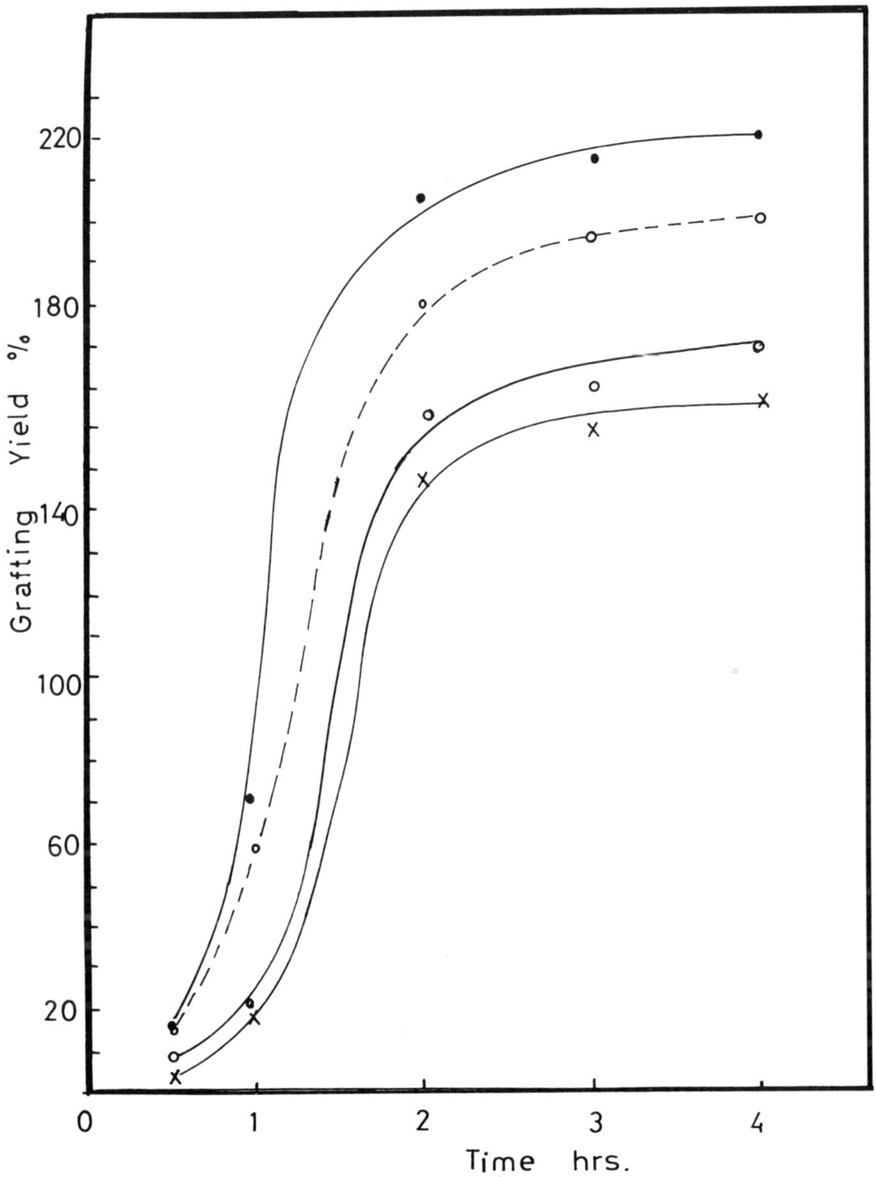

Figure 9. Grafting rate onto bagasse treated with ammonium hydroxide: ○ - - - ○, 10:1; ●—●, 20:1; and with sodium hydroxide: ×—×, 2%; ○—○, 4%.

Literature Cited

1. Tirzina, J.; Issled, Obl.; Khim, Drev. Tezisy Dokl. Konf. Moledykh Uch. 1975, 151, 41 (Russ.); Chem. Abstr. 1976, 84, 180747q.
2. Straforelli, J. B., Ph.d. Thesis, University of British Columbia, Vancouver, B.C., 1973.
3. Kubota, H.; Ogiwara, Y. J. Appl. Polym. Sci. 1969, 13, 1569.
4. Erdélyi, J. Zellstoff und Papier. 1970, 243.
5. Hornof, V.; Kokta, B. V.; Valade, J. L. J. Appl. Polym. Sci. 1976, 20, 1543.
6. Phillips, R. B.; Kobayashi, A. J.; Brown, W.; Stannett, V.T. "Pollution Abatement by More Effective Lignin Utilization: Grafting to Lignin and Lignin-Containing Pulps"; North Carolina State University, 1970.
7. Hornof, V.; Kokta, B. V.; Valade, J. L. J. Appl. Polym. Sci. 1975, 19, 1573.
8. Mansour, O. Y.; Schurz, J. Svensk Papperstidn. 1973, 76, 258.
9. Mansour, O. Y.; Nagaty, A. J. Polym. Sci., Polym. Chem. Ed. 1975, 13, 2785.
10. Mino, G.; Kaizerman, S.; Rasmussen, E. J. Polym. Sci. 1959, 38, 393.
11. Mansour, O. Y.; Seif El-Dien, M. Paperi ja Puu 1972, 54, 189.
12. Dudzik, Z.; Cvetanovîé. Proceedings of the Fourth International Congress on Catalysis, Moscow, 1968, Adadémiai Kiadó, Budapest, 1971, p 175.
13. Parkes, G. D. "Mellor's Modern Inorganic Chemistry"; Longmans, Green and Co.: London, New York and Toronto, 1951, p 721.
14. Mansour, O. Y.; Nagaty, A.; Salama, M. A.; Abdel Mouty, F. Paperi ja Puu (in press).
15. Mansour, O. Y.; Moustafa, A. B. J. Polym. Sci., Polym. Chem. Ed. 1975, 13, 2795.

RECEIVED January 22, 1982.

The Xanthate Method of Grafting

Part IX: Grafting of Acrylonitrile onto High-Yield Hardwood Pulp

B. V. KOKTA, R. C. LO, and C. DANEAULT

Centre de Recherche en Pâtes et Papiers, Université du Québec à Trois-Rivières, C.P. 500, Trois-Rivières, Québec, Canada G9A 5H7

Hardwood pulp was copolymerized with acrylonitrile as a monomer using the xanthate grafting method. Grafting has been initiated using a redox system of ferrous ion and hydrogen peroxide. The effect of operating conditions on grafting was investigated. Among the factors studied, the most important were initial pH, the concentration of hydrogen peroxide and the pulp/monomer ratio. Grafting parameters has been compared to that of softwood pulps. In addition, the effect of different residual concentration of lignin in pulp was examined. Sodium chlorite served as the delignification agent. The results showed minimum grafting at 6-8% lignin level. Finally, the optimum set of reaction conditions, which gives almost 100% grafting efficiency at very high conversion level has been proposed.

The xanthate method of grafting, discovered by Faessinger and Conte (1) is one of the most promising for industrial application. In our laboratory we have been working on the optimization of xanthate method. In order to develop an efficient method we have tried to reduce the homopolymer formation to a minimum and consequently to increase grafting efficiency. Maximum monomer conversion was another important objective in the developing of a cheap grafting method. We have shown that for low lignin pulps (2), for high yield pulps (3,4), as well as for mechanical (5) and thermomechanical pulps (6,7), the xanthate method could be optimized by properly adjusting the reaction conditions like the hydrogen peroxide concentration, the initial pH and the pulp/monomer ratio. The pulps used in the above mentioned references (2-7) were softwood ones (spruce and balsam fir).

It was shown that, in general, the paper properties could be improved by the grafting of acrylamide onto low strength pulps (like high yield and mechanical pulps) (8). It is well known that hardwood pulp fibers are considerably shorter (about 1 milli-

0097-6156/82/0187-0269$6.00/0
© 1982 American Chemical Society

meter) compare to softwood fibers (about 3 millimeters). In addition, hardwood fibers, being less compressable than the softwood ones give generally low strength pulps. Therefore, it seems logical that grafting on hardwood pulps would improve their mechanical properties (9). Unfortunately very little have been published on grafting hardwood pulps using the xanthate method with the exception of work done by Young (10). He has found, that grafting efficiencies in case of aspen could be close to 53% for both the styrene and acrylonitrile monomers. Conversion values obtained were 76.5% and 17% for acrylonitrile and styrene, respectively.

The objective of this work was to optimize grafting parameters for hardwood pulp in general and aspen in particular. The monomer used was acrylonitrile.

Experimental

a) Materials:

Pulp: the high yield bisulfite pulp of aspen, yield 93%, brightness 46.6%, C.S.F. 325, lignin 14.8%.

Monomer: acrylonitrile (Eastman Chemical Grade) was purified by distillation; the central fraction was collected and stored in a refrigerator in dark bottles.

All other chemicals employed in the work were analytical grade and were used as supplied by manufacturers.

b) Delignification of pulps:

Sodium chlorite was used as the bleaching agent at pH 4. The concentration of chlorite required to obtain a given degree of delignification was determined by preliminary experimentation.

In a bleaching operation, the required quantity of pulp (150 g air-dried) was desintegrated and placed into a polyethylene bag. The calculated amount(from 2 to 12%) of $NaClO_2$ dissolved in 300 ml water was subsequently added and the mixture was homogenized by trituration. The pH was adjusted to 4 with sulfuric acid, and the mixture was once again homogenized and the pH readjusted if necessary. The bag, placed in a controlled-temperature bath was maintained at 50°C for 120 minutes. After bleaching, the pulp was filtered, washed 5 times with 1000ml aliquots of water and then subjected to alkaline extraction by sodium hydroxide (1.5% by weight in respect to dry pulp) at 80°C during 1 hour. The extracted pulp was thoroughly washed until pH 7 and then vacuum filtered to approximately 20% of consistency. All pulps were preserved in a refrigerator.

The pulps prepared by this procedure as well as the non bleached high yield bisulfite pulp have been characterized by the standard methods of the Canadian Pulp and Paper Association (CPPA Technical Section) as shown in Table I.

c) Procedures used:

The technique of pulp pre-conditionning as well as the polymerization conditions have been described in a previous paper (11). Specific conditions used in this work were as follows:

Pulp conditioning: Pulp 9.0 ± 0.02 g (oven-dried weight); partial mercerization at 25°C, 45 min. in 300 ml 0.75 N NaOH; (solubility in NaOH:3.5%); xanthation, 2 hrs in presence of CS_2 va-

TABLE I

Properties of Pulps Prepared by Chlorite Delignification.

Pulp	Kappa[a]	Lignin %	Copper index[b]	Carboxyl index[c]
BT-0	86.3	14.8	2.2	16.4
BT-1	74.5	12.8	2.1	22.4
BT-2	62.0	10.7	1.9	24.8
BT-3	52.8	9.1	2.0	24.5
BT-4	37.2	6.4	1.8	23.7
BT-5	31.8	5.5	1.8	23.2
BT-6	25.4	4.4	1.5	22.5

[a]CPPA standard G.18.

[b]CPPA standard G.22.

[c]TAPPI standard T237 su.63.

pors, 25°C; washing, 300 ml of acidified (pH:4.5) distilled water; ion exchange, 2 min in 150 ml of 0.004% $(NH_4)_2Fe(SO_4)_2$; washing, 250 ml of distilled water (pH: 6.5).

Copolymerization: Monomer, acrylonitrile, 9.09 g; impregnation 15 min, 25°C; surfactant (Tween-80), 0.9 g; distilled water, 210 ml; H_2O_2 (conc. 30%) 5 ml; H_2SO_4 to pH: 5; reaction time, up to 19 hrs; atmosphere, nitrogen; reaction temperature, 25°C; quenching of polymerization: washing of product with distilled water and 1% of $K_2S_2O_5$, 5 min; washing with 1000 ml of distilled water.

Extractions: The apparent level of grafting, polymer loading and conversion was determined by Soxhlet extraction using dimethyl formamide. The oven dry weight was corrected for pulp lost in the mercerization step. Grafting parameters were defined as follows:

Polymer loading, %: (A-B)/(B) x 100
Grafting efficiency, %: (A-B)/(D-B) x 100
Degree of conversion, %: (D-B)/(C) x 100

where A is weight of product after copolymerization and extraction, B is weight of pulp, C is weight of monomer charged; and D is weight of product after copolymerization.

Results and Discussion

1) Effect of Operating Conditions on Grafting Efficiency and Conversion.

a) Dependence of grafting on initial pH.

Dependence of grafting on initial pH is illustrated in Figure 1. Grafting efficiency is not much affected by pH, even though the optimum value seems to be between pH 4 and 7. This is a little lower than the values obtained for mechanical softwood pulp, for which maximum grafting efficiency was in the proximity of pH 6-9 (4). In case of softwood kraft semibleached pulp (2% lignin), grafting efficiency was directly proportional to pH up to the value of pH 10 when it reached 83% (2).

As far as conversion is concerned, there is a sharp increase from 5% conversion at pH = 1 to 70% conversion at pH=4 where the conversion stays constant up to pH=7. This compares well to the values for softwood mechanical pulp for which the conversion was highest at pH=5 (67.8%), (4). On the other way, conversion values for softwood semibleached kraft pulp were highest at pH=2 (87%) and than steadily decreased to the 55% value at pH=8 and to 0% at pH=10 (2). From the results shown above, it seems apparent that the effect of pH on grafting is similar for high lignin pulps either hardwood as in this study or softwood mechanical pulp (4).

Considering that both grafting efficiency as well as conversion in this study shows optimal values for pH 4 - 7, we have used pH 5 which favors the attachment of ferrous ion on partial cellulose xanthate and other ion exchange groups present. (The term "partial cellulose xanthates" stands for xanthates bound statistically to any component of the pulp (cellulose, hemicellulose, lignin). The importance of the presence of ferrous ions on grafting is shown in Diagrams I-II-III, which show reaction mechanism sug-

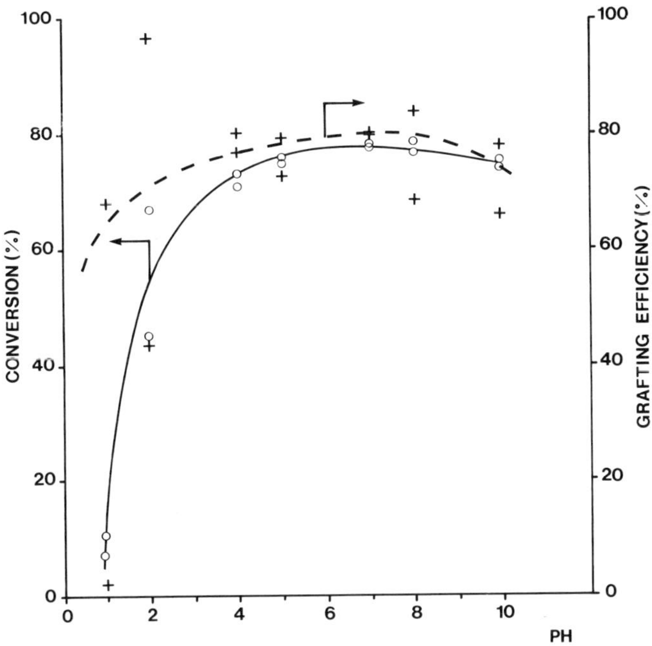

Figure 1. Dependence of conversion (○) and grafting efficiency (+) on pH.

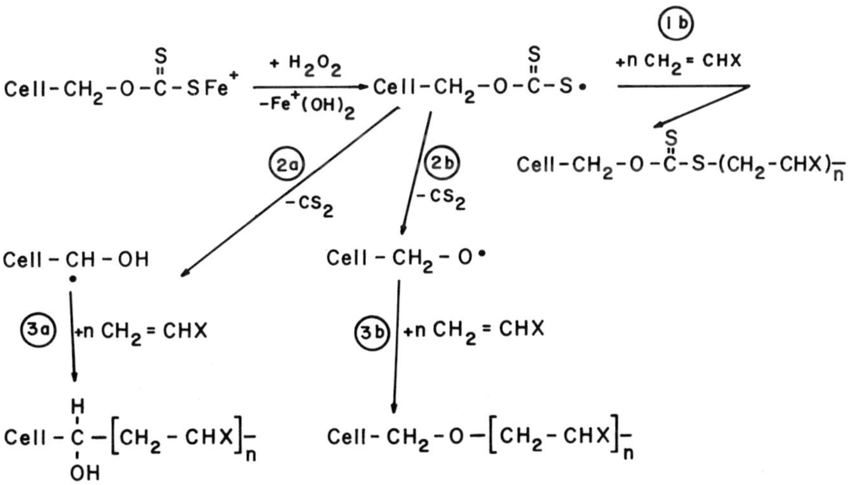

Diagram I. Activation of cellulose xanthate by hydrogen peroxide.

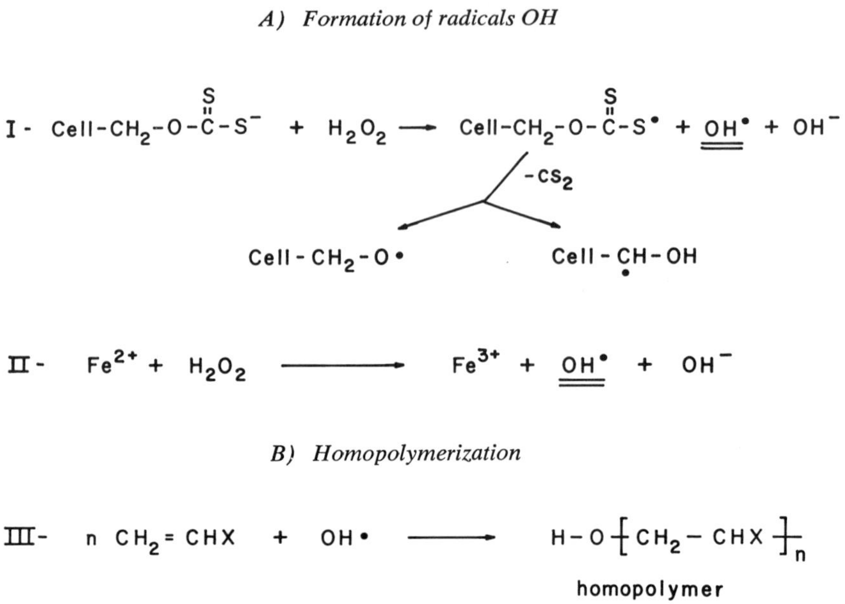

Diagram II.

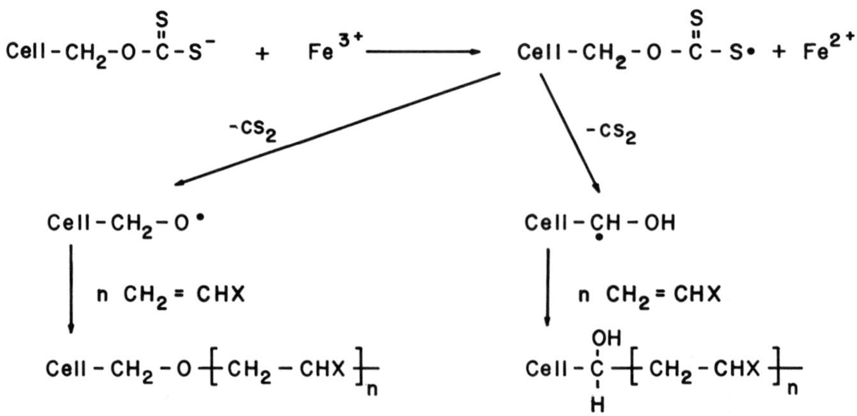

Diagram III. Activation of cellulose xanthate by ferric ion.

gested by Teichmann (12) and Krassig (13). Even though the grafting reaction could be carry out without ferrous ion present (diagram II) as was shown by Dimov and Pavlov (14) in case of pure cellulose, the presence of lignin makes grafting extremely slow in the absence of ferrous ion. The ferrous ion, when oxidized to ferric ion appears to contribute to an increased grafting efficiency as shown in diagram III (15).

b) Dependence of grafting on time.

Dependence of grafting efficiency and monomeric conversion on reaction time are shown in Figures 2 and 3. It is shown that a rapid polymerization takes place during the first 2 hrs of the reaction; the rate then decreases, and after about 8 hrs the reaction almost ceases. It is interest to note that grafting efficiency is directly proportional to time during the first 2 hrs of the reaction and after that remains constant at about the 85% level. It seems that the homopolymer formation (see diagrams II-III) is remakably suppressed at pH 5 at which there is not very much ferrous ion in solution to form free OH radical and polymerization takes place mostly on partially xanthated cellulose.

c) Dependence of grafting on hydrogen peroxide. (Figure 4)

It is evident that in the interval studied there is an increase of grafting efficiency (from 80 to 90%) and increase in conversion with H_2O_2 concentration up to 3g of peroxide (14.28g/1). Further increase of peroxide does not have much influence. This is in agreement with our previous work with softwood mechanical pulp (4).

The higher concentration of peroxide leads to a greater number of grafted chaines with lower molecular weights (2).

d) Dependence of grafting on monomer quantity.

The next set of experiments has been done to establish the effect of monomer concentration on grafting. Figure 5 shows the dependence of total conversion and grafting efficiency on the quantity of monomer used. While the monomer conversion is increasing up to the 90% level when monomer/pulp ratio reaches one and then stays constant, the grafting efficiency decreases at the same time. This indicates that the homopolymerization is favored by the presence of more monomer. This phenomenon is probably due to a greater rate of incidence of chain transfer by monomer. As was indicated in our previous work with softwood kraft pulp (2), which showed similar effect of monomer concentration on grafting of acrylonitrile, the possible transfer of growing graft chains to monomer causes a formation of homopolymer in the system unless an immediate transfer back to the backbone material can take place. The latter process is more probable when there is a higher pulp monomer ratio and better grafting efficiency is therefore expected at low monomer concentration.

In addition, at higher pulp/monomer ratio there is relatively higher amount of potentially active sites per quantity of monomer used; therefore the copolymerization reaction in the proximity sites is more probable than the homopolymerization in solution.

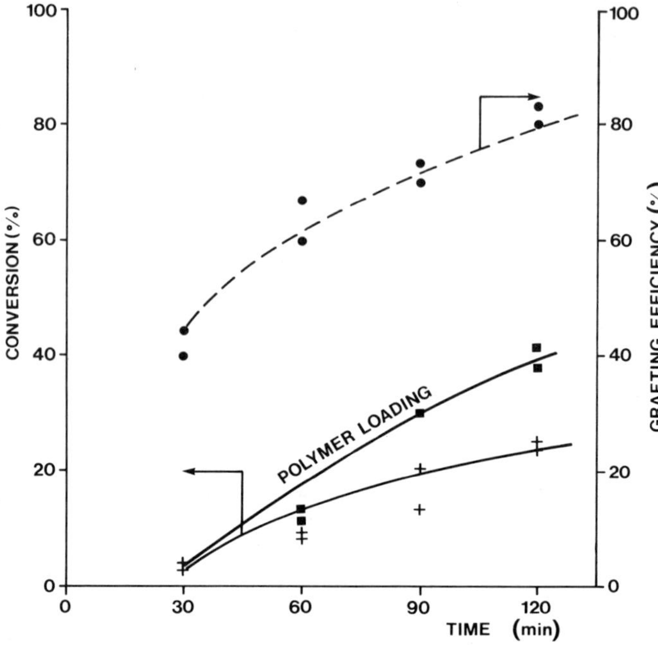

Figure 2. Dependence of conversion (+), grafting efficiency (●), and polymer loading (■) on reaction time. Conditions: BT-2 pulp and 10.7% lignin.

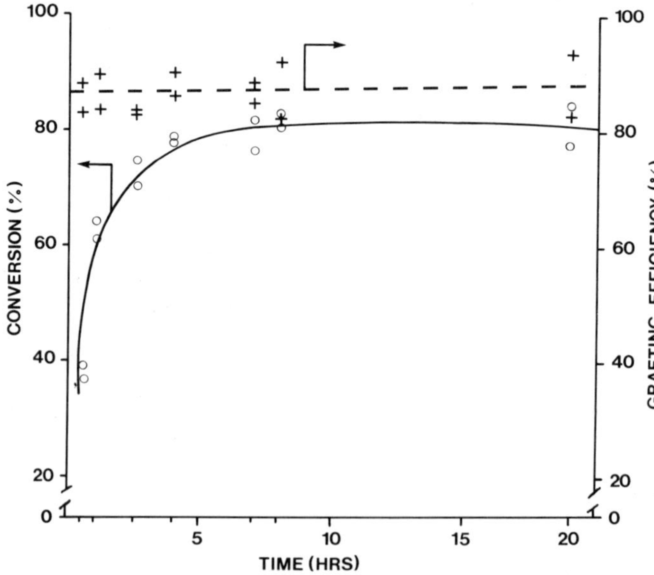

Figure 3. Dependence of conversion (○) and grafting efficiency (+) on reaction time.

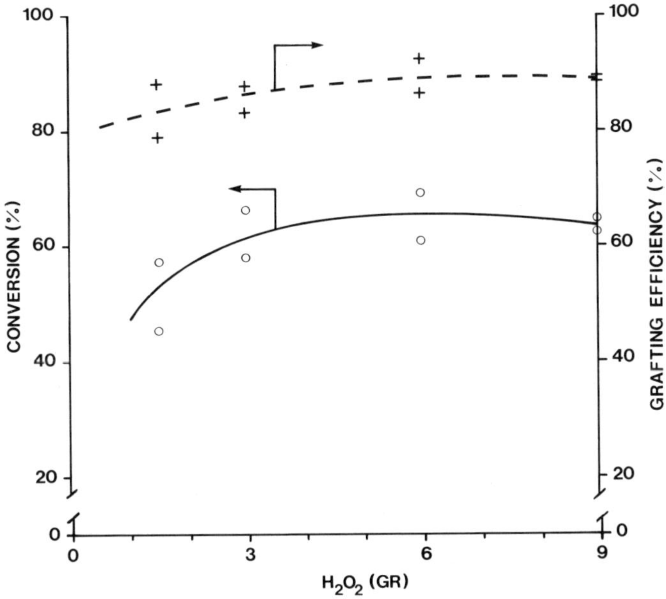

Figure 4. Dependence of conversion (○) and grafting efficiency (+) on quantity of hydrogen peroxide.

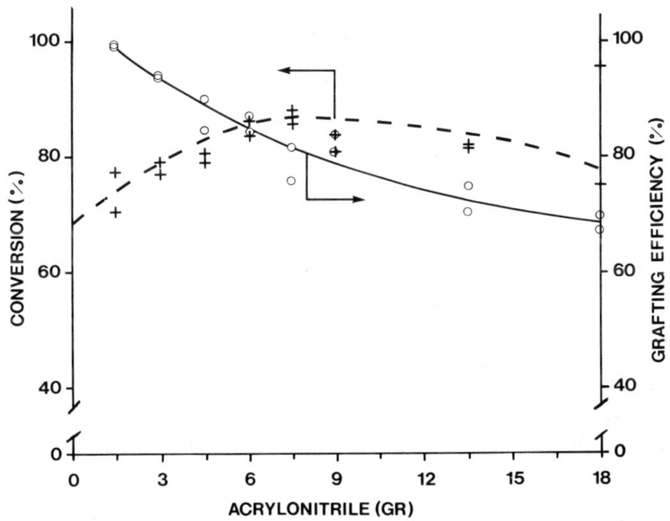

Figure 5. Dependence of conversion (+) and grafting efficiency (○) on quantity of monomer.

e) Dependence of grafting on temperature.

A negligible effect of temperature on grafting is shown in Figure 6. It is obvious that the temperature does not have any effect on grafting efficiency and a very small effect on conversion. These results are not surprising in view of the similar results in our previous work (11). It seems that the positive effect which the increase of temperature exerts on the initiation and propagation of polymerization compensates for the negative effect of saponification (increased rate of saponification of the xanthate groups in acidic medium before the radical formation as indicated in diagram I).

Fortunately, grafting at 25°C is advantageous from the energetic point of view.

2) Effect of Lignin on Grafting Efficiency, Conversion and Polymer Loading.

a) Effect of Bleaching on Delignification

In order to find out how the residual lignin affects grafting parameters, high yield hardwood pulp of aspen was bleached with several different concentrations of sodium chlorite in order to prepare pulp with decreasing lignin content. As in our previous work with softwood pulp (3) sodium chlorite was selected as the bleaching agent in order to minimize cellulose degradation. The resulting pulps were characterized by standard methods and their properties are listed in Table I.

The lignin content of the pulps varies from 14.8% down to 4.4%. Even though the viscosity of pulp was not measured, it was shown in our previous work that the method of bleaching used does not degrade cellulose (3).

b) Effect of Lignin on Conversion.

It is evident from Figure 7, that high bisulfite hardwood pulp behaves in a similar fashion as high bisulfite softwood pulp (16). The plots of conversion versus lignin content indicate the presence of a minimum. This minimum conversion is observed at approximately 8-10% of lignin and occurs quite regularly and independently of the time of reaction. At the shorter reaction time used (0.5 hr), only 2.5% conversion is observed. Polymerization increases in 1 hr reaction time to 10% conversion and in case of 18 hrs polymerization fairly high conversions are recorded throughout the range of lignin content employed. The presence of a small minimum is nonetheless notable. In case of softwood high yield sulfite pulp, the minimum of conversion was situated at about 4% higher (e.g., 12-15%) than the hardwood range, (16).

c) Effect of Lignin on Polymer loading.

Figure 8 shows the dependence of polymer loading as a function of lignin content. Once again, it is possible to observe the presence of a minimum on the polymer loading-versus-lignin curves, which is situated at approximately 8-12% lignin as compared to 11-16% lignin for softwood equivalent pulp (16).

d) Effect of lignin on grafting efficiency.

Figures 9 and 10 show that there is also a minimum on the grafting efficiency-versus-lignin curves, which is situated this

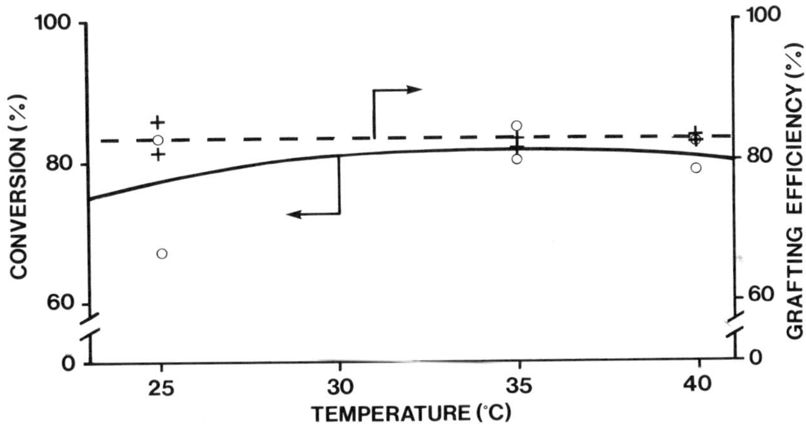

Figure 6. Dependence of conversion (○) and grafting efficiency (+) on temperature.

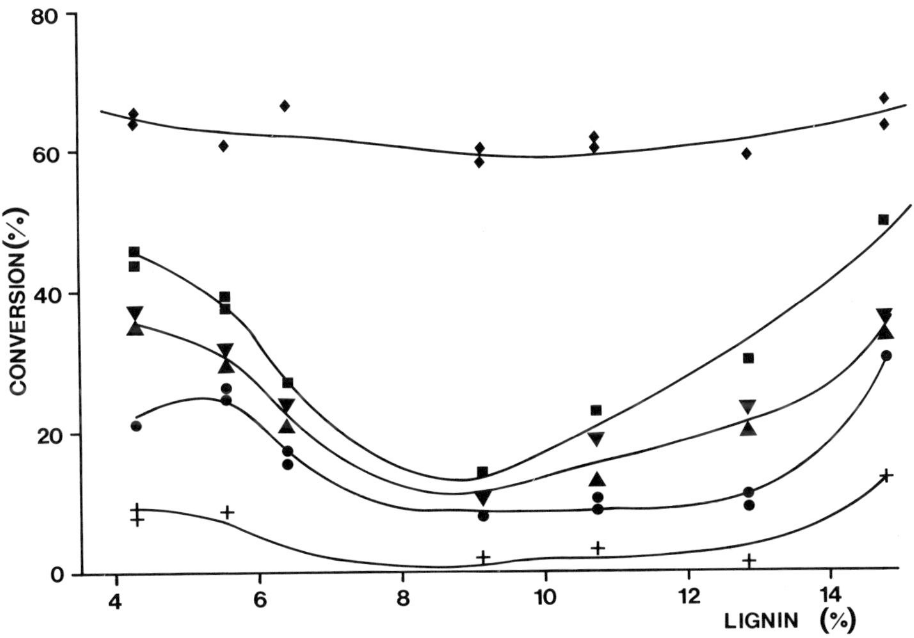

Figure 7. Effect of lignin on conversion. Key: ◆, 18 h; ■, 2 h; ▲, 1.5 h; ●, 1 h; +, 0.5 h.

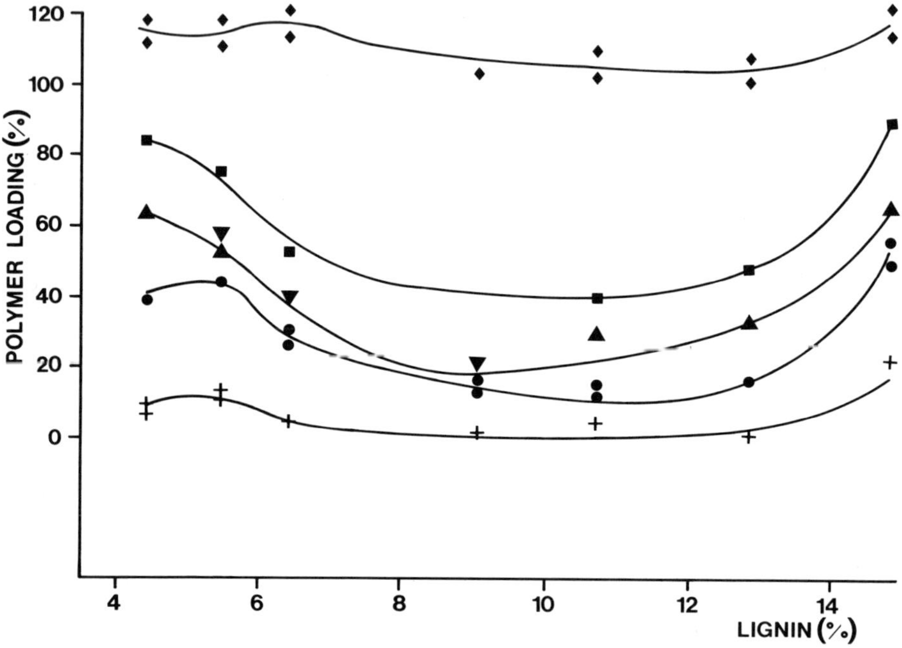

Figure 8. Effect of lignin on polymer loading. Key: ♦, 18 h; ■, 2 h; ▼, 1.5 h; ●, 1 h; +, 0.5 h.

time at approximately 10-13% lignin. This minimum which is very pronounced at 30 min of reaction becomes almost inexistant in the case of 18 hrs reaction.

The mechanism that might explain the presence of minimum is not yet completely clear. It is certain, that lignin participates in the copolymerization reaction as we have shown in the two recent publications, (17,18) in which lignin was copolymerized either with styrene (17) or acrylic monomers (18). It seems probable that the composition of lignin as well as cellulose and holocellulose remaining after different levels of bleaching may be chemically and topochemically different (see Table I). Because of the complexity of the lignin structure and to the different character of lignocellulose substrates used in our work, it is difficult to make any general conclusion. As was suggested previously (16), any correlation based on macroscopic parameters such as the concentration of lignin are bound to have but a limited significance. As in our previous work, it was observed that the minimum coincides with the maximum in the carboxylic index, (Table I) which relates to the cation exchange capacity. A similar correlation between the graftability of sulfite pulp and its exchange capacity has been also noted in our previous work (16). In conclusion, even though there is the definite correlation between carboxylic group index and minimum in respect to graftability in our system, a detailed work with model compounds would be required to futher elucidate the mechanism of lignin participation in the copolymerization reaction.

Conclusion

It has been shown that the xanthate method could be successfully applied in order to graft hardwood high yield bisulfite pulp made of aspen. The optimum grafting condition, which would give almost 100% grafting efficiency (i.e.,100% formation of true graft and no homopolymer) and rather high monomer conversion of 80% are as follows:

Pulp/monomer:	6 (weight ratio)
H_2O_2:	8g/1
Initial pH:	5
Monomer:	acrylonitrile
% Lignin:	14%
Pulp Yield:	90%
150 ml 0.004%	$(NH_4)_2Fe(SO_4)_2$

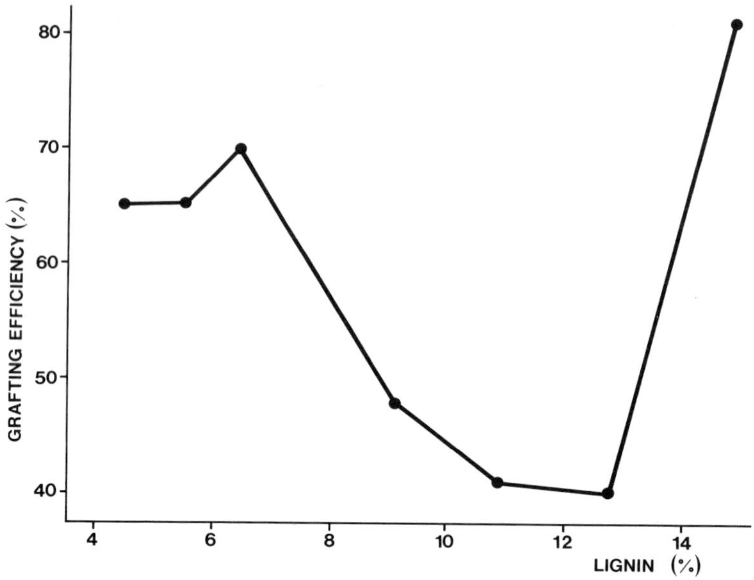

Figure 9. Effect of liginin on grafting efficiency. Reaction time is 30 min.

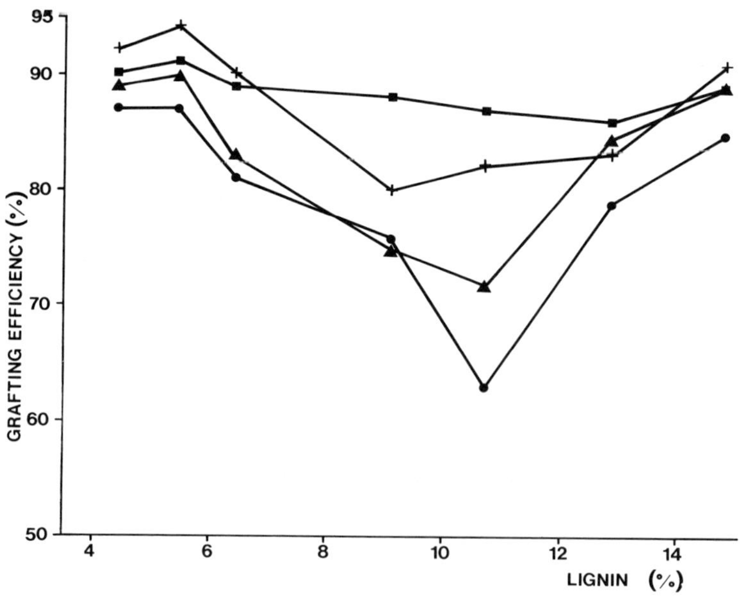

Figure 10. Effect of lignin on grafting efficiency. Key: +, 18 h; ■, 2 h; ▲, 1.5 h; ●, 1 h.

References

1. R.W. Faessinger and J.S. Conte, Scott Company, Belgian Patent 646, 284 (oct. 8, 1964); British Patent 1,059,641 (feb. 22, 1966); N.S. Patent 3,330,784 (sept. 11, 1967); German Patent 1,468,965 (march 6, 1969).
2. Hornof, V., Kokta B.V. and Valade J.L., J. Appl. Polym Sci., 1975, 19, 545-556.
3. Hornof V., Kokta B.V. and Valade J.L., J. Appl. Poly. Sci., 1975, 19, 1573-1584.
4. Hornof V., Kokta B.V., and Valade J.L., J. Appl. Poly. Sci., 1977, 20, 477-485.
5. Hornof V., Daneault C., Kokta B.V. and Valade J.L., J. Appl. Poly. Sci., 1977, 21, 2991-3002.
6. Kokta B.V., Valade J.L., and Daneault C., Pulp and Paper Can., 1979, 80 (9),TR59-65.
7. Kokta B.V., Valade J.L. and Daneault C., Pulp and Paper Can. 1981, 82 (3),TR5-14.
8. Neimo L., Ehrnrooth E., and Sihtola H., Polym. News, 1974, 2 (1/2),22.
9. Quère J., Phillips R.B., Guiroy G., Compte rendu du Centre Technique du Papier, 1973, Grenoble, France, 83-87.
10. Young R.A., Agric J., Food Chem., 1977, 25 (4),738-742.
11. Kokta B.V. and Valade J.L., Tappi, 1972, 55, 366.
12. Teichmann H.E., Brickman W.J., Faessinger R.V., Mayer C.G. and Krässig H.A., Tappi, 1974, 57 (7),61-64.
13. Krassig H.A., J. Macromol Sci. Chem., 1976, A10, 759-768.
14. Dimov K. and Pavlov P., J. Polymer Sci., 1969, A-1, 7, 2775.
15. Samoilov I.V., Livshits R.M., Morin B.B., Rogovin Z.A., U.S. Patent 3, 878, 144, 1975.
16. Hornof V., Kokta B.V., and Valade J.L., J. Appl. Polym. Sci., 1976, 20, 1543-1554.
17. Chen R., Kokta B.V. and Valade J.L., J. Appl. Polym. Sci., 1979, 24, 1609-1618.
18. Chen R., Kokta B.V. and Valade J.L., J. Appl. Polym. Sci., 1980, 25, 2211-2220.

RECEIVED December 17, 1981.

18

Graft Copolymerization of Lignosulfonate with Methacrylic Acid and Acrylate Monomers

RUBIE CHEN and BOHUSLAV V. KOKTA

Centre de Recherche en Pâtes et Papiers, Université du Québec à Trois-Rivières, Trois-Rivières, Québec, Canada G9A 5H7

Lignosulfonate from sulfite spent liquor was grafted, after purifying through ultrafiltration with acrylic monomers: methyl acrylate, methyl methacrylate and methacrylic acid. Copolymerization was redox initiated by peroxide-ferrous ions in aqueous and non-aqueous media under nitrogen atmosphere. The effects of reaction medium, reaction time, lignin/monomer ratio, pH, and peroxide charge, on the copolymerization were studied. Among the media employed, the magnitude of conversion of acrylate monomer was found to be the highest in water, lower in dioxane, and the lowest in methanol; while for acrylic acid the medium preference of obtaining better conversion was in the order of: water, methanol and dioxane. The best reaction conditions in water were: 3% peroxide, 1:3 lignin/monomer without any buffering, under which conversion would reach 88-96% and the grafting efficiency 35-62%.

Since the prices of petroleum products have soared, valorization of natural resources has intensively attracted the attention of researchers in different fields. To fully utilize wood, which is a renewable raw material, one must make better use of its 25-30% lignin content which forms the principal dissolved material in the spent liquor of chemical pulping processes. The annual production of sulfite pulp in Canada is approximately 2.6 million tons (data based on 1979), thus, at least one million tons of lignosulfonate is produced, representing a potential resource of raw material for other uses. However, among 35 sulfite mills in Canada, only two or three of them possess a recovery system for spent liquor, that means a very small portion of lignosulfonate has been recovered. The recovered lignin has been

0097-6156/82/0187-0285$6.00/0
© 1982 American Chemical Society

used, taking advantage of its colloidal properties, as the agents for emulsification, deflocculation, chelation, adhesion, drilling mud additive, etc. Besides, research has been oriented to the utilization of lignin in polymer domains, e.g., as a moiety of either thermosetting or thermal plastics.

The graft of lignin with vinyl monomers has been studied in the last two decades, Koshijima and Muraki published a series of reports on radical grafting of vinyl monomers onto hydrochloric acid lignin, induced by radiation (1, 2, 3) or by chemical initiators (4). Nam et al. (5, 6) investigated the grafting of methacrylate onto lignosulfonate, using hydrogen peroxide as an initiator. Phillips and his coworkers (7, 8, 9) studied the radiation-induced graft copolymerization of styrene onto hydrochloric acid lignin as well as various kraft softwood lignins. Naveau (10) prepared methacrylic derivatives of acid-hydrolyzed lignin through esterification. Chernyavskaya and Berlin (11) worked on the grafting of methyl acrylate onto hydrolyzed wood lignin using peroxide initiator. Simionescu and his coworkers (12, 13) studied the radiation grafting of hydrochloric acid poplar lignin and reed lignin with vinyl monomers. Graft copolymerization of ozone-activated hydrochloric acid lignin with styrene was reported by Katuscak et al. (14, 15).

Through the awareness of that, in their studies, most of the above mentioned researchers used lignins specifically laboratory -prepared, the main objective of this work was to derive lignin-plastic copolymers from lignin available commercially; the other objectives involved the kinetic study and the optimization of reaction conditions. Reported in the present paper are the graft copolymerizations of calcium lignosulfonate with methyl acrylate, methyl methacrylate, and with methacrylic acid; the effects of several reaction variables (medium, reaction time, pH, lignin/monomer ratio, and initiator charge) on the copolymerization results are discussed.

EXPERIMENTAL

Materials. The lignin material used in this study was a commercial calcium salt of lignosulfonate (LS) (Toranil B, St. Regis Paper Co.), isolated from the spent liquor of softwood sulfite pulping. It was purified, in our laboratory, through ultrafiltering a 2% water solution (using a Millipore immersible molecular separate kit) and drying the filtrate under vacuum. The material as received contained 8.8% methoxyl group (specified by the supplier) and its lignosulfonate content was 86.8% with reference to the purified material (determined in our laboratory by ultraviolet absorption at 280 nm). Vinyl monomers, namely, methyl acrylate (MA), methyl methacrylate (MMA) and methacrylic acid (MAA) were distilled under vacuum; distillation was performed over cupric sulfate and through a copper-ring packed column to remove stabilizers. The central cut of each distillation was collected in a dark bottle and stored in a refrigerator. Methanol was also purified by vacuum distillation. The other chemicals were of reagent grade and used without further purification.

Copolymerization. The detail experimental procedures of copolymerization have been reported elsewhere (16), however, the procedures will be described briefly here and are shown schematically in Figure 1. Graft copolymerization was redox initiated using hydrogen peroxide and ferrous chloride and was conducted under nitrogen atmosphere with gentle agitation. After reaction, extractions were made to separate the resulting copolymer from the vinyl homopolymer and the unreacted lignosulfonate. In the LS-acrylate systems, acetone was used to extract homopolymers, and water to extract the unreacted lignosulfonate. In the LS-(methacrylic acid) system, because water could dissolve the unreacted lignosulfonate and the two reaction products (i.e., homopolymer and grafted copolymer), ethanol and methanol were used to extract the homopoly(methacrylic acid) and the copolymer, respectively. Wherever no specification is given, the reaction conditions used in this work were fixed as follows: CaLS = 1 g; monomer = 5ml measured at 25°C (MA = 4.79 g, 5.57×10^{-2} mol; MMA = 4.70 g, 4.70×10^{-2} mol; MAA = 5.07 g, 5.90×10^{-2} mol); ferrous chloride = 15 mg; hydrogen peroxide = 30 mg; reaction medium = 20 ml. When using water as a reaction medium in the LS-MA (or MMA) system, a small quantity of emulsifier (Tween 20, 1 g /100 ml of medium) was employed.

The copolymerization results were interpreted in terms of the parameters defined below:

$$\text{Total monomer conversion, } \% = [(TS - WL)/M] \times 100$$
$$\text{Grafting efficiency, } \% = [1 - HP/(TS - WL)] \times 100$$
$$\text{Degree of lignosulfonate reacted, } \% = [(WL - LU)/WL] \times 100$$

where WL stands for the lignosulfonate used in the reaction; M, the monomer; TS, the total solid obtained after reaction; HP, homopolymer; and LU, lignosulfonate unreacted. All these symbols are of the same weight unit.

RESULTS AND DISCUSSION

Reaction medium. When lignosulfonate was subjected to graft copolymerization with vinyl monomers, the extent of copolymerization due to the effect of medium varied from one monomer to another. In a LS-styrene system (16), it was found that methanol was a better medium than water under certain given conditions; while in a LS-acrylonitrile system (17),the contrary was true, i.e., water better than methanol. This contradiction was thought due to the fact that styrene has electropositive (i.e., electron-releasing) substituent while acrylonitrile has electronegative (i.e., electron-attracting) substituent. In the present study,

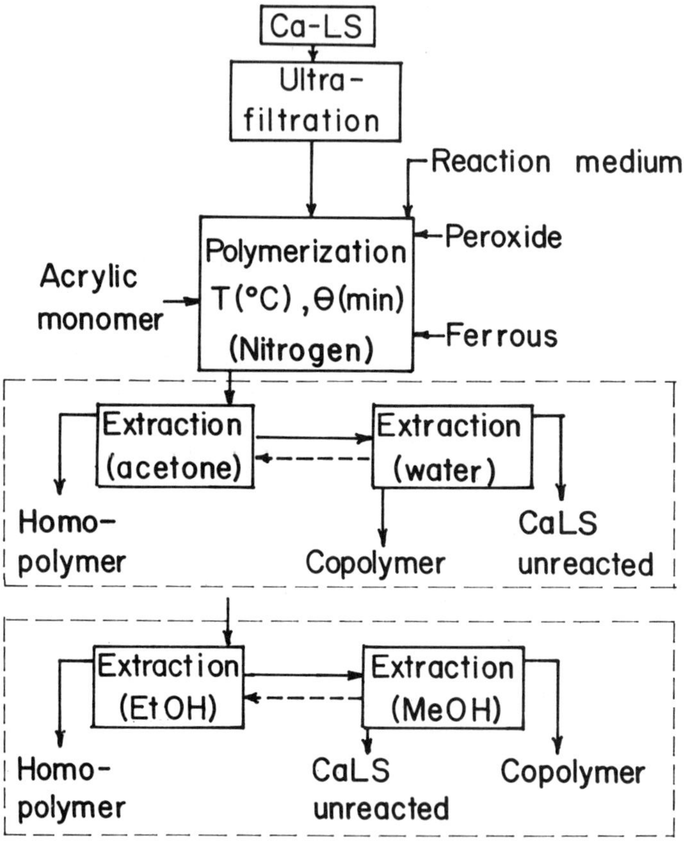

Figure 1. Schematic diagram of experimental procedures.

copolymerization of lignosulfonate and acrylate or acrylic acid was examined in several media commonly used in lignin research work. The resulting kinetic curves are given in Figures 2 and 3, respectively, for the LS-(methyl methacrylate) and LS-(methacrylic acid) systems. It can be seen that after 2 hours of reaction the total conversion of methyl methacrylate in the former system attained 88% in water and 60% in dioxane; though the initial conversion rates were nearly the same in both media. The lowest conversion, 3%, occurred in methanol. In the second system (Figure 3), the total conversion of methacrylic acid reached 96, 61 and 30% in water, methanol, and dioxane, respectively, with initial rate decreasing in that order. The medium which brought about higher conversion was found to bring about also higher grafting efficiency. For example, the grafting efficiencies of methyl methacrylate onto lignosulfonate were 30, 19 and 8%, respectively, in water, dioxane and methanol. The medium preference of the LS-(methyl acrylate) system was similar to that of the LS-(methyl methacrylate).

Reaction time. As already seen in Figures 2 and 3, most of the reactions were rapid at first then slowed down. Figure 4 shows typically the effect of reaction time on copolymerization, taking the LS-(methacrylic acid) system in aqueous medium as an example. About 80% of the methacrylic acid monomer polymerized after 20 min of reaction and the final 96% conversion reached within one hour. The grafting efficiency appeared to increase along with the conversion; from 35% at 5 min to its plateau (about 60%) at one hour of reaction. However, not so much change in the degree-of-lignosulfonate-reacted was found during the same period, i.e., from 43% at 5 min to 50% at 1 hour, implying that the number of active sites on lignin macromolecules probably became fixed as soon as copolymerization started and this number was not altered as the reaction proceeded. Figure 5 gives another example showing the LS-(methyl acrylate) system which attained its final states at not longer than 2 hours of reaction.

Ratio of LS/monomer. In the previous work (16), it was reported that an unavoidable inhibiting effect, owing to the quinonoid structure of lignosulfonate, on the polymerization was observed when studying the graft of styrene onto lignosulfonate by redox initiation; the induction period was proportional to the concentration of lignosulfonate, in other words, proportional to the LS/monomer ratio. However, in this study, no such evidence was observed in the LS-acrylic systems, though in theory this kind of inhibiting effect could also be imposed to these systems; Figures 6 and 7 show the influence of the LS/monomer ratio on copolymerization. In general, there was no adverse influence on the monomer conversion when the LS/monomer ratio (in weight) was augmented from 0.1 to 1.0. In the systems involving methyl acrylate and methacrylic acid, total conversions were practically sta-

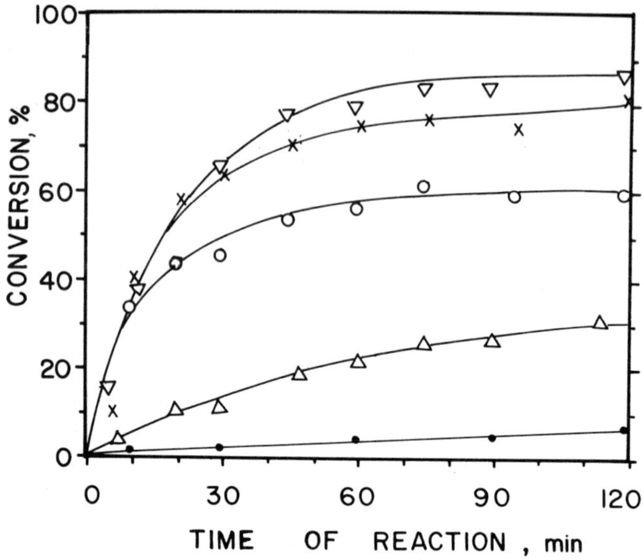

Figure 2. Variation of the total monomer conversion with reaction time for the LS–(methyl methacrylate) system at 50°C in different media: ▽*, water;* ×*, water–dioxane 1:1;* ○*, dioxane;* △*, water–methanol 1:1; and* ●*, methanol.*

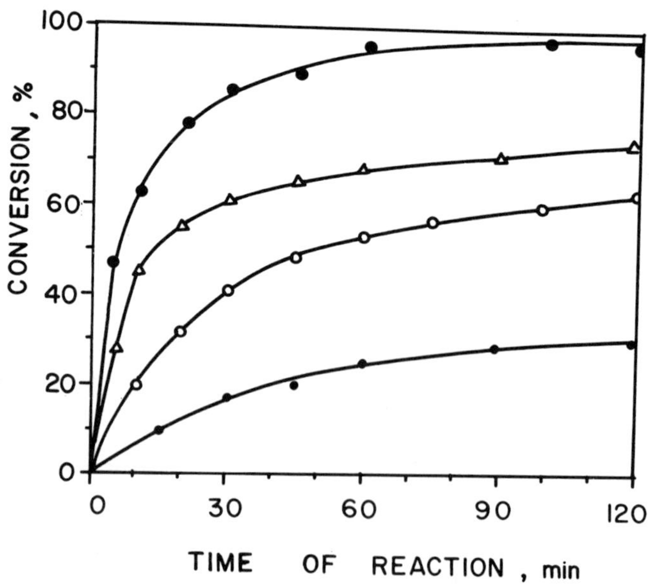

Figure 3. Variation of the total monomer conversion with reaction time for the LS–(methacrylic acid) system at 30°C in different media: ●*, water;* △*, dimethylformamide;* ○*, methanol; and* •*, dioxane.*

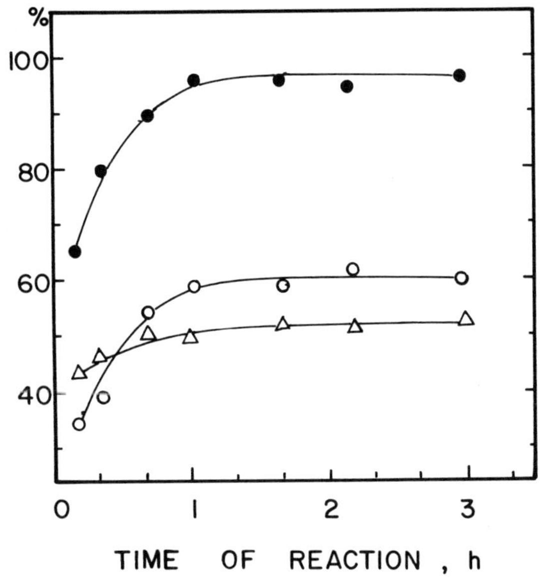

Figure 4. Influence of reaction time on copolymerization of lignosulfonate and methacrylic acid in water medium at 30°C: ●, total monomer conversion; ○, grafting efficiency; △, degree of LS reacted.

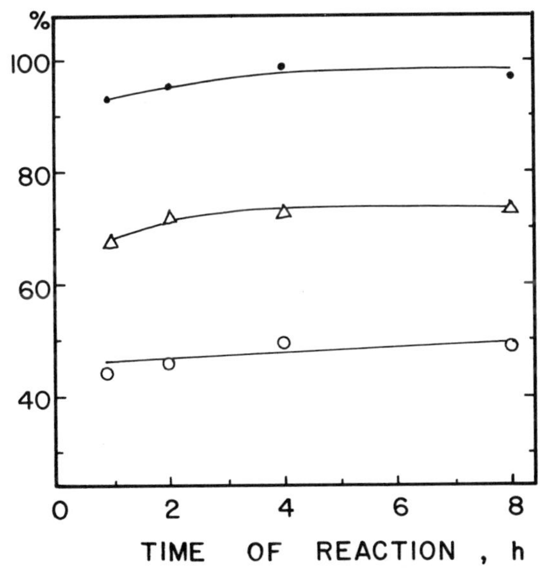

Figure 5. Influence of reaction time on copolymerization of lignosulfonate and methyl acrylate in water medium at 30°C: ●, total monomer conversion; △, degree of LS reacted; ○, grafting efficiency.

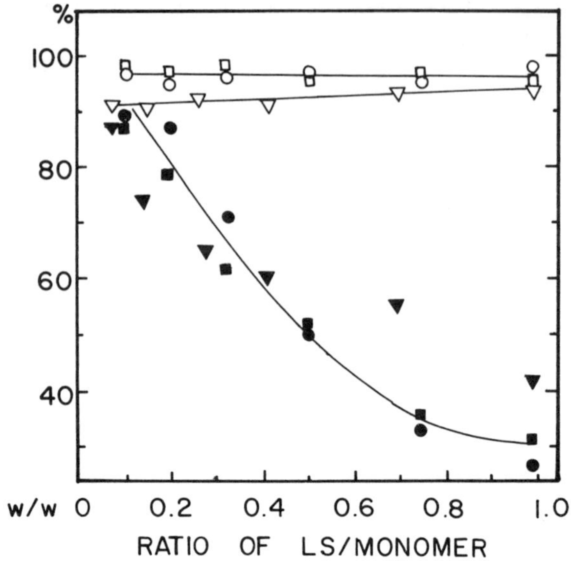

Figure 6. Effect of the LS/monomer ratio on the total monomer conversion (□, ▽, ○) and the degree of LS reacted (■, ▼, ●) for the systems: □/■, LS–(methyl acrylate); ▽/▼, LS–(methyl methacrylate); and ○/●, LS–(methacrylic acid). Conditions: water medium; 30°C; 120 min reaction time; and LS = 1 g.

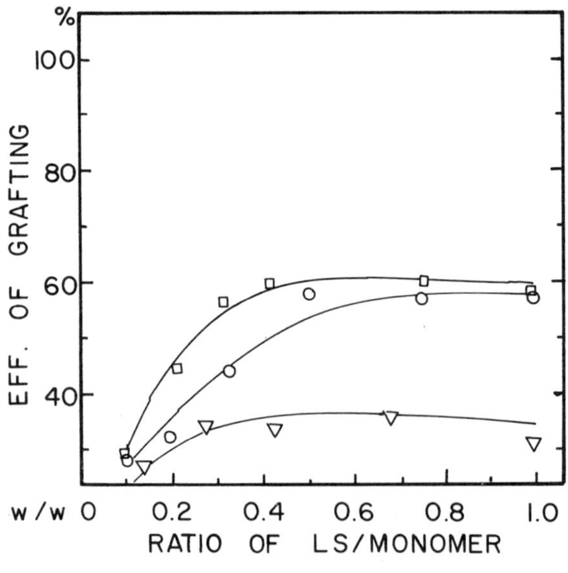

Figure 7. Effect of the LS/monomer ratio on the grafting efficiency for the systems: □, LS–(methyl acrylate); ▽, LS–(methyl methacrylate); and ○, LS–(methacrylic acid). Conditions: water medium; 30°C; 120 min reaction time; LS = 1 g.

bilized at more than 95 percent; while in the LS-(methyl methacrylate) system, the conversion seemed to increase slightly from 90 to 94% in the same range of LS/monomer ratio. It is interesting to note (in Figure 7) that the grafting efficiency was improved to a certain extent by increasing the LS/monomer ratio then it remained unchanged. Both the LS-(methyl acrylate) and LS-(methacrylic acid) systems reached their plateaus of around 60% efficiency at LS/monomer = 0.5; whereas the LS-(methyl methacrylate) system attained a plateau of 35% at LS/monomer = 0.3. These two limits, i.e., 0.5 and 0.3, suggested that the involvement of lignosulfonate in the graft copolymerization would reach a saturation point in each system.

As for the degree-of-lignosulfonate-reacted (Figure 6), a negative influence was found in all the three systems. It decreased almost proportionally to the LS/monomer ratio. In the range of the ratio studied the largest decrease was from 90% to 26% in the LS-(methacrylic acid) system and the smallest change was from 86% to 42% in the system of LS-(methyl methacrylate). At this moment and without further detail investigation, it is unable for the authors to explain the undesirable decrease in the degree-of-lignosulfonate-reacted, except that a speculation is made that the relative reactivity of lignosulfonate might diminish with the LS/monomer ratio.

Initial pH. The solubility of lignosulfonate in water increases with increasing pH,and the hydrodynamic volume of lignin molecule expands correspondingly. Furthermore, the degree of dissociation of calcium lignosulfonate is larger at higher pH, and lignosulfonic molecule could thus attract electrically more water molecules to its surrounding. These water molecules would become more or less obstacles to the approaching polyvinylic radicals. As a consequense, an increase in the initial pH produced a negative influence on copolymerization, which is evidently shown in Figures 8 and 9.

For the LS-(methyl acrylate) system (Figure 8), the variation of conversion with pH was quite moderate in the range between 2 and 10, producing a drop of not more than 4% (from 98.2 to 94.4%). Grafting efficiency did not change significantly, and its value scattered around 60%; while the degree-of-lignosulfonate-reacted decreased from 62% to about 50%. More significant variation occurred in the LS-(methacrylic acid) system (Figure 9) where the conversion was lower in the alkaline than in the acid side and attained a minimum of 80.2% at near neutral pH. This kind of variation was similar to what had been reported for the homopolymerization of methacrylic acid (18). Grafting efficiency varied with pH in a similar way as conversion did; its minimum occurred at pH = 7 but the efficiencies at both the high and low pH values (i.e., 12 and 2) were almost identical. The degree-of-lignosulfonate-reacted decreased from 53% to near 40% as pH increased from 2 to 8 and did not change with further increases in pH.

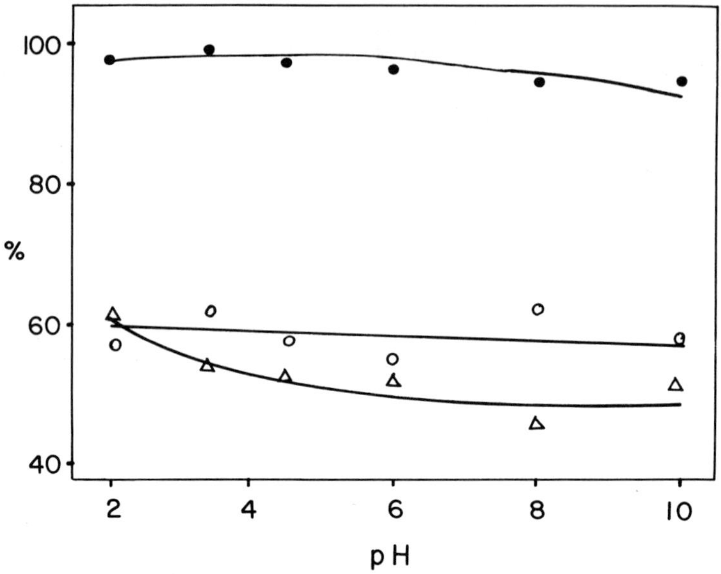

Figure 8. Effect of pH on copolymerization of lignosulfonate and methyl acrylate. Conditions: water medium; 30°C; 120 min reaction time. Key: ●, total monomer conversion; ○, grafting efficiency; △, degree of LS reacted.

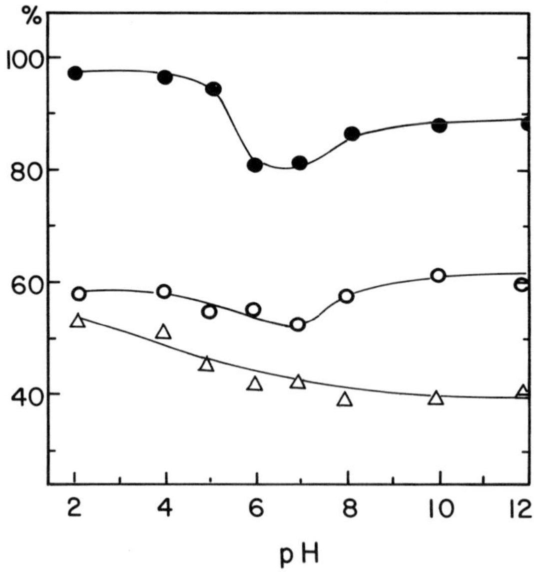

Figure 9. Effect of pH on copolymerization of lignosulfonate and methacrylic acid. Conditions: water medium; 30°C; 120 min reaction time. Key: ●, total monomer conversion; ○, grafting efficiency; △, degree of LS reacted.

It is known that the colour of lignosulfonate is darker at higher pH, thus the copolymer produced at pH 2 was tan and at pH 10 was dark brown. It should be pointed out that, without any buffering, the pH values of the LS-acrylate and LS-(methacrylic acid) systems were 4.4 and 2.4, respectively.

Peroxide charge. Though there are stable free radicals in lignin preparations (19, 20, 21), the concentration of these radicals is far from sufficient to provoke copolymerization with vinyl monomers; otherwise, these radicals may inhibite copolymerization. By means of redox activation, using the Fe^{+2}-H_2O_2 pair, copolymerization was found to occur readily under suitable conditions, which was due to the formation of a greater number of active phenolic radicals in the lignin molecule. It is well known that the ferrous-H_2O_2 pair involves reactions where ferrous ions are oxidized to produce HO• and HO_2• (22) in a cyclical equilibrium:

$$Fe^{+2} + H_2O_2 \rightarrow Fe^{+3} + HO^- + HO\cdot$$

$$Fe^{+2} + HO\cdot \rightarrow Fe^{+3} + HO^-$$

$$Fe^{+3} + H_2O_2 \rightarrow Fe^{+2} + H^+ + HO_2\cdot$$

It is the HO• which induces the vinyl radicals a part of which then grafts onto lignin.

$$HO\cdot + nCH_2{=}CHR \rightarrow HO{-}(CH_2{-}CHR)_n{-}CH_2{-}\dot{C}HR \left\langle \begin{array}{l} \nearrow \text{Homopolymer} \\ \searrow \text{Copolymer} \end{array} \right.$$

It was thought (6) that the freshly oxidized ferric ion might competitively react with the lignosulfonate to generate the lignin macroradical, thus making lignin itself act as an initiator:

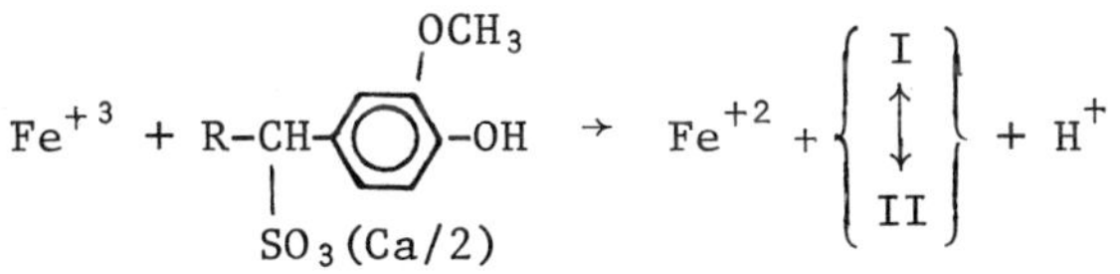

where

I II

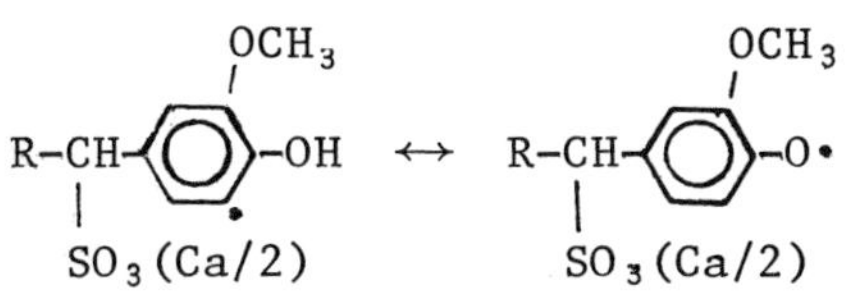

These reactions could happen in this work only when sufficient amount of peroxide was present. Thus the quantity of peroxide used played an important role in the control of copolymerization. Figure 10 demonstrates the influence of peroxide charge on copolymerization for the LS-(methyl methacrylate) system. As shown, there was no distinguishable variation in the grafting efficiency in the range studied; the values obtained were situated between 20 and 30%. Meanwhile, the total monomer conversion and the degree-of-lignosulfonate-reacted initially increased with peroxide charge. When the quantity of peroxide augmented from 10 to 30 mg, both the conversion and the degree-of-lignosulfonate-reacted increased about two times of their original value (i.e., from 43 to 90% and from 30 to 61%, respectively). After that, further addition of peroxide resulted in only a small gain in conversion and no further augmentation in the degree-of-lignosulfonate-reacted. This implyed that a limilted amount of peroxide reacted directly or indirectly (e.g., via oxidized ferric ions) with the lignin macromolecules to generate active sites for grafting. Beyond that limit, it was believed that the excess of peroxide radicals worked essentially with vinyl monomers to produce only homopolymers, supported by the variation of the molecular weight of homopolymers (obtained in the same reaction) with peroxide charge -- the molecular weight of homopoly(methyl methacrylate) decreased with increasing peroxide charge.

Copolymer. The appearence and properties of a particular copolymer obtained depended to a large degree on the type of monomer with which the copolymer was prepared. LS-(methyl acrylate) and LS-(methyl methacrylate) copolymers were brownish amorphous solids. The former was rubbery and moderately hard; and the latter was plastic like and harder than the former. At room temperature, both were insoluble in ordinary solvents and quickly swelled in water to about six times in volume. At elevated temperature, the lignin backbone of these two copolymers could be degraded through acid or alkaline hydrolysis.

LS-(methacrylic acid) copolymer was brownish brittle solid which was soluble in water, methanol, dimethyl formamide, and dimethyl sulfoxide at room temperature. Figure 11 shows the reduced viscosity, η_{sp}/C, of this copolymer and that of a methacrylic acid homopolymer (for comparison purpose) in methanol at 25°C. From this figure, an abrupt increase in the copolymer viscosity can be observed at very low concentration, suggesting that this copolymer might form microgel (probably owing to the particles having extremely high molecular weight or crosslinks) in methanol solution.

CONCLUSION

Commercially available calcium lignosulfonate can be readily and effectively grafted with methyl acrylate, methyl methacry-

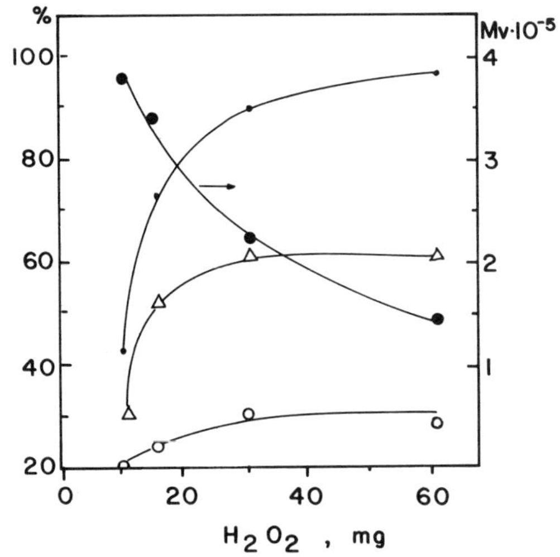

Figure 10. Effect of peroxide charge on copolymerization of lignosulfonate and methyl methacrylate. Conditions: water medium; 50°C; 120 min reaction time. Key: •, total monomer conversion; △, degree of LS reacted; ○, grafting efficiency; ●, molecular weight of MMA homopolymer.

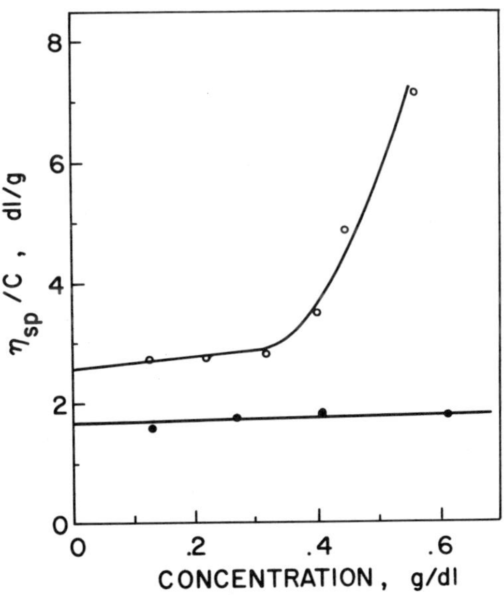

Figure 11. Reduced viscosity versus concentration at 25°C. Key: ○, LS–(methacrylic acid) copolymer; and ●, homopoly(methacrylic acid) in methanol.

late, and with methacrylic acid, using Fe^{+2}-H_2O_2 initiation. Increase in peroxide charge increased the monomer conversion and decreased the molecular weights of the acrylic homopolymers while the grafting efficiency and the degree-of-lignosulfonate-reacted approached certain limits. The best reaction conditions in water medium were: 3% peroxide, and lignin/monomer 1:3.

ACKNOWLEGEMENTS

The authors gratefully acknowledge the financial assistance received from the Natural Sciences and Engineering Research Council of Canada and from the Ministry of Education of the Province of Québec. The work described in this article forms part of the research program of the Centre de Recherche en Pâtes et Papiers at Université du Québec à Trois-Rivières.

LITERATURE CITED

1. Koshijima, T.; Muraki, E.; J. Japan Wood Res. Soc., 1964, 10(3), 110.
2. Koshijima, T.; Muraki, E.; Preprints IUPAC Internatl. Symp. Macromol. Chem. (Tokyo), 1966, Session 3, 60.
3. Koshijima, T.; Muraki, E.; J. Polymer Sci. A1, 1968, 6, 1431.
4. Koshijima, T.; Muraki, E., J. Japan Wood Res. Soc., 1967, 13(8), 355.
5. Nam, C.; Koshijima, T.; Muraki, E.; J. Polymer Sci. A1, 1971, 9, 855.
6. Nam, C.; Maku, T.; Saito, M.; Koshijima, T.: Cellul. Chem. Technol., 1974, 8, 263.
7. Phillips, R.B.; Brown, W.; Stannett, V.T.; J. Appl. Polymer Sci., 1971, 15, 2929.
8. Phillips, R.B.; Brown, W.; Stannett, V.T.; J. Appl. Polymer Sci., 1972, 16, 1.
9. Phillips, R.B.; Brown, W.; Stannett, V., J. Appl. Polymer Sci., 1973, 17, 443.
10. Naveau, H.P., Cellul. Chem. Technol., 1975, 9, 71.
11. Chernyavskaya, S.B.; Berlin, A.A.; Khim. Drev., 1978, No. 1, 96.
12. Simionescu, C.; Cernatescu-Asandei, A.; Stoleru, A.; Cellul. Chem. Technol., 1975, 9(4), 363.
13. Simionescu, C.; Anton, I.; Cellul. Chem. Technol., 1969, 3(4), 387.
14. Katuscak, S.; Mahdalik, M.; J. Appl. Polymer Sci., 1973, 17, 1919.
15. Katuscak, S.; Hrivik, A.; Mahdalik, M.; Paperi Puu, 1971, 53, 519.
16. Chen, R.; Kokta, B.V.; Valade, J.L.; J. Appl. Polymer Sci., 1979, 24, 1609.
17. Chen, R.; Kokta, B.V.; Valade, J.L.; J. Appl. Polymer Sci., 1980, 25, 2211.

18. Ito, H.; Shimizu, S.; Suzuki, S.; Kogyo Kagaku Zashi, 1958, 58, 194; Chem. Abstr., 1955, 49, 14375d.
19. Steelink, C., Adv. Chem. Ser., 1966, 59, 51.
20. Katuscak, S.; Hrivik, A.; Macak, K.; Paperi Puu, 1972, 54(4a), 201.
21. Hon, D.N.-S.; Glasser, W.G.; Tappi, 1979, 62(10), 107.
22. Haber, F.; Weiss, J., Proc. Roy. Soc. (London), 1934, A147, 332.

RECEIVED December 17, 1981.

CHEMICAL MODIFICATION OF LIGNOCELLULOSIC FIBERS

19

Chemical Modification of Wood with Thioacetic Acid

SATISH KUMAR and S. C. AGARWAL

Forest Research Institute and Colleges, Dehra Dun, India

Acetylated wood possesses unique properties such as dimensional stability and resistance to biological degradation. Acetylation with thioacetic acid is a new approach in which shortcomings observed in acetylation with acetic anhydride or acetyl chloride are overcome. Reactions with thioacetic acid can be carried out in vapor-phase and excess reagent recovered with xylene. The only by-product released is H_2S gas, which is recycled. High levels of acetylation sufficient to protect against biodegradation and up to 66 percent antishrink efficiency can be achieved by using suitable catalysts. Wood moisture content up to 7 percent has no adverse effect on acetylation. Preliminary tests indicated the acetylated wood to be resistant to a brown rot fungus even at low acetyl contents. Limited studies on strength properties indicated a gain in such properties with acetylation.

Wood is a complex polymer composite made up of cellulose, hemicelluloses and lignin. Because of its organic origin, wood is easily degraded by a variety of micro-organisms, insects, and marine organisms. Sunlight (UV radiation), rain, heat, and abrasion affect physical and chemical changes in wood. Under changing conditions of relative humidity, wood swells or shrinks making it dimensionally unstable and unsuitable for some precise applications. The hygroscopic nature and biological deterioration potential of wood has been traced to the presence of a large number of free hydroxyl groups in all the three major components. These hydroxyl groups act as reactive sites favoring adsorption of water vapor. Most enzyme systems attach themselves to these hydroxyl groups before metabolising the wood substrate to easily digestible simple compounds. Certain wood species are known to be resistant to degradation by organisms, as these contain toxic compounds in their heartwood. Treatment with wood preservatives

0097-6156/82/0187-0303$6.00/0
© 1982 American Chemical Society

also renders wood unpalatable to such organisms. Such protection of wood is based on the theory that these toxic compounds are solubilised in the enzymatic fluids secreted by the organisms and immobilize the latter through their toxicity.

Brief Review of Past Work on Chemical Modification of Wood

Chemical reaction between reactive sites in wood components and a chemical reagent to form a non-polar bond between the two is defined as chemical modification. This definition excludes all impregnation treatments which do not form covalent linkages such as polymer inclusions, some coatings, heat treatments, etc. (1). Chemical modification reactions, developed primarily to impart dimensional stability to wood, were found to provide a high resistance against biological and physical deterioration. Protection of wood through chemical modification has been based on the theory that the enzyme must forge a direct contact with the substrate. Fungal enzymes like most others, being specific in nature, require that the substrate molecule or a portion of it must fit the active sites of enzyme in a lock and key type relationship (2). Chemical modification alters the specific configuration of wood components, due to substitution of hydroxyl groups. The blocking of hydroxyl groups also changes the hydrophilic nature of wood, causing exclusion of water from the substrate, which is a biological necessity of many organisms. Both these effects are probably equally responsible for imparting high decay resistance to chemically modified wood (1), as in most cases neither the reacting compounds nor the modified wood are toxic in nature. Electron microscopic studies have revealed that the reductions in fungal secretions and affinity for the substrate are the most probable reasons for decay resistance (3). Chemically modified woods thus offer a viable alternative to conventional preservation techniques, which although highly effective against organisms do carry a certain degree of pollution hazards.

Rowell (1) reviewed the various types of chemical modification reactions and reagents used for the same and listed their advantages and disadvantages. Chemical modification reactions may be classified by the type of carbon-oxygen-carbon bond of which ethers, acetals and esters are of major importance. Ether type linkages are formed by methylation reactions with dimethyl sulfate or methyl iodide (4, 5) and reactions with alkyl or allyl chlorides (6-10), β-propiolactone (11), acrylonitrile (12, 13) and epoxides (14-18). The most promising reactions have been found with epoxides, such treatments being highly effective with no side effects. Epoxides are, however, very expensive and relatively higher levels of treatment are required due to occurrence of polymerisations reactions and degree of substitution can not be determined from weight gains (18).

Acetals are obtained by a two stage reaction with aldehydes

resulting in cross-linking (19, 20). Although high decay resistance is obtained at very low levels of chemical loading (21), the reagents give a bad smell and the acid catalysed reactions impair the mechanical properties (19, 22, 23). Milder catalysts are being investigated for minimal effect on strength properties (20, 24).

Esterification of wood can be carried out through acetylation (4, 8, 25, 26, 28), crotonylation (9), propylation (28), butylation (10) and phthaloylation (27). Reactions with isocyanates also yield a desirable wood product, but isocyanates are highly toxic, expensive and reduce toughness (29). None of the other esterification reactions showed any advantage over acetylation.

Acetylation of Wood. The acetylation process was developed as early as 1865 for production of cellulose acetate, but its application to wood was conceived only in 1928 (30). Acetylation of wood is the most widely studied reaction of all the chemical modification reactions. The most common acetylating reagent studied is acetic anhydride although trifluoroacetic anhydride, ketene and acetyl chloride have also been used. Suida (31) obtained the first patent on acetylation of wood using acetyl chloride. Reaction with acetyl chloride releases HCl, a strong mineral acid, which causes degradation of wood. Recent studies on acetylation of wood pretreated with lead acetate, with acetyl chloride in vapor phase revealed that, although a higher acetylation level could be so achieved, there was not much advantage in using acetyl chloride over acetic anhydride as subsequent reactions with lead acetate produced an equivalent amount of acetic acid (32).

A large number of catalysts and swelling agents have been studied for the acetylation reaction. The best acetylation conditions have been reported in vapor phase treatment with uncatalysed acetic anhydride in xylene solutions at 100-130°C (28). Acetylation reaction between wood and acetic anhydride proceeds as follows:

$$\text{Wood-OH} + (CH_3CO)_2O = \text{Wood-O-}\overset{\overset{\displaystyle O}{\|}}{C}CH_3 + CH_3COOH$$

The main disadvantage of this reaction is that 50% of the reagent is lost in the form of acetic acid adding to treatment costs. Acetylated wood continues to smell of acetic acid, corrodes metal fasteners and may cause hydrolysis of wood in the long run, if acetic acid is not completely removed. Moreover, ester bonds are labile to acid hydrolysis and the continuous presence of acetic acid in the wood is liable to reverse the reaction. Commercial systems to recover unreacted reagents and the by-products based on refrigeration system have been developed in Russia, but the cost involved is so far the limiting factor (33).

Acetylation of Wood with Thioacetic Acid

Reaction between thioacetic acid and wood proceeds as follows:

$$\text{Wood-OH} + CH_3COSH = \text{Wood-O}\overset{\overset{O}{\|}}{C}CH_3 + H_2S\uparrow$$

The only by-product is H_2S gas, which is recycled for the manufacture of thioacetic acid (34) making the process a closed circuit one and thus non-polluting (35).

Experimental

Two species, chir (*Pinus roxburghii* Sargent.) size 7.5 x 2.75 x 1.25 cm^3 and mango (*Mangifera indica* Linn.) size 7.5 x 1.25 x 1.25 cm^3 were studied to establish different reaction variables. Four replicates were taken for each treatment and experiments were carried out with oven-dried samples except in case of those where effect of moisture content was studied. Samples were placed in a reaction chamber and reaction was carried out under reflux conditions at 93 $\pm$ 1°C, the boiling point of thioacetic acid. Experimental set-up for vapor phase acetylation is shown in Figure 1.

To study the rate of reaction and obtain adequate weight percent gain (WPG), reaction periods varying from 1 to 8 hours were used. Effect of sample thickness on overall acetylation levels was examined on 7.5 cm long mango samples of different cross-sections (6 x 6 mm^2, 12.5 x 12.5 mm^2, 20 x 20 mm^2, 25 x 25 mm^2, and 30 x 30 mm^2). These samples were acetylated in vapor phase for 3 and 6 hours.

Five sets (4 samples each set) of mango samples were conditioned to five different moisture levels and conditioned samples were acetylated in vapor phase for 3 hours to study the effect of wood moisture content on WPG.

Pretreatment with various swelling agents and catalysts, e.g., dimethyl formamide, 4 percent urea-ammonium sulphate (1:1), pyridine, 4 percent potassium acetate, 4 percent ammonium chloride, 4 and 6 percent trichloriacetic acid and 1, 2 and 4 percent monochloroacetic acid was done to evaluate the performance of these chemicals in acetylation reactions using thioacetic acid. Effect of time on acetylation in the presence of pyridine and monochloroacetic acid was studied with mango.

WPG obtained in wood samples was calculated as follows:

$$WPG = \frac{Wa - Wo}{Wo} \times 100$$

Where Wo is the oven-dried weight of untreated wood sample and Wa is the oven-dried weight of the same sample after acetylation.

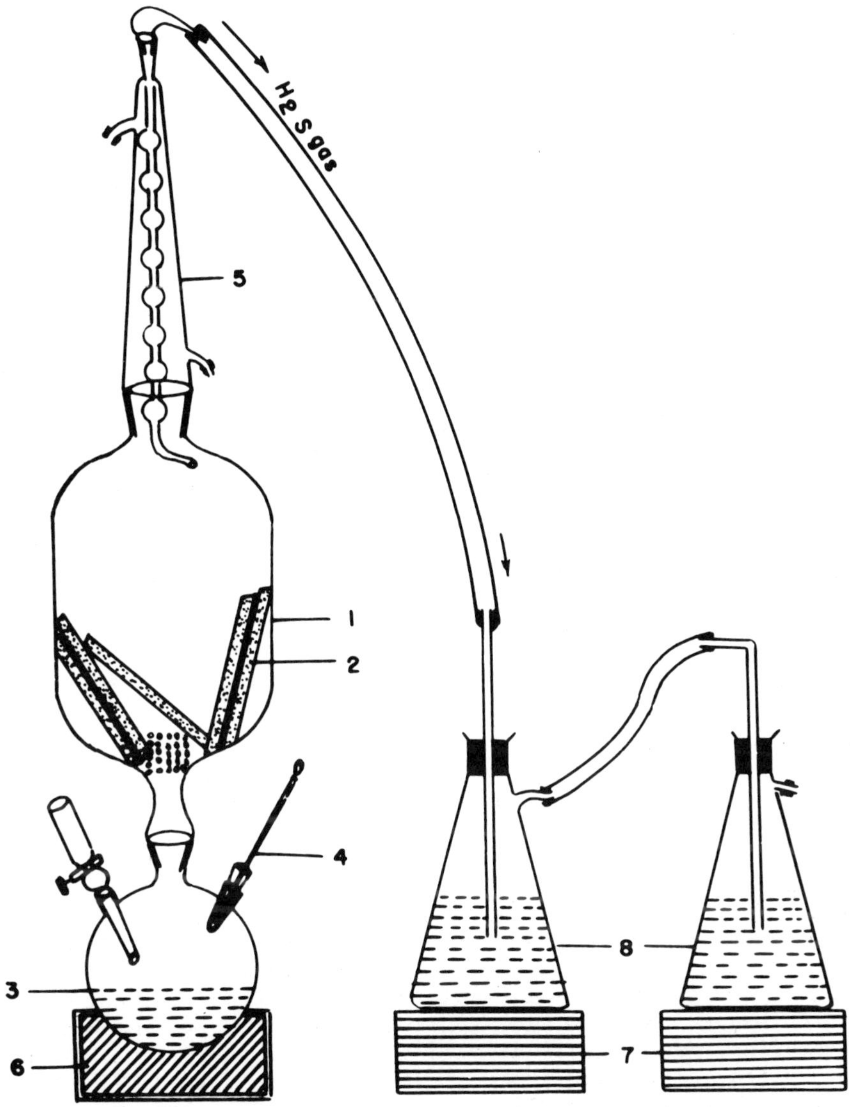

Figure 1. Vapor phase acetylation assembly. Key: 1, reaction chamber; 2, wood samples; 3, thioacetic acid; 4, thermometer; 5, water condenser; 6, heating mantle; 7, magnetic stirrer with heater; and 8, acetic anhydride and catalysts.

Acetyl content of untreated as well as treated samples was determined following the standard ASTM method (36).

Dimensions of the samples were measured with a dial gauge (to an accuracy of 0.001 cm) before treatment (oven-dried), immediately after treatment (saturated with the reagent) and again after oven-drying at 100 $\pm$ 2°C. Fully swollen dimensions were also measured after impregnating the samples with water under vacuum. In chir, only tangential swelling and shrinkage were evaluated. The antishrink efficiency (ASE) imparted due to acetylation was calculated as below (37):

$$ASE = \frac{(\text{Percent swelling in untreated sample}) - (\text{Percent swelling in treated sample})}{(\text{Percent swelling in untreated sample})} \times 100$$

Results and Discussion

Acetylation is a single site reaction and one acetyl group is required for every hydroxyl group reacted. Since there is no crosslinking or polymerization of the reagent, the weight gain can be directly translated into acetyl groups added. Goldstein et al (28) observed that acetyl content obtained in ponderosa pine was higher than the observed WPG, as net WPG represents the weight gained minus the extractives leached out. Figure 2 shows that a linear relation exists between WPG and acetyl content as obtained by chemical analysis. A similar relation has been also reported by Peterson and Thomas (3), who reported a single linear relation for loblolly pine (a softwood) and yellow poplar (a hardwood). Our data yielded two distinct relations for the two woods. These differences probably arise from the leaching of extractives as well as the difference in acetyl contents of untreated wood. Before acetylation, chir had an acetyl content of 1.27% whereas mango contained 4.44%.

Effect of Time and Sample Thickness. Table I depicts the acetylation level obtained in mango at various time intervals. It may seem that even after 6 hours of reaction, WPG is insufficiently low to impart decay resistance to wood. The rate of vapor phase reactions is determined by the rate of permeation of reagent vapors into the wood. It has been observed that even longitudinal permeability of wood decreased by 2-10 times during acetylation (38). This probably is the reason for slow increase in WPG. In chir also a WPG of 7 percent and 10.9 percent was observed in four and eight hours acetylation respectively. In an earlier study using only 0.25 cm thick samples a WPG of 16.67% was obtained in four hours (35). Similar limitations were observed by Tarkow et al (37), who concluded that 3 mm thickness was the most optimum to obtain uniform acetylation. The effect of cross-section on acetylation is quite clear in Table II.

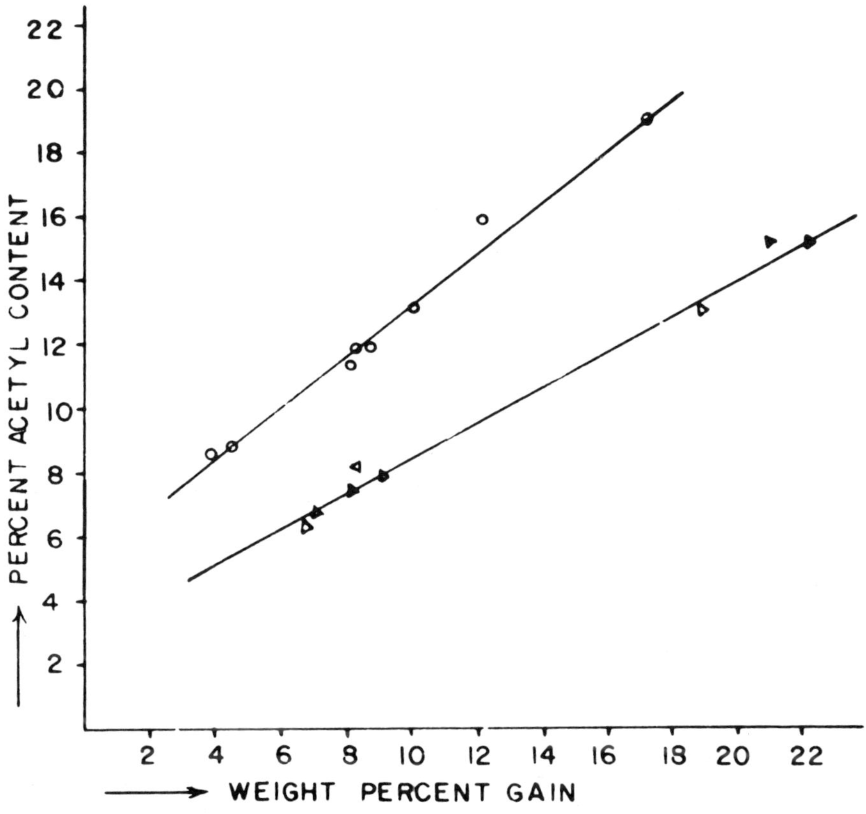

Figure 2. Relationship between weight percent gain (WPG) and acetyl content. Key: ○, mango; △, chir.

Table I. Effect of Time on Vapor Phase Acetylation of Mango

Treatment time (hours)	Vapor phase acetylation			Vapor phase acetylation followed by extraction with xylene for 1 hour	
	Immediate WPG	WPG (net)	Acetyl content %	WPG (net)	Acetyl content %
1	34.7	3.7	8.58	3.4	8.52
1.5	40.6	4.4	8.79	4.1	8.88
2	43.8	8.1	11.33	7.5	11.16
3	48.4	8.7	11.93	8.2	11.85
4	83.2	8.0	12.48	7.6	12.41
6	83.6	8.4	12.90	7.7	12.79

Vapor phase treatments are generally preferred over liquid phase as the former can be carried out under controlled conditions to regulate the absorption of the reagent. It was, however, observed that it was not possible to regulate the over absorptions of thioacetic acid even in vapor phase treatments. When dilute solutions of thioacetic acid in xylene were used for reaction, the reaction did not proceed satisfactorily. Substantial amounts of thioacetic acid condensed into wood samples during treatment leading to nearly 80% WPG immediately after the treatment. This unbonded thioacetic acid not only upsets the economy of the process but also continues to emit a foul smell for long periods. Moreover, prolonged contact of thioacetic acid may have adverse effect on wood strength as it is believed to be a depolymerization agent for lignin (39). Extraction of unreacted thioacetic acid was therefore carried out with xylene vapors immediately after the reaction was terminated. The extraction had only a slight effect on the net WPG as may be seen in Table I. Material balance analysis showed that it was possible to recover all the unreacted reagent in one hour in the small samples used. The quantities of thioacetic acid consumed could be related to the bonded acetyl groups. Earlier workers recommended a reaction period of 8-20 hours to get proper WPG with acetic anhydride (28,33,37). As longer reaction periods usually become uneconomical, suitable catalysts need to be tested for higher WPG.

Table II. Effect of Sample Thickness (Mango) on WPG

Cross-section (cm^2)	WPG	
	3 hour treatment	6 hour treatment
0.6 x 0.6	11.64	-
1.25 x 1.25	8.70	8.40
2.0 x 2.0	7.61	9.59
2.5 x 2.5	7.39	9.42
3.0 x 3.0	7.42	9.46

Effect of Wood Moisture. It has been recommended that for acetylation with acetic anhydride, the moisture content of wood should be about 2% as excess moisture is likely to react with acetic anhydride and produce acetic acid (37). Goldstein et al. (28) observed that raising the moisture to 22% considerably slowed the reaction and each 1 percent of moisture in wood would lead to hydrolysis of about 5.7% acetic anhydride. Low moisture contents are not possible to attain in commercial treatment of wood. With ketene gas it has been possible to acetylate wood with as high as 20% moisture content with WPG about 25% (40). Thioacetic acid is only partially stable in cold water and dissociates at higher temperatures. The presence of moisture in wood could thus be critical in treatments with thioacetic acid also. Results of mango treated at 5 different moisture levels are depicted in Figure 3. As may be seen a moisture content up to 7.5% has no adverse effect on WPG. At 10% moisture content the WPG decreased to 4.6. With further increase in moisture, there was a gradual decrease in WPG attained.

Effect of Various Swelling Agents and Catalysts. Because of the immobile nature of the wood-matrix, a swelling agent is considered necessary for easy transport of reagent to the reaction sites. A wide range of swelling agents and catalysts have been investigated to improve the acetylation reactions. Thioacetic acid, itself, is a good swelling agent affecting nearly 13.5% swelling with respect to oven-dry volume of wood. This probably is the reason that rate of reaction with thioacetic acid is slightly higher than with acetic anhydride. Since WPG obtained in case of thioacetic acid were very low even after 8 hours reaction, various swelling agents and catalysts suggested by earlier workers were studied with chir to see their effect on acetylation level. Among the various pretreatments tested, mono-

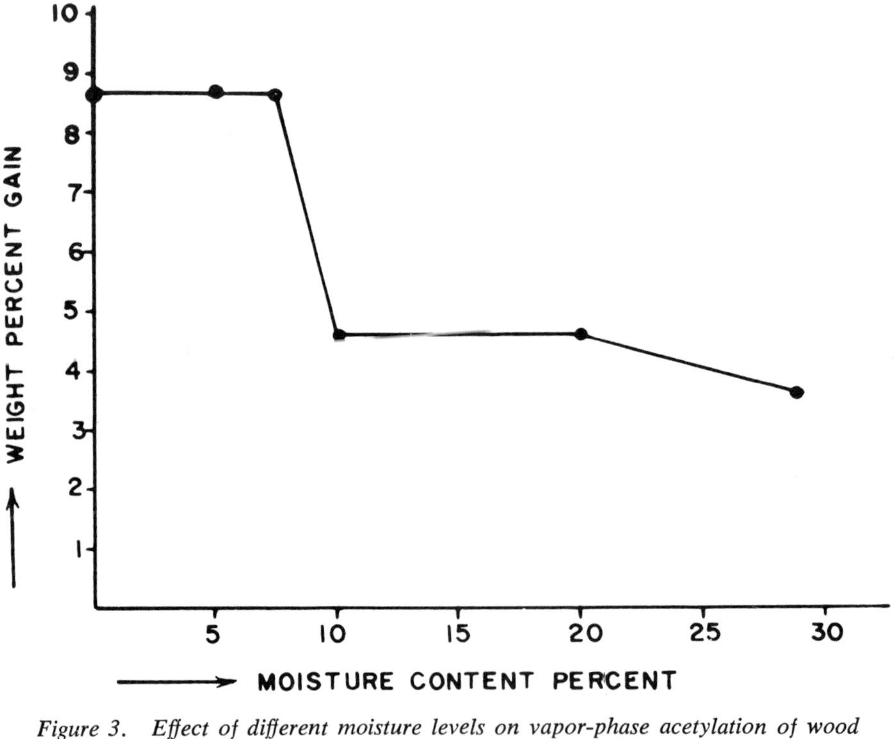

Figure 3. Effect of different moisture levels on vapor-phase acetylation of wood with thioacetic acid. Reaction time is 3 h.

chloroacetic acid (2% and above) gave the highest WPG. Sufficient WPG were also obtained with 4% trichloroacetic acid (Table III) (41).

A detailed study with respect to acetylation time was made in case of monochloroacetic acid and it was observed that an acetylation time of 2 hours was sufficient to obtain a WPG of 16.78%. In the case of mango, only pyridine and monochloroacetic acid were tried. While pyridine had very little effect, monochloroacetic acid improved WPG to over 17% in six hours treatment (Table IV).

Properties of Acetylated Wood

Decay & Termite Resistance. Cellulose is the principal food of termites in wood. Termites also digest wood through enzymes present in the intestinal protozoa. Acetylated wood has been found resistant to fungal, termite and even shipworm attack (28, 37). Even low WPG (7.5-7.8) severely retarded decay with most decay fungi. WPG around 15-17 were found most effective (3,28). Termite resistance tests under laboratory conditions gave threshold value near 19 WPG (28). Acetylated veneers (WPG 19.2) in ground contact gave average life of 17.5 years compared with 2.7 years for untreated controls (42). Wood samples with WPG around 12 were found to be light to moderately attacked by termites (*Microcerotermes beesoni* Snyder) when tested according to forced feeding test developed by Sensarma (43). Limited studies on chir samples treated with thioacetic acid gave weight losses around 5.5% at WPG of 10-11 with *Poria monticola*, a brown rotter indicating acetylation of cellulose even at lower WPG. Losses came down to approximately 3% at WPG around 19 (44).

During acetylation with acetic anhydride, at acetyl levels around 12 percent, about 22 percent of the hydroxyls in spruce wood are reported to be substituted. These hydroxyls are the most accessible ones and are therefore believed to be mainly in the lignin and the hemicelluloses (45). Lignin has been found to be the most easy to acetylate compared with cellulose in other lignocellulosic materials like jute also (46). At acetylation level of 13.5% as much as 86.4% of the lignin, 21.6% of the hemicellulose and only 9.3% of the cellulose were acetylated in pinewood (47). Low acetyl levels can impart resistance against white rot fungi, these fungi preferentially attack the lignin and acetylation with thioacetic acid appears to have an edge over acetic anhydride in providing resistance against brown rotters as well. However, according to Peterson & Thomas (3), free hydroxylsin lignin macromolecules are very infrequent and resistance to white rot is achieved through blocking of cellulose as the lignin degrading catalysts of the white rot fungi derive their energy from simultaneous metabolism of carbohydrates. This hypothesis is, however, contradicted by the fact that in pinewood at acetyl level of 13.5%, nearly 52.1% of the acetyl groups introduced were

Table III. Effect of Various Swelling Agents and Catalysts on WPG and Shrinkage-Swelling Behavior in Chir. (4 Hour Vapor-Phase Treatment with Thioacetic Acid.) (41)

Pretreatment	WPG	Acetyl content %	Tangential swelling % (O.D. to fully swollen)	Tangential shrinkage % (fully swollen to O.D.)	ASE %
Control (untreated)	-	1.27	6.32	5.94	-
No catalyst	7.08	6.75	4.61	4.39	25.8
Dimethyl formamide	8.10	7.46	4.66	4.45	28.1
Urea-ammonium sulphate	9.13	7.91	4.44	4.26	28.3
Pyridine	10.39	9.33	4.38	4.32	29.4
Potassium acetate	6.83	6.30	4.30	4.12	30.6
Ammonium chloride	8.32	8.20	3.99	3.28	36.4
Trichloroacetic acid(4%)	18.92	13.05	2.40	2.35	60.7
Trichloroacetic acid(6%)	18.89	12.92	-	-	56.3
Monochloroacetic acid (1%)	12.29	-	-	-	-
Monochloroacetic acid (2%)	22.05	-	-	-	-
Monochloroacetic acid (4%)	22.39	15.30	2.03	1.99	66.6

Table IV. Vapor Phase Acetylation of Mango Samples Pretreated with Pyridine/Monochloroacetic Acid

Treatment time (hours)	Pretreatment					
	Nil		Pyridine (24% retention by O.D. wt. of wood)		Monochloroacetic acid (2.5% retention by O.D. wt. of wood)	
	WPG	Acetyl content (%)	WPG	Acetyl content (%)	WPG	Acetyl content (%)
1	3.7	8.58	3.9	8.78	8.6	11.69
3	8.7	11.93	10.1	13.12	12.1	15.88
6	8.4	12.90	9.8	14.47	17.2	19.05

associated with lignin indicating sufficient availability of hydroxyls for reaction. Thus unavailability of the lignin hydroxyls to fungal enzymes due to acetylation appears to be the more plausible reason for decay resistance against lignin consuming fungi.

Effect on Hygroscopicity and Dimensional Stability. Water adsorption in wood is attributed mostly to hydroxyl groups located on polysaccharide chains. Lignin is considered to be the least hygroscopic wood component while hemicelluloses contribute to a larger portion of water adsorbed. Nearly 1/3 of the cellulose in wood is amorphous in nature, which is responsible for water adsorption. The available information on molecular association of various components of wood indicates that lignin and hemicelluloses form a three dimensional interpenetrating matrix surrounding the cellulose microfibrils. Lignin and hemicelluloses being easily accessible are thus likely to be first substituted and a very high percentage of lignin is reported to be substituted at acetyl levels around 13.5% (47). At slightly lower acetyl content (13%) it has been observed that the hygroscopicity of chir is reduced by over 50% (Table V).

Adsorption of water is accompanied with swelling of wood. Since cellulose constitutes almost 50% of the wood substance, its fractional contribution to sorption is the highest. Swelling of wood is thus manifested by the adsorption of water by hydroxyl groups located on surface of micells and amorphous regions of cellulose, crystalline zones being impenetrable. Reduction in

Table V. Effect of Acetylation on Hygroscopicity of Chir at Various Relative Humidities.

Relative-humidity (%)	Moisture content of control (%)	Reduction in hygroscopicity at acetyl content level	
		6.75%	13.00%
28	5.6	28.6	41.1
48	9.5	38.9	50.0
60	12.0	41.7	52.1
72	14.5	39.3	52.8 5

shrinkage and swelling is usually represented by ASE obtained due to treatment (37). Tangential swelling and shrinkage along with ASE obtained at various WPG/acetyl levels have been listed in Table III. Studies have indicated that ASE by acetylation is achieved through substitution of hydroxyls as well as bulking of the cell wall. A WPG of 20-25% has been reported to yield an ASE of about 70% in spruce (37). Figure 4 shows the relation between acetyl content and ASE as obtained by acetylation with acetic anhydride in the presence of pyridine (37) and thioacetic acid with various catalysts and swelling agents. As has been reported earlier, at about 13.5% acetyl content, most of the lignin and a higher fraction of hemicellulose is acetylated while nearly 90% of the cellulose remains unreacted (47). Electron microscopic studies on acetylated pine also reveal that at lower WPG, esterification was very low in earlywood tracheids. A gradient existed across the cell wall thickness, esterification being very low in S_1-S_2 region compared with S_2-S_3 region in the tracheid cell wall (3). According to Tarkow *et al.* (37) at this acetyl level the ASE is only 35%. It appears that in the two component systems used by Tarkow *et al.* (37), pyridine swells the wood structure and acetic anhydride is not able to penetrate through the lignin-hemicellulose network to acetylate or bulk the cellulose at the same rate as pyridine may penetrate. Since thioacetic acid itself is swelling in nature, it accords higher dimensional stability even at lower acetyl contents probably due to its penetration to the cellulose structure also.

Effect on Strength Properties. The few studies conducted on the effect of acetylation on strength properties indicate that acetylation with acetic anhydride had no immediate adverse effect

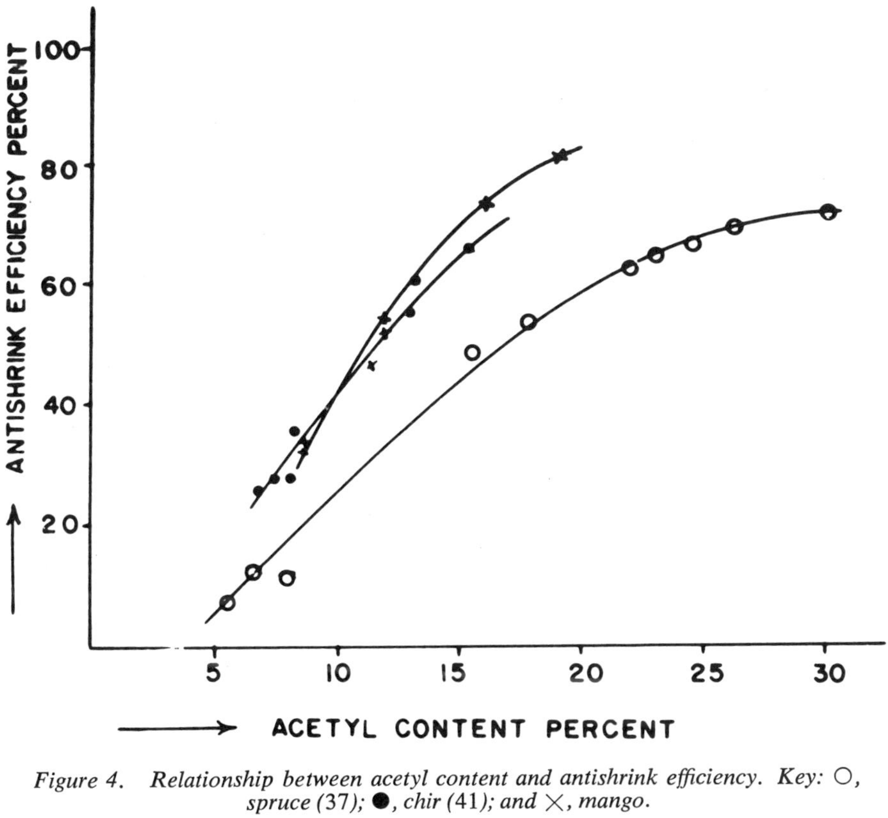

Figure 4. Relationship between acetyl content and antishrink efficiency. Key: ○, *spruce (37);* ●, *chir (41); and* ×, *mango.*

on most strength properties. Modulus of elasticity and modulus of rupture in bending, modulus of elasticity under tension and toughness showed improvement in the few woods studied (37). Goldstein et al. (28) also reported an increase in wet compressive strength and impact strength after acetylation.

Limited tests on the effect of acetylation with thioacetic acid on strength properties of mango indicated that MOR in static bending improved by approximately 50% and maximum crushing strength in compression by 25% at a WPG of 8.5. Impact strength determined by Charpy test also showed an improvement of about 50%. This improvement in mechanical property is due to increase in density of wood. Even density, determined on oven dry weight and oven-dry volume basis, increased by about 3.9% at a WPG of 8.7, despite the fact that volume under oven-dry conditions undergoes considerable increase (about 4.7%) due to bulking effect. A WPG of 8 results in an increase of about 4.6% in wood density (O.D. wt. swollen volume basis).

Summary and Conclusions

D D During chemical modification of wood with thioacetic acid, the only by-product formed is H_2S gas. This easily escapes from wood favoring a forward reaction. The liberated H_2S is recycled to produce thioacetic acid, making the process a closed circuit one. Since there is no crosslinking or polymerization of the reagent, acetylation level can be directly gauged from the WPG. The unreacted reagent can be easily recovered from wood by solvent extraction with xylene. Thioacetic acid is a good swelling agent for wood, which facilitates penetration of the reagent deep into the wood structure. Even thicker samples can be acetylated during vapor phase treatment. Another advantage of using thioacetic acid is that wood containing moisture content up to 7 percent can be acetylated, whereas in acetylation with acetic anhydride wood moistures beyond 2 percent result in reagent losses. Among a wide range of catalysts tested, monochloroacetic acid and trichloroacetic acid yielded the best results.

Limited tests on strength properties of wood acetylated with thioacetic acid showed that MOR in static bending, impact strength, maximum crushing strength in compression and density improved. It is, therefore, evident that thioacetic acid has either no adverse effect on wood components and their interlinks or the improvements affected by modification far exceed these effects, so as to be of little consequence and ending in net gains.

Wood acetylated with thioacetic acid showed resistance to decay and termites at low WPG around 12. The available information on pattern of substitution of hydroxyls during acetylation with acetic anhydride suggests substitution of lignin hydroxyls at low acetylation levels. Resistance to micro-organisms, particularly those consuming cellulose, even at low acetylation levels suggests preferential substitution of cellulose hydroxyls with thioacetic acid. This is also supported by the high ASE (70-75%) obtained at low WPG (about 15%).

Detailed studies on the pattern of substitution of hydroxyl group located on various wood components will lead to better understanding of various properties of modified wood. More effective catalysts also need to be investigated to obtain higher acetylation levels in shorter periods.

Literature Cited

1. Rowell, R. M. Proc. Amer. Wood Preserv. Assoc. 1975, 71, 41-51.
2. Lehninger, A. L. "Biochemistry". Worth Publishers Inc., New York, 1970, p. 83.
3. Peterson, M. D.; Thomas, R. J. Wood and Fiber 1978, 10(3), 149-63.
4. Narayanamurti, D.; Handa, B. K. Das Papier 1953, 7, 87-92.
5. Rudkin, A. W. Australian J. Appl. Sci. 1950, 1, 270-283.
6. Kenaga, D. L.; Sproull, R. C.; Esslinger, J. Southern Lumberman 1950, 180, 45.
7. Kenaga, D. L.; Sproull, R. C. J. Forest Prod. Res. Soc. 1951, 1, 28-32.
8. Risi, J.; Arseneau, D. F. Forest Prod. J. 1957, 7, 210-213.
9. Ibid., 245-246.
10. Ibid., 261-265.
11. Goldstein, I. S. U.S. Patent 1960, 2,931,741.
12. Baechler, R. H. Forest Prod. J. 1959, 9, 166-171.
13. Kenaga, D. L. U.S. Patent 1963, 3,077,417.
14. McMillin, C. W. Forest Prod. J. 1963, 13, 56-61.
15. Liu, C.; McMillin, C. W. U.S. Patent 1965, 3,183,114.
16. Aktiebolag, M. D. U.S. Patent 1965, 1,408,170.
17. Barnes, H. M.; Choong, E. T.; McIlhenny, R. L. Forest Prod. J. 1969, 19, 35-39.
18. Rowell, R. M.; Gutzmer, D. I. Wood Science 1975, 7, 240-46.
19. Tarkow, H.; Stamm, A. M. J. Forest Prod. Res. Soc. 1955, 3, 33-37.
20. Stevens, M.; Schalck, J.; Raemdonck, J. V. The International J. of Wood Preserv. 1979, 1(2), 57-68.
21. Stamm, A. J.; Baechler, R. H. Forest Prod. J. 1960, 10, 22-26.
22. Stamm, A. J. Tappi 1959, 42(1), 39-44.
23. Burmester, A. Holzforschung 1967, 21(1), 13-20.
24. Vihavainen, T.; Piispanen, K.; Mansikkamaki, P. Intl. Res. Group on Wood Preservation 1980 Doc. No. IRG/WP/3146.
25. Stamm, A. J.; Tarkow, H. U.S. Patent 1947, 2,417,995.
26. Stamm, A. J.; Harris, E. E. "Chemical Processing of Wood" Chem. Pub. Co. Inc. 212, 5th Avenue, New York 1953, p.253.
27. Risi, J.; Arseneau, D. F. Forest Prod. J. 1958, 8, 252-255.

28. Goldstein, I. S.; Jeroski, E. B.; Lund, A. E.; Nielson, J.F.; Weaver, J. M. Forest Prod. J. 1961, 11, 363-370.
29. Baird, B.R. Wood and Fiber 1969, 1, 54-63.
30. Hess, K. Berichte 1928, 61, 1460.
31. Suida, H. Austrian Patent 1930, 122,499.
32. Singh, S. P.; Dev, L.; Kumar, S. Intl. Jour. Wood Preservation 1980 (in press).
33. Otlesnov, Yu.; Nikitina, N. Proc. of the Latvian Agricultural Academy (LSKhA) 1977, 130, 50-53.
34. Clarke, H. T.; Hartman, W. W. J. Amer. Chem. Soc. 1924, 46, 1731.
35. Singh, S. P.; Dev, I.; Kumar, S. Wood Science 1979, 11, 268-270.
36. American Society for Testing Materials ASTM Standard, 1972, D871-72 (15), 274-277, Philadelphia.
37. Tarkow, H.; Stamm, A. J.; Erickson, E.C.O. USDA Forest Service FPL 1950, 1593.
38. Kumar, S.; Singh, S. P.; Sharma, M. J. Timb. Dev. Assoc. (India) 1979, 25(3), 5-9.
39. Glasser, W. Comments received on paper (Ref. 35) 1978.
40. Karlson, I.; Shvalbe, K. For. Prod. Abstr. 1980, 3, 2943.
41. Singh, S. P.; Kumar, S. J. Timb. Dev. Assoc. (India) 1981, 27(3), 30-35.
42. Davidson, H. L. USDA Forest Service Research Note 1977, FPL 02.
43. Sensarma, P. K. "Wood Destroying Termites of India"; PL 480 Project Report, Forest Research Institute & Colleges, Dehra Dun, India 1976.
44. Singh, S. P.; Sharma, R. P.; Kumar, S. Unpublished data, Forest Research Institute, Dehra Dun, India.
45. Kirk, T. K. Chemistry and Biochemistry of Decay, in "Wood Deterioration and its Prevention by Preservative Treatments"; Syracuse Wood Science Series 5 (Ed. W.A. Côté), Syracuse University Press, 1973; 167.
46. Callow, H. J. J. Textile Institute 1952, 43, T 423.
47. Truksne, D. Forest Products Abstracts 1980, 3 (11), 2946.

RECEIVED December 24, 1981.

20

Thermoplasticization of Cellulose and Wood by Graft Copolymerization and Acylation

NOBUO SHIRAISHI—Kyoto University, Department of Wood Science and Technology, Kyoto 606, Kyoto, Japan

TUTOMU AOKI and MISATO NORIMOTO—Kyoto University, Wood Research Institute, Uji 611, Kyoto, Japan

MASAKAZU OKUMURA—Harima Chemicals Inc., Kokogawa 675, Hyogo, Japan

A homogeneous grafting using an organic cellulose solvent as a reaction medium makes cellulose thermally meltable, while heterogeneous graftings examined do not convert cellulose or wood into a meltable material. When thermally unmeltable acetylated wood as well as acetylated-propionylated wood with low propionyl contents are further chemically modified by grafting, the thermally meltable properties are rendered even by very low graft add-ons. These modified cellulose and wood with thermally meltable properties can be molded to form transparent sheets at adequate temperatures and pressing time not causing thermal degradation, under pressures usually used for the compression molding of plastics.

Wood is a thermally insensitive material. Generally accepted concepts that (a) cellulose is a crystalline polymer, (b) lignin has a three-dimensional network molecular structure with very high molecular weights, and (c) chemical bondings are formed between wood components as are found in lignin-carbohydrate complexes, lead us to have a confidence that wood is not a thermally meltable material. The annual rings can still be seen in the cross section of charcoal obtained after heat treatment of wood above carbonization temperatures. This demonstrates clearly that wood, by its own inherent nature, dose not melt with heat treatment. Consequently, once wood is ground into powder with the size of less than 5 mm and loses its fibrous properties, it can no longer be used as molded materials or boards without effective use of binders, adhesives or synthetic polymers. This is quite different from the properties of thermoplastic synthetic polymers which can be molded to any shape even from the finely ground powder.

It is then of interest to give thermoplastic property to wood or cellulose in order to obtain similarly workable materials as

0097-6156/82/0187-0321$8.00/0
© 1982 American Chemical Society

from fossil resources. Though not so workable as petroleum- or coal-based plastics, thermal properties of grafted cellulosic polymers have been reported. For example, Arthur et al. (1) enhanced thermoelasticity of cellulose by grafting with main interest of lowering the softening temperature. On the other hand, Yoshimura (2), for instance, prepared cellulose-based materials with transition temperatures higher than the starting cellulose derivatives through crosslinking by grafting. No reports have been made on converting cellulose or wood into a thermally meltable material by grafting.

The present paper reports novel methods to prepare thermally meltable cellulose and wood-based materials and their properties. Two methods are emphasized: (a) a grafting and (b) acylation combined with grafting.

Experimental

Materials. Commercial cotton (Pakistani cotton) cut into 5 mm length and Whatman cellulose powder CF-11 was used as cellulose samples after Soxhlet extraction with alcohol-benzene for 7 h. Wood meal (mainly 40 - 80 mesh) from Makanba (Betula Maximowicziana Regel) was used after washing with cold water. Methyl methacrylate and styrene were purified by the conventional methods. Other reagents used were analytical reagent grade.

Preparation of Cellulose-Methyl Methacrylate Graft Copolymer. Cotton cellulose (1 g) was allowed to stand with 1 % aqueous solution of ammonium persulfate (0.5 g) for 30 min at room temperature. Methyl methacrylate (MMA) (20 ml) and methanol (20 ml) were added to the mixture. After degassed, the flask was sealed and shaken at 60 °C for 1 to 4 h. The graft product was isolated by pouring into a large excess acetone followed by Soxhlet extraction with acetone for 48 - 72 h.

Preparation of Cellulose-Styrene-Sulfur Dioxide Graft Copolymer (Homogeneous Grafting). The procedure for preparation of cellulose-styrene copolymer in a SO_2-diethylamine (DEA)-dimethylsulfoxide (DMSO) medium was described in our previous paper (3). The grafting proceeds homogeneously throughout the reaction.

Preparation of Wood-Methyl Methacrylate Graft Copolymer. Wood Meal (0.5 g), an acetic acid-sodium acetate buffer solution with pH 4.6 (5 ml), a ferrous sulfate aqueous solution (5 ml; $FeSO_4$ $1x10^{-4}$ mole), MMA (8 ml), and a hydrogene peroxide aqueous solution (5 ml; H_2O_2 $1x10^{-3}$ mole) were mixed in this order. After degassed, the flask was sealed and shaken at 50 °C for 1 to 8 h. The reaction mixture was poured into a large excess of methanol. To remove homopolymer, the product was then extracted with acetone for 36 - 72 h.

Preparation of Wood-Styrene-Sulfur Dioxide Graft Copolymer. Procedure for preparation of wood-styrene-sulfur dioxide graft copolymer was almost the same as described in our previous paper (3), except that wood was grafted instead of cellulose and that the reaction proceed heterogeneously.

Acetylation and Mixed Acetylation-Propionylation of Wood. Wood meal (15 g) was pretreated with an acetic acid (26.3 ml)-acetic anhydride (4.2 ml) mixture for overnight. Acylating agents were precooled to about -25 °C and added to the pretreated mixture. The compositions of the acylating agents were acetic anhydride (8.91 ml)-acetic acid (48 ml) for acetylation, and acetic anhydride (62.2 ml)-propionic anhydride (36.5 ml)-acetic acid (48 ml) for the mixed acetylation-propionylation. Perchloric acid (0.19 ml) was used as a catalyst and premixed with the acylating agents. The reaction mixture was gently stirred. The temperature of the mixture rose to 25 °C at reaction time of about 1 h, after which the reaction temperature was kept constant at 25, 35, or 45 °C. The total reaction time was 6 h. When the reaction was completed, perchloric acid was neutralized by adding an acetic acid solution of potassium carbonate (1.5 ml; K_2CO_3 0.16 g). The reaction mixture was then poured into a large excess of deionized water or methanol, and the solid product was repeatedly washed by water or methanol.

Preparation of Acylated Wood-Styrene Graft Copolymer. To dried acetylated wood or acetylated-propionylated wood (2.5 g), pyridine (20 ml) and styrene (20 ml) were added. After passing through nitrogen for 1 min, the flask was stoppered, dipped slowly in liquid nitrogen to freeze the content, and sealed. The grafting was achieved by irradiating the mixture at $2x10^4$ r/h with γ-radiation from a cobalt 60 source to a definite total dose. The graft product with the homopolymer was obtained by precipitating in a large excess of methanol, washing by methanol, and drying. The product was, further, extracted with benzene to remove the homopolymer. The dose rate was low enough not to cause the degradation of wood.

Measurement of Thermal Softening and Melting. About 20 mg samples of both untreated and chemically modified cellulose and wood were used for this measurement. Thermal softening and melting were observed as the collapse of a column of powder under a constant pressure of 3 kg/cm^2 in a heated glass capillary tube. A thermomechanical analyzer (Sinku Riko Co. Ltd., TM 3000) was used as measuring device. The measurement was conducted from 20 °C to 400 °C at a programmed heating rate of 1 °C/min.

Results and Discussion

Thermoplasticization of Cellulose by Heterogeneous Graft Copolymerization. As the first trial to obtain a thermoplastic cellulose graft copolymer, a heterogeneous grafting was examined. A catalyzed grafting method using ammonium persulfate as the catalyst was employed for copolymerizing MMA to cotton cellulose. The results are shown in Table I. The reaction proceeds readily to a high graft add-on by this method.

Table I. Heterogeneous Graft Copolymerization of MMA onto Cellulose Using Ammonium Persulfate as a Catalyst

Sample	Polymerization Time (hr)	Weight Increase (%)	Grafting Efficiency (%)
PC-1	1	125	20.3
PC-2	2	178	16.4
PC-3	3	374	18.8
PC-4	4	413	18.5

Thermoplastic properties of these grafted cellulose samples were examined and the results are shown in Figs. 1 and 2. In Fig. 1, deformation of the sample (Δ) under a constant load at a programmed heating rate versus temperature (T) is shown. The curve for untreated cellulose shows one softening region where the deformation occurs only to a Δ value of 0.67. Other curves for the grafted materials show two specific regions around 130 °C and 340 °C where the deformations occur sharply. The deformation of all the grafted cellulose samples reaches the Δ value of 1.0, indicating a complete flow of the material. This phenomenon can be ascribed to either melting of all the components or thermoplastic flow of the material with remaining solid parts.

To identify the mechanism, the samples were submitted to high pressure (50 kg/cm^2) at around 300 °C to mold films of the grafted cellulose and moldability of the composites was examined. Unaltered cellulose fibers were observed in the films (Photo. 1). This can be interpreted that the grafted celluloses which have more than 125 % polymer add-ons do not melt thoroughly, and the complete flow observed in the thermodiagram in Fig. 1 is due to thermoplastic flow in which PMMA is melted and cellulose remain solid.

In Fig. 2, plots of the deformation rate (dΔ/dT) against T for the samples in Table I are shown. Two distinct peaks around 130 and 340 °C are in good accord with the conclusions drawn from Fig. 1. A subpeak appears about 300 °C.

To interpret these peaks, thermomechanical behavior of homopolymethyl methacrylate (PMMA) with different molecular weights in

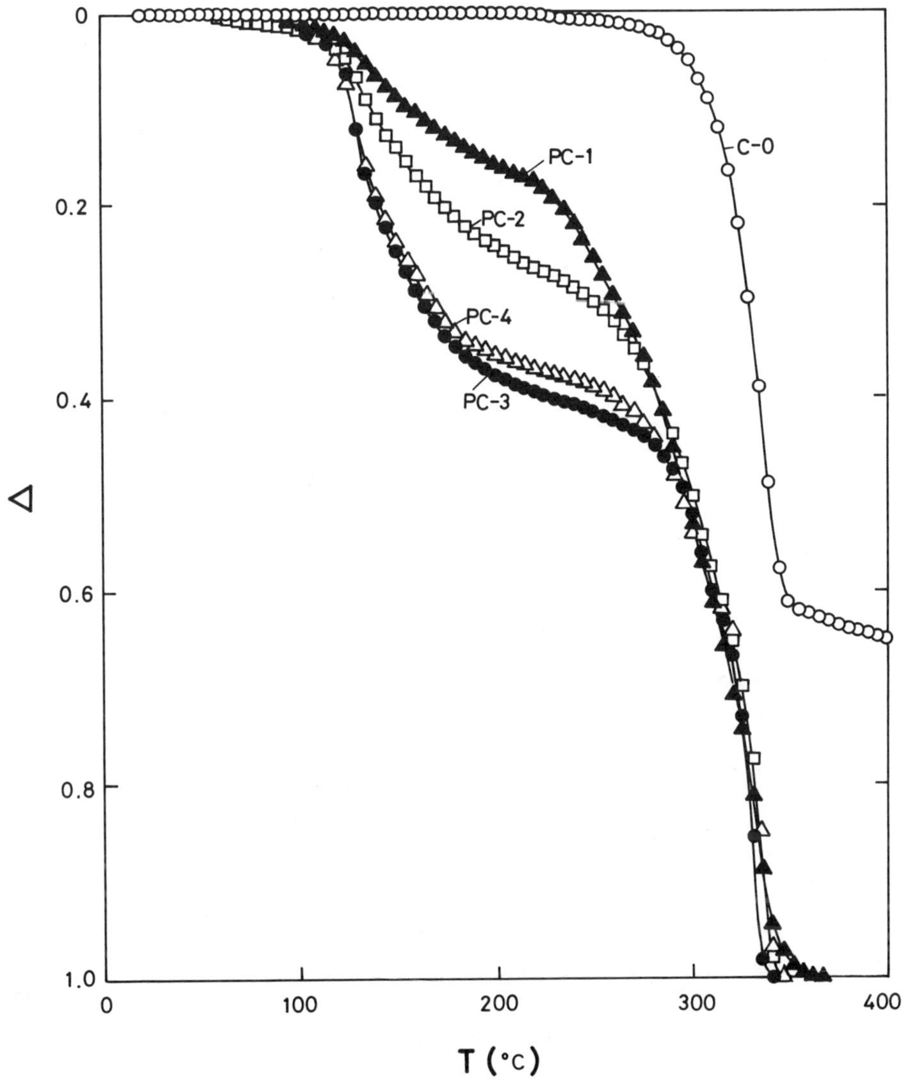

Figure 1. Plots of the deformation vs. temperature. Key: C–O, untreated cellulose and PC-1—PC-4, cellulose–PMMA composites prepared by heterogeneous grafting (see *Table I*).

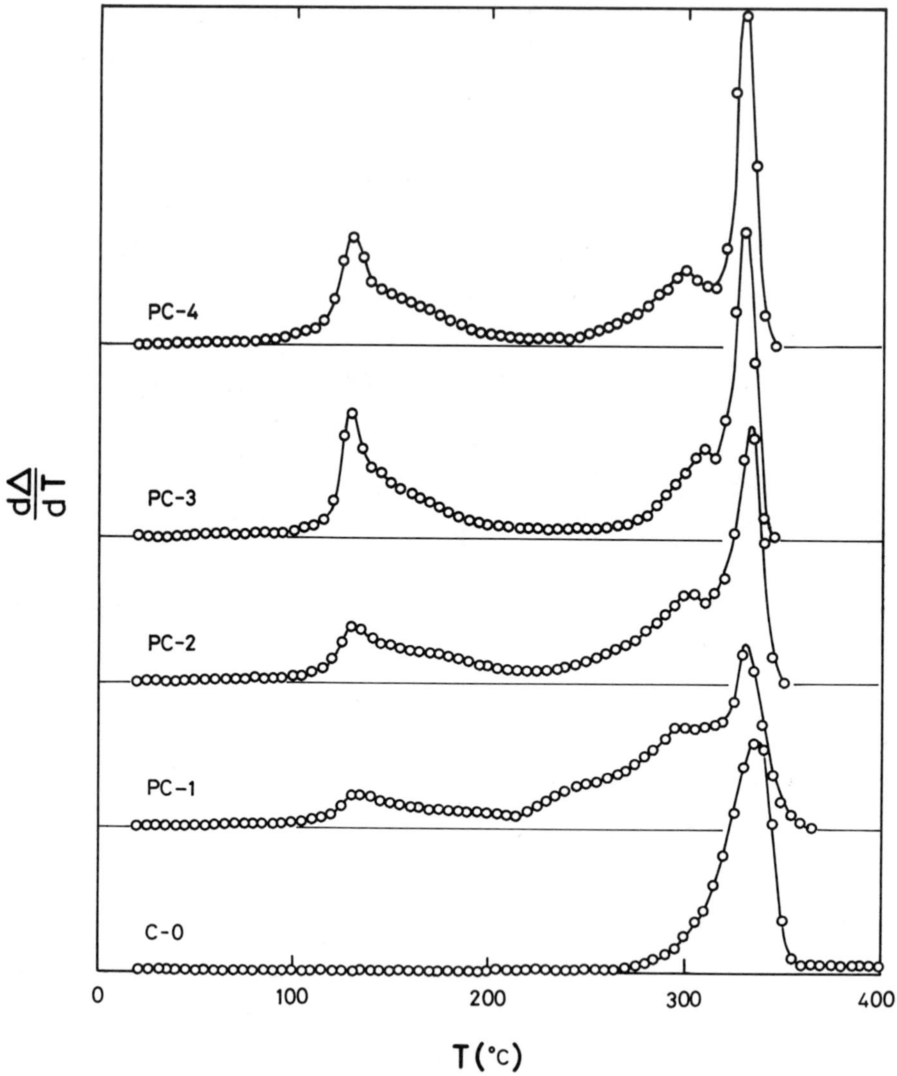

Figure 2. Plots of the deformation rate vs. temperature. Key: C–O, untreated cellulose and PC–1—PC–4, cellulose–PMMA composites prepared by heterogeneous grafting (see *Table I*).

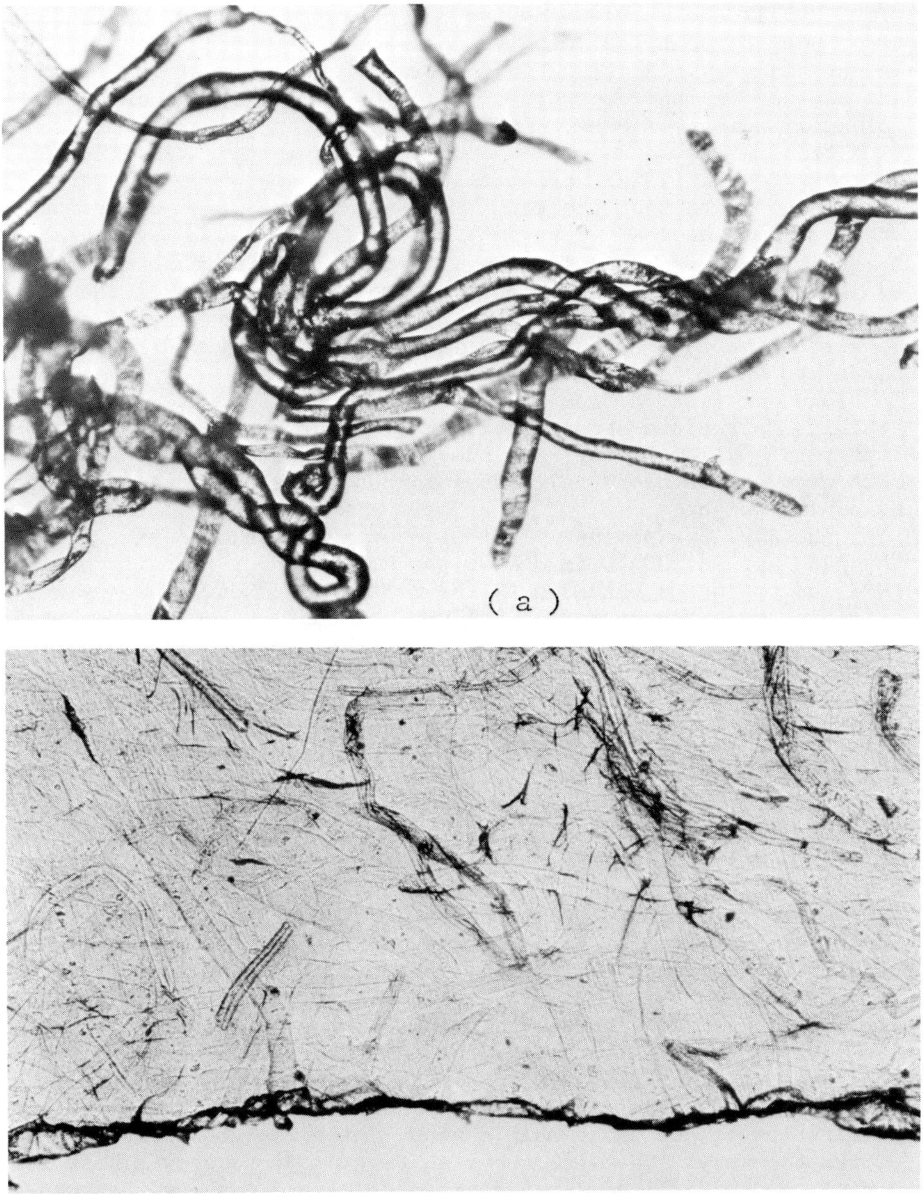

Photo 1. Cellulose sample used. Top: A part of molded film from the cellulose–PMMA composite, PC–3. Molding conditions: temperature 300°C; time, 2 min; pressure, 50 kg/cm². Bottom: under a microscope.

the range 10^4 to 10^6 were examined, and the results are shown in Fig. 3. All the curves show two regions with considerable deformation; one at about 130 °C which does not shift with molecular weight changes is ascribed to the glass-rubber transition of PMMA, and another in the temperature range 160 - 350 °C depends on the molecular weight of PMMA. Independence of the glass-rubber transition temperature and dependence of apparent melting temperatures on the molecular weight are recognized with amorphous polymers. It is noted here that the term "melting" means a flow occurring under a constant load (3 kg/cm^2). The flow is caused by whole molecular motions, in which the slipping of the main chains plays a significant role and the lower the molecular weight of the polymer is, the easier the slipping of the main chains takes place. Thus, the second peak appearing in the range 160 - 350 °C is concluded to be due to the melting of the sample.

Another finding from Fig. 3 is that PMMA extracted from the cellulose structure shows a melting behavior at the highest temperature (350 °C) among the PMMA samples. This indicates that the PMMA formed within the cellulose amorphous region has a very high molecular weight.

The Fig. 3 explains well the interpretation of peaks in Fig. 2. The peak around 130 °C is due to the glass-rubber transition of PMMA and the one around 340 °C the melting of PMMA. The thermal softening peak for untreated cellulose around 340 °C is overlapping the melting peak of PMMA for the grafted cellulose. The wide envelope with the sub-peak (ca. 300 °C) for the grafted cellulose appearing between the two main peaks (150 - 320 °C) may be due to interaction between cellulose and PMMA within the composite. Effects of the interaction on the thermomechanical behavior have been discussed by various investigators (4 - 9).

The grafted cellulose samples prepared by the heterogeneous grafting thus do not show melting behavior regardless of their high polymer add-ons. This can be attributed in part to the fact that no uniform introduction of grafted branches along the cellulose main chain is achieved by the heterogeneous grafting method employed in this report; the grafting can not be expected to occur in the crystalline region of cellulose.

To confirm this partly, the grafted cellulose with decrystallized cellulose structure were prepared by dissolving the heterogeneously-prepared composites in a non-aqueous cellulose solvent, the paraformaldehyde (PF)-DMSO system, and regenerating them in methanol. This treatment actually resulted in a permanent decrystallization of cellulose main chains, which was confirmed by X-ray diffractometory. That is, the resultant products show almost completely decrystallized X-ray diagrams. Figure 4 shows thermomechanical diagrams of the decrystallized composites with those for untreated cellulose and decrystallized cellulose. The diagrams of untreated and decrystallized cellulose reveal that the decrystallization treatment lowers the thermal softening peak by 70 °C. This is considered to be caused by destruction of the crystalline struc-

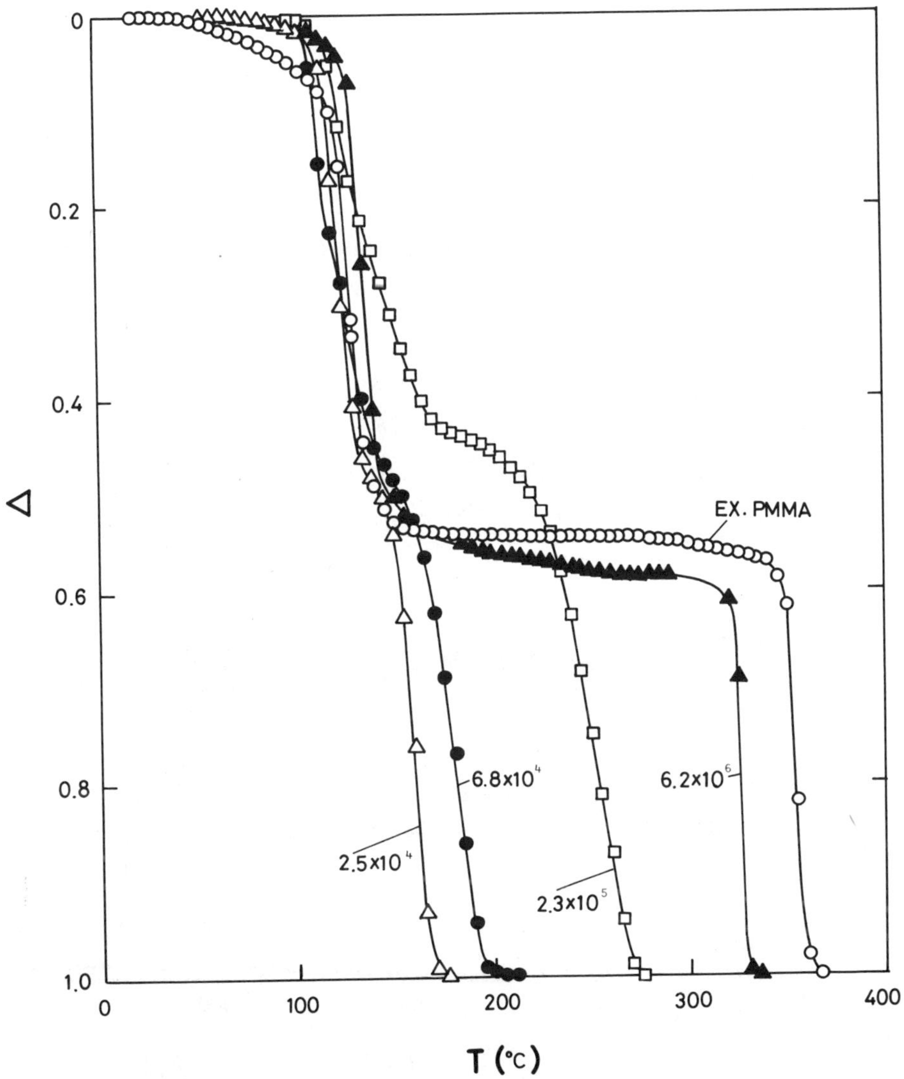

Figure 3. Thermomechanical behavior of various homo-PMMA with different molecular weights. (PMMA is extracted from the cellulose structure of the cellulose–PMMA composite prepared by the heterogeneous grafting).

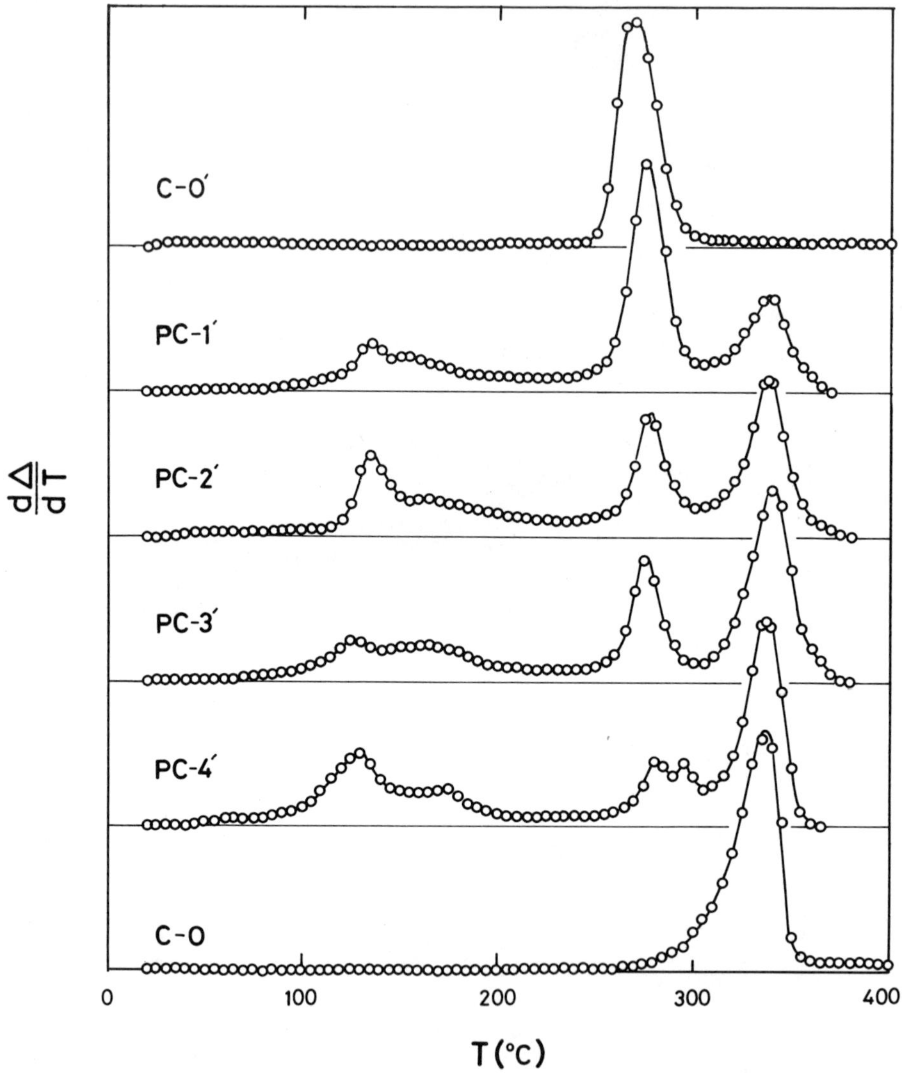

Figure 4. Plots of the deformation rate vs. temperature for PC-1′—PC-4′, the decrystallized cellulose–PMMA composites; C –O′, decrystallized cellulose; and C–O, untreated cellulose. Decrystallization was achieved by using a PF–DMSO solution.

ture of cellulose and loosening of the cohesive structure of cellulose. This change in the diagram of cellulose by the treatment evidences that the separation of the peak for cellulose from one for PMMA at 340 °C is realized by the decrystallization. Both the peaks appeared at the same temperature (340 °C) in Fig. 2.

Although an appreciable thermoplasticization was realized by the decrystallization treatment, the treatment did not give cellulose graft products which melt upon heating. This lack of melting was confirmed by microscopic observations of films prepared with the decrystallized composites.

<u>Thermoplasticization of Cellulose by Homogeneous Copolymerization</u>. A homogeneous grafting was carried out to prepare thermoplasticized cellulose using a non-aqueous cellulose solvent as a reaction medium. Cellulose was first dissolved in the SO_2-DEA-DMSO solution and then styrene was graft-copolymerized onto cellulose by a cobalt 60 γ-ray mutual irradiation method. This method was found to introduce uniformly polysulfone branches with low molecular weight ($\overline{Mn}$ = 3.2 - 3.7 x10^3) along the cellulose main chain and the number of the branches introduced per unit cellulose chain was found to be large (6.4 - 10.6) (10). This grafting method gave polysulfone, because of the conditions of polymerization in the presence of SO_2. The results of the grafting are shown in Table II.

Typical thermomechanical diagrams for the grafted products are shown in Fig. 5 together with homo-polysulfone. Although the grafted cellulose sample having a true grafting value of 21.0 % does not show a melting behavior, the sample with 104.6 % true grafting shows a melting behavior very clearly. The sample with 86.8 % true grafting was also found to be meltable. The true melting for the latter two cases is confirmed as follows: (a) the sharp melting occurs at 126 °C for the homo-polysulfone, which is about 70 °C lower than that for the polysulfone-grafted cellulose (more than at least 86 % grafting). This implies that the complete flow found for the grafted cellulose is not caused by a mere thermoplastic flow; (b) the molded films of the grafted products with enough add-ons are homogeneous and transparent. The results obtained indicate that cellulose can melt upon heating when the true grafting

Table II. Homogeneous Graft Copolymerization of Styrene onto Cellulose Using a SO_2-DEA-DMSO Solution as a Reaction Medium

Sample	Polymerization Time (hr)	Weight Increase (%)	Grafting Efficiency (%)
SC-1	10	58.3	21.0
SC-2	60	111.4	86.8
SC-3	80	131.9	104.6

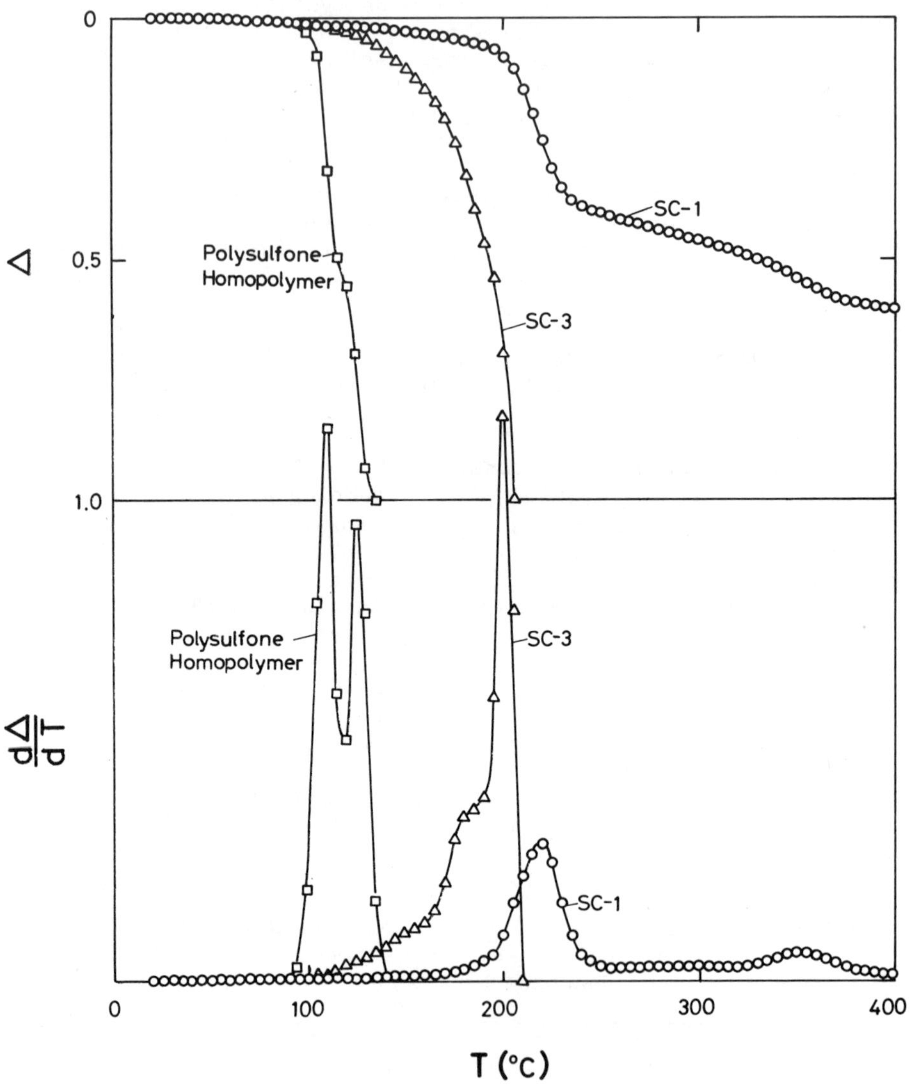

*Figure 5. Thermomechanical diagrams of a polysulfone homopolymer and SC–1 and SC–3, the cellulose–polysulfone composites prepared by the homogeneous graft copolymerization of styrene onto cellulose (*See *Table II).*

exceeds at least 86.8 %. The results are of interest because they show that cellulose can be converted into a thermally meltable material even by grafting, provided that the introduction of branch polymers is achieved uniformly with high density.

Thermoplasticization of Wood by Heterogeneous Graft Copolymerization (Redox Initiation). A redox grafting method using ferrous sulfate-hydrogene peroxide as the initiator was conducted to copolymerize MMA to wood meal. The results of grafting are shown in Table III.

Thermoplastic properties of these grafted wood samples are shown in Figs. 6 and 7. Figure 6 shows that the deformation of grafted wood samples with more than 101 % polymer add-on attains a Δ value of 1.0. The deformation of the grafted samples having polymer add-ons less than 86.8 %, however, does not reach the Δ value of 1.0. It can be concluded that the complete flow is realized with the grafted wood having more than 100 % of weight increase. It was confirmed by microscopic observations of the corresponding molded films that this phenomenon, the complete flow, is caused by a thermoplastic flow.

Another finding obtained from Fig. 6 is that the grafting even with a low degree of the polymer add-on can alter the thermoplastic properties of wood. Wood samples, though grafted only to a lower degree than that to cause a thermoplastic flow of the composite as a whole, have thermal softening temperatures lower than untreated wood. The degree of the thermal softening appearing at lower than 300 °C becomes greater with an increase in the polymer add-on.

From Fig. 7, it can be concluded that the existence of small amounts of PMMA in the cell wall of wood shifts the thermal softening peak to a low temperature by about 60 °C without changing the profile of the curve. As pointed out, this has been interpreted in terms of the interaction of wood components with PMMA (4 - 9). For the composite with more than PMMA content of 50 %, the peaks due to thermoplasticity of PMMA appear clearly, though overlapping the thermal softening curve for wood. A peak at about 130 °C is attributable to the glass-rubber transition of PMMA and the other peak at about 340 °C to the melting of PMMA.

Thermoplasticization of Wood by Graft Copolymerization in Decrystallized State. We have reported that wood can effectively be decrystallized without a weight-loss by treating with a non-aqueous cellulose solvent, the SO_2-DEA-DMSO solution (11). Thus, use of the non-aqueous cellulose solvent as a reaction medium for the graft-copolymerization of monomers to wood was expected to result in products with branch polymers more uniformly distributed. The results obtained by the homogeneous grafting of cellulose (10) were expected to support this idea.

From this point of view, a grafting of wood in a decrystallized state was carried out using the SO_2-DEA-DMSO system as the reaction medium. Wood meals were first contacted with the cellulose

Table III. Redox Graft Copolymerization of MMA onto Wood Using Ferrous Sulfate - Hydrogen Peroxide as an Initiator

Sample	Polymerization Time (hr)	Weight Increase (%)	Grafting Efficiency (%)
PW-1	1	10.6	32.1
PW-2	2	21.1	40.0
PW-3	3	42.1	34.9
PW-4	4	55.3	22.7
PW-5	5	75.1	19.8
PW-6	6	86.8	13.0
PW-7	7	101	10.8
PW-8	8	132	9.39

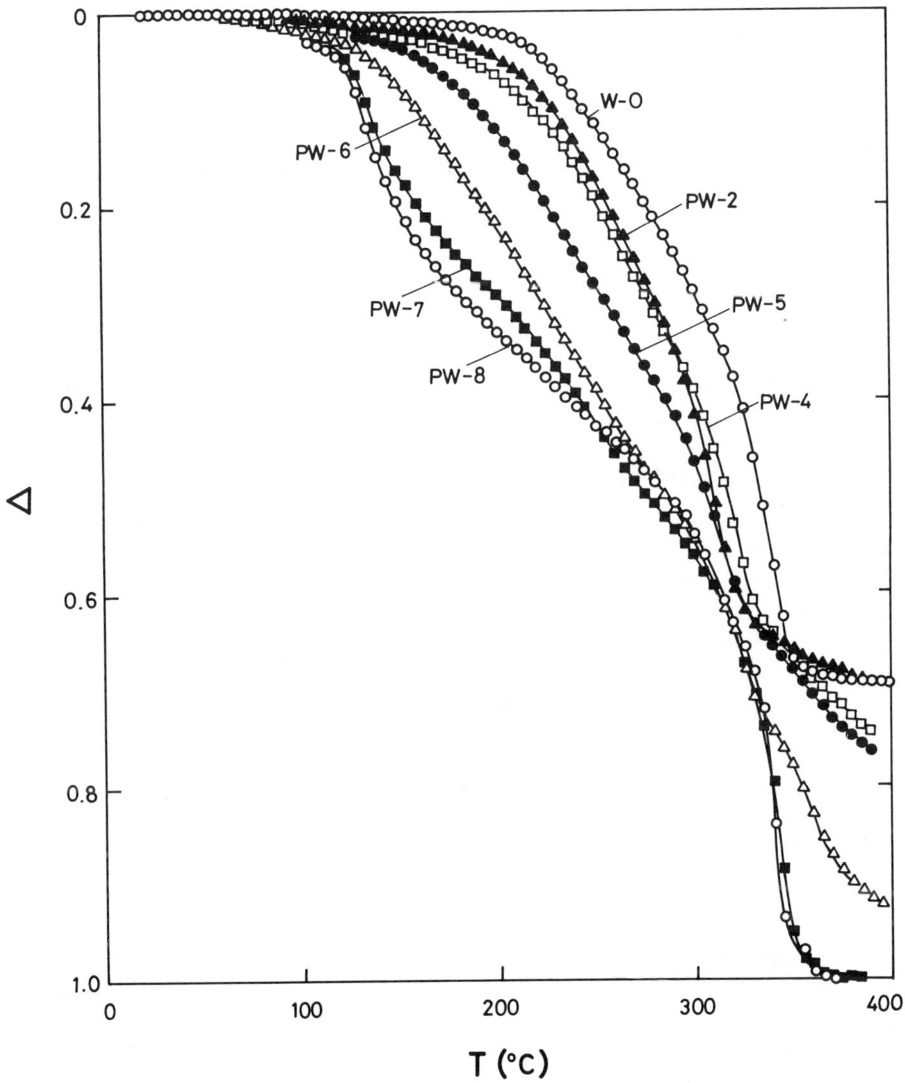

Figure 6. Plots of the deformation vs. temperature for W–O, untreated wood and PW–2—PW–8, the wood-PMMA composites prepared by the redox grafting method (See *Table III*).

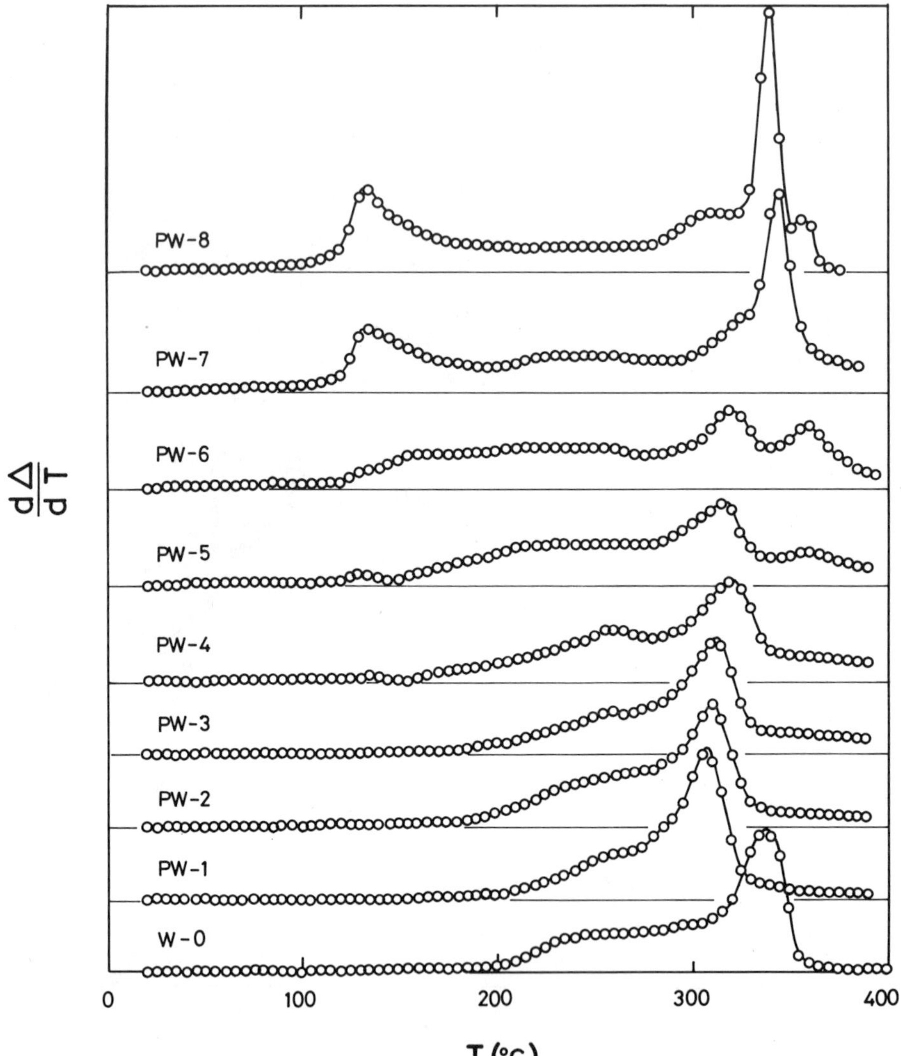

Figure 7. Plots of the deformation rate vs. temperature for W–O, untreated wood and PW–2—PW–8, the wood–PMMA composites prepared by the redox grafting method (See *Table III).*

solvent, and then styrene was added to the mixture and graft-copolymerized.

Although the cellulose solvent was used as the medium in this grafting, the reaction proceeded heterogeneously, because the presence of lignin prevented complete dissolution of wood. The presence of lignin is also known to retard the graft-copolymerization chemically. These may explain partly why the grafting by this system proceeded very slowly compared with the corresponding homogeneous grafting of cellulose as shown in Fig. 8.

Typical thermomechanical diagrams for these grafted wood are shown in Figs. 9 and 10. A complete flow is not observed in Fig. 9, because the polymer add-ons are low. It can be learned from Figs. 9 and 10, however, that the thermal softening temperature of wood is lowered considerably by the grafting with very low degrees of the polymer add-on (even less than 10 %).

In Fig. 10, it is found that curves for the grafted wood samples show two peaks. Both of the peaks shift to lower temperature with an increase in the polymer add-on. As shown in Fig. 5, cellulose grafted with polysulfone melts around 200 °C. The lower temperature peak for the grafted wood in Fig. 10 shifts toward the temperature at which the polysulfone-grafted cellulose shows a complete flow and enlarges its peak height with increasing polymer add-ons. Hence, this peak is considered to be due to the thermal behavior of the polysulfone grafted cellulose within the wood cell wall. The higher temperature peak for the grafted wood samples, ascribable to the thermal softening of unaltered parts of wood, also, not only shifts to lower temperature but decreases its peak height with an increase in the polymer add-ons. The latter is explainable by the gradual decrease in the unaltered parts of wood by grafting. The former can be related to the formation of meltable graft polymers within wood. Because the graft products with enough polymer add-ons can show melting behavior at a lower temperature range than the thermal softening temperature of wood, the grafted products within wood are considered to act as an external plasticizer to lower the thermal softening temperature of the unaltered parts of the grafted wood.

Comparison of the results in Fig. 10 with those in Fig. 7 makes it clear that the differences in monomer species, distribution of grafted branches, and the degree of decrystallization of wood result in quite different thermograms.

<u>Thermoplasticization of Wood by Introducing Higher Aliphatic Acyl Groups</u>. As shown in the previous sections, wood is not easily converted into a thermally meltable material by grafting, although cellulose is found to be convertable to the meltable material by grafting using a non-aqueous cellulose solvent as the reaction medium. It should be noted that in order to obtain the thermally meltable wood the grafting should be conducted so as to get products with homogeneously introduced graft-side-chains and, at the same time, with high polymer add-ons. Considering these two factors,

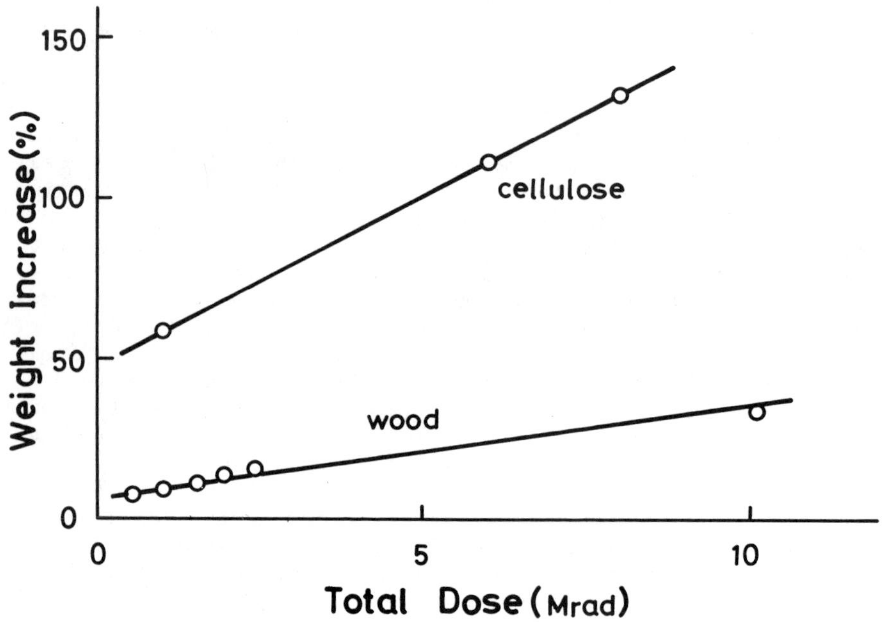

Figure 8. Total dose–weight increase curves for the graft polymerization of styrene onto cellulose and wood using a SO_2–DEA–DMSO solution as a reaction medium.

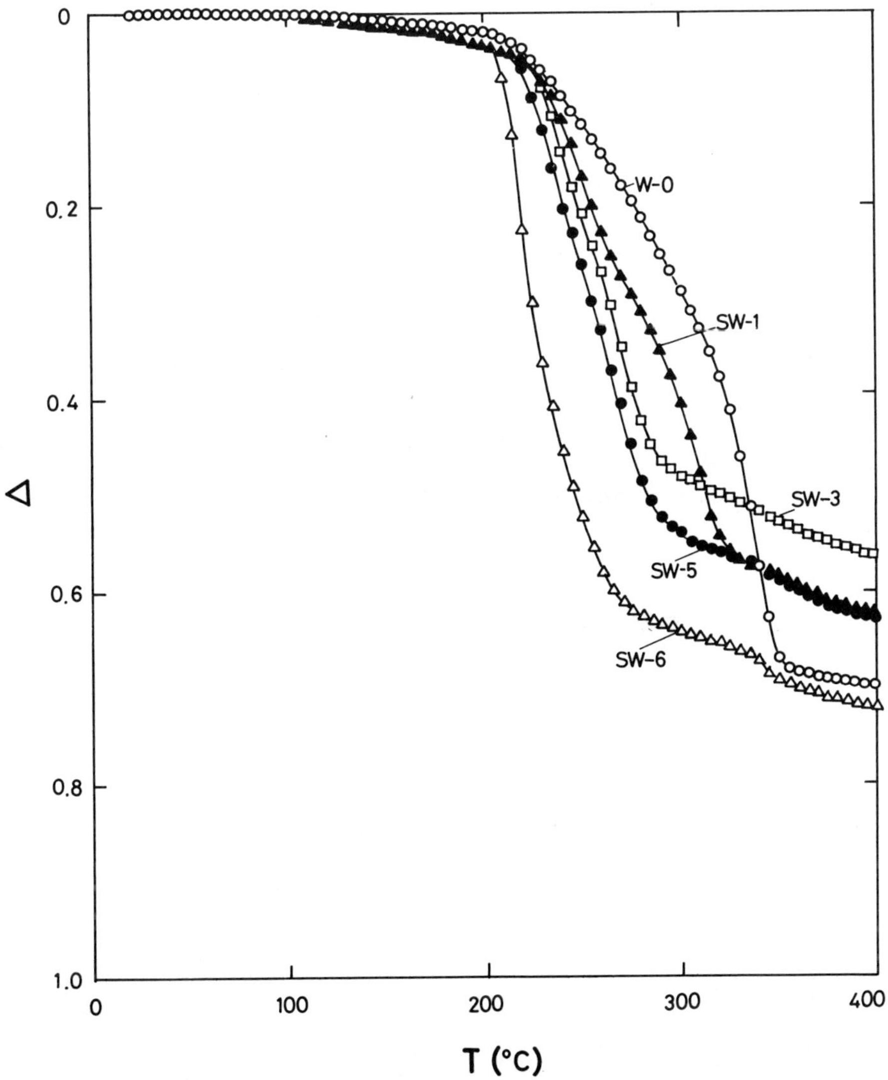

Figure 9. Plots of the deformation vs. temperature for W–O, untreated wood, and SW–1, –3, –5, and –6, the wood–polysulfone composites prepared by the graft copolymerization using a SO_2–DEA–DMSO solution as a reaction medium. The weight increases are SW–1, 7.9%; SW–3, 11.4%; SW–5, 14.9%; and SW–6, 33.0%.

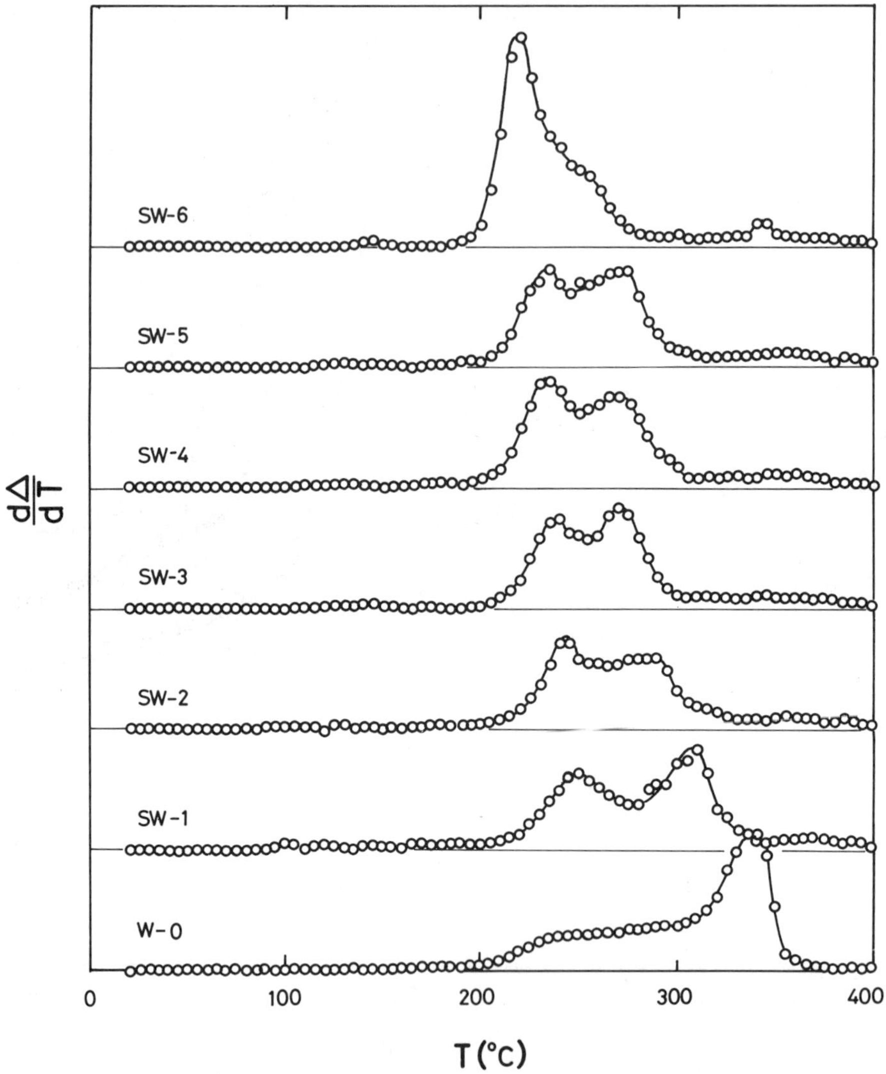

Figure 10. Plots of the deformation rate vs. temperature for W–O, untreated wood and SW-2—SW-6, the wood–polysulfone composites prepared by the graft copolymerization using a SO_2–DEA–DMSO solution as a reaction medium. The weight increases are SW–1, 7.9%; SW–2, 9.4%; SW–3, 11.4%; SW–4, 12.8%; SW–5, 14.9%; and SW–6. 33.0%.

the grafting procedures are considered to be a rather difficult method for thermoplasticization of wood compared with other chemical modification such as acylation and etherification.

We have reported that when wood is acylated with a series of higher aliphatic acid chloride, caproyl to stearoyl chloride, in a non-aqueous cellulose solvent, an N_2O_4-dimethylformamide (DMF)-pyridine solution, the resultant modified wood is given thermally sensitive properties showing clear melting behavior in their thermomechanical diagrams (12, 13). In this method, a very high degree of acylation was not always required to provide the thermally meltable property to wood, and when almost one third of hydroxyl groups in wood were acylated, the products became the thermoplasticized materials. This suggested that the use of cellulose solvents in wood modification results in the uniform introduction of certain chemical groups along the cellulose chain (14)

It was desirable, however, to investigate the possibilities of preparing thermoplastic wood using more common reaction procedures than the above method. If wood can be converted into a thermoplastic material by some simple and economical methods it will have a great value.

As the first step, we have tried to acylate wood in a trifluoroacetic anhydride (TFAA)-higher aliphatic acid system at 30 or 50 °C (TFAA method) and in a higher aliphatic acid chloride-pyridine-DMF system at 100 °C (Chloride method) (15). Both the methods resulted in thermally meltable products. An example of the thermomechanical diagram for the products is shown in Fig. 11. In this figure, the diagram for a lauroylated wood sample prepared by the TFAA method is compared with that for untrated wood. The lauroylated wood shows thermal behavior with a sharp drop caused by complete flow of the sample at 195 °C.

In order to examine whether or not the flow is ascribed to melting of the lauroylated wood, we tried to mold sheet from it by hot-pressing at 140 °C under a pressure of about 150 Kg/cm^2. Transparent sheets were molded from lauroylated wood meals as shown in Photo. 2. This result indicates clearly that the flow behavior observed in the thermomechanical diagram is attributable to melting. The molding temperature used, however, is considerably lower than the melting temperature found in Fig. 11. On this temperature difference, we have recently reported that the apparent melting temperature of acylated wood sample is lowered with an increase in the applied pressure (16); amorphous polymers like esterified wood with higher aliphatic acyl groups do not have melting points being distinctly definable, but have apparent melting temperatures (flow temperatures). The flow temperature varies with measuring conditions such as the pressure applied and the heating rate.

The apparent melting temperature obtained for various acylated wood prepared by both the TFAA and the Chloride methods by using the thermomechanical analyzer under a pressure of 3 Kg/cm^2 are shown in Table IV. The acylated wood samples prepared by the TFAA method show somewhat lower apparent melting temperatures compared with those prepared by the Chloride method.

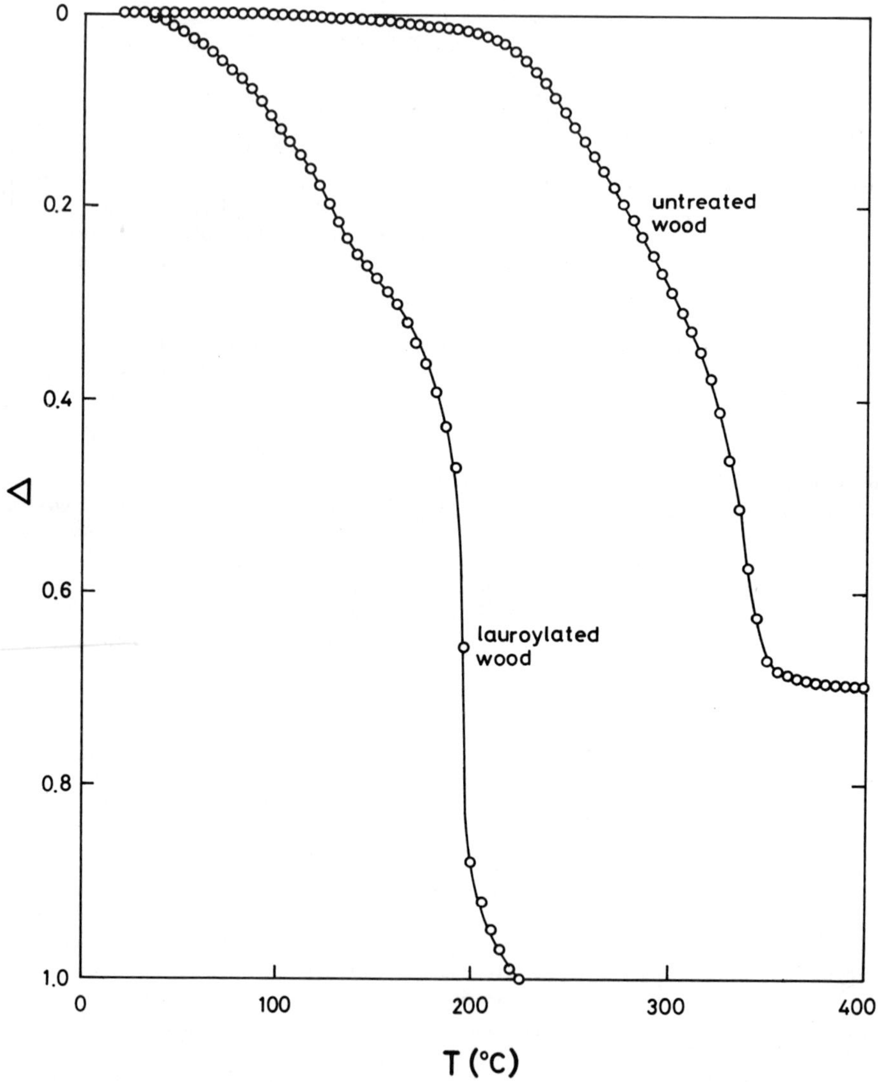

Figure 11. Thermomechanical diagrams of untreated wood and lauroylated wood.

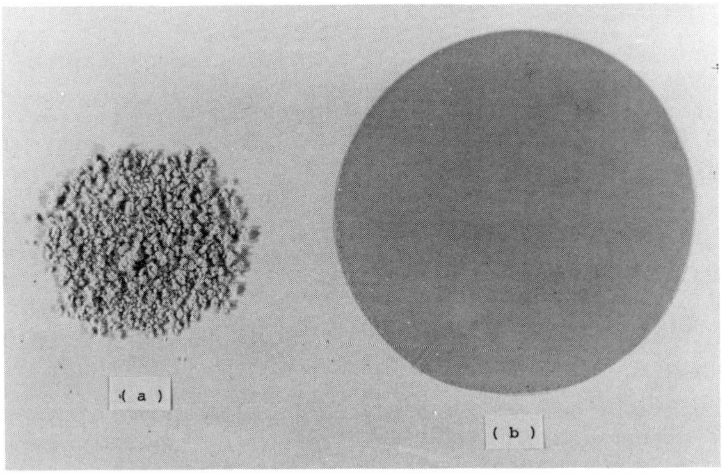

Photo 2. Lauroylated wood meals, a; and a sheet prepared from the lauroylated wood meals by compression molding, b. Molding conditions: temperature, 140°C; time, 2 min; pressure, 150 kg/cm².

Table IV. Apparent Melting Temperature of Various Higher Aliphatic Acid Esters of Wood Prepared by the TFAA or the Chloride Method

Sample (ACYL)	MELTING TEMP. (°C)	
	TFAA	CHLORIDE
BUTYRYL	300	310
VALEROYL	235	305
CAPROYL	250	260
CAPRYLYL	210	245
CAPLYL	205	290
LAUROYL	195	240
MYRISTOYL	200	-
PALMITOYL	195	295
STEALOYL	-	220

Thermoplasticization of Wood by Introducing Lower Aliphatic Acyl Groups Followed by Grafting. Thermoplasticity of acetylated wood is found to be dependent on the method of preparation and the degree of substitution (15, 17). Acetylated wood samples prepared by the TFAA method showed a clear melting phenomenon at 320 °C under a pressure of 3 Kg/cm^2 (15). Wood samples acetylated by the Chloride method or by a method utilizing the acetic anhydride-pyridine or triethylamine-DMF system (25 °C) did not undergo complete flow, while a considerable thermoplasticity was provided (15). Although wood samples fully acetylated by a procedure utilizing the acetic anhydride-acetic acid-sulfuric acid system did not show clear melting, their partially saponified samples gave thermomechanical diagrams with a sharp drop corresponding to complete flow (17). Acetylated wood samples prepared by a procedure using the acetic anhydride-acetic acid-perchloric acid system did not show clear melting, either. Thermal properties of the acetylated wood were enhanced by mixed esterification with other acyl groups. That is, wood esters containing either propionyl or butyryl groups in addition to acetyl revealed meltable properties, if the mixing ratio was appropriate (17).

With these results in mind, we tried to enhance the thermal properties of the esterified wood samples with lower aliphatic acyl groups by grafting. In the first trial, wood samples acylated with the acetic anhydride-propionic anhydride-acetic acid-perchloric acid system were grafted with styrene. The thermomechanical diagrams of the products are shown in Fig. 12 with that of the starting material. In this case, the thermomechanical behavior was examined with the graft products without extraction of the homopolymer. This is based on an idea that the grafting can be compared with the corresponding polymer blend in obtaining molded sheets or films with good qualities. It was also found in this experiment that the extraction of the homopolymer did not change the meltable properties of the acylated-grafted products, giving similar results as shown in Fig. 12 for the non-extracted material.

It is learned from the figure that the grafting can convert the unmeltable acetylated-propionylated wood sample into meltable materials, and the apparent melting temperature decreases with an increase in the amounts of polymer deposites. The further striking finding is that the degree of grafting enough to cause such a drastic change in the thermoplastic property of the esterified wood is very small. Even the graft products with total weight increase of less than 10 %, which are prepared by irradiation to a total dose of less than 0.2 Mrad, behave as thermally meltable materials.

Secondly, thermally unmeltable acetylated wood was tried to be converted to a meltable material. Three kinds of acetylated wood samples, which has been prepared by the acetic anhydride-acetic acid-perchloric acid system at different temperatures of 25, 35 and 45 °C, were all converted into thermally meltable materials by the grafting. An example is shown in Fig. 13. In this figure,

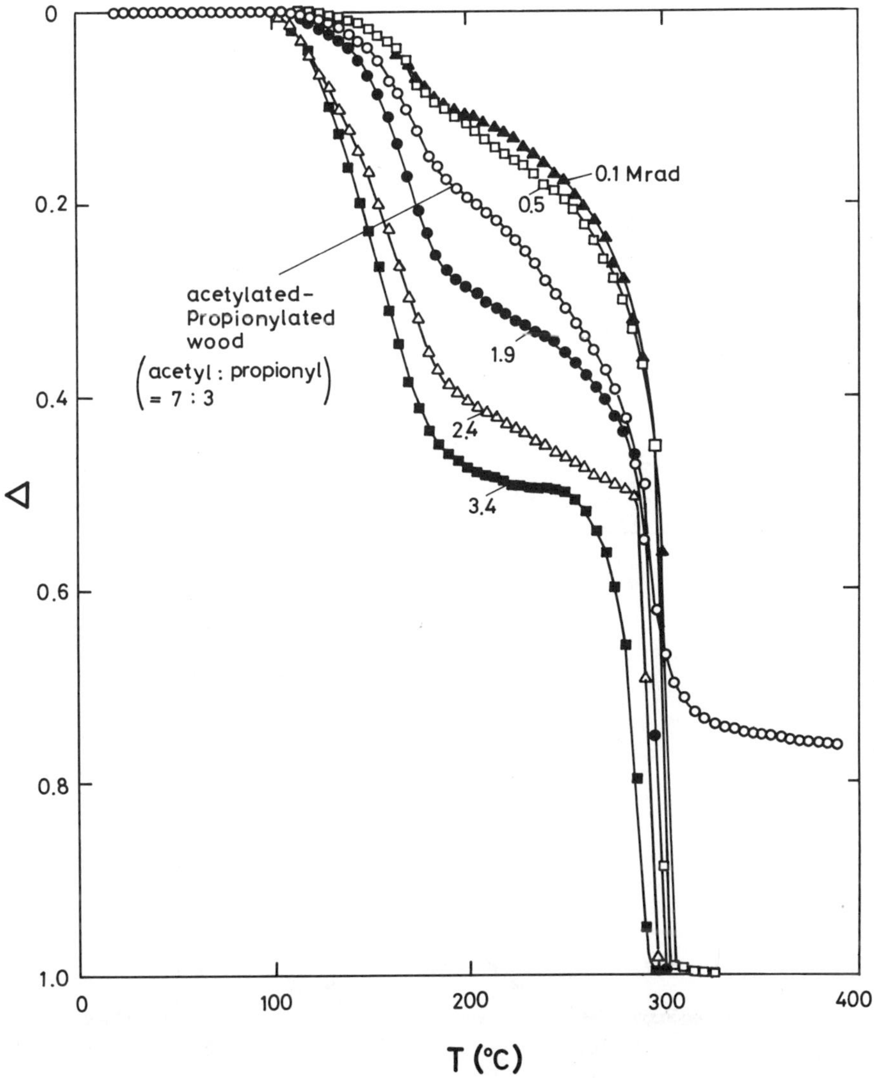

Figure 12. Thermomechanical behavior of an acetylated–propionylated wood sample and the acetylated–propionylated wood–polystyrene composites prepared by the γ-ray induced graft copolymerization in a pyridine medium. Numerical values on the curves represent the total dose of the irradiated γ-ray. Key (total irradiation, resultant weight increase based on the weight of wood): ▲, 0.1 Mrad, 4.2%; □, 0.5 Mrad, 12.2%; ●, 1.9 Mrad, 49.1%; △, 2.4 Mrad, 66.0%; and ■, 3.4 Mrad, 87.1%.

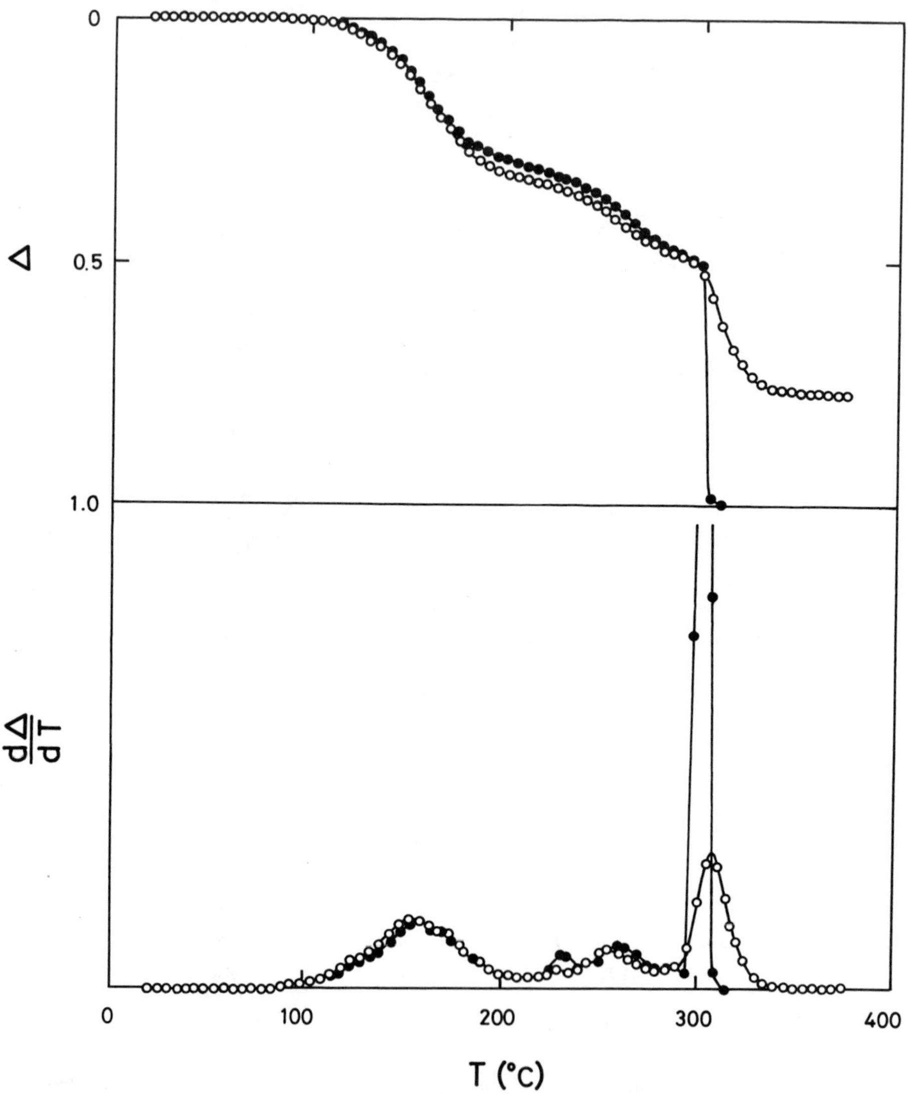

Figure 13. Thermomechanical behavior of ○, the acetylated wood and ●, the acetylated wood–polystyrene composite prepared by the γ-ray induced graft copolymerization in a pyridine medium. Conditions: total dose, 2 Mrad; resultant weight increase, 76.7%.

a thermomechanical diagram for the grafted product of acetylated wood is compared with that for the corresponding acetylated wood. It is clearly seen that the grafting is effective for preparing a thermally meltable material from the acetylated wood. Although irradiation to a total dose of 2 Mrad was applied for the grafting in this case, the low degree of irradiation as low as 1 Mrad or less has almost the same effect for the thermoplasticization of acetylated wood.

These results obtained can be interpreted, at least, in terms of the effect of external plasticization caused by the deposition of polystyrene within wood cell wall in addition to the insufficient internal plasticization previously provided by the acylation.

The above results shown in Figs. 12 and 13 can be discussed in connection with the application of the thermoplasticized wood. The thermoplasticized wood can be used as material for molding, and as one way of utilization, can be used as blend composites with synthetic polymers. If this blending is made by grafting as shown above, two benefits can at least be pointed out: (a) the thermoplasticity of wood materials is enhanced. (Better results can be obtained with esterified wood.) (b) the compatibility of the plasticized wood with synthetic polymers increases by the grafting. These factors are considered to be advantageous for preparing molded composites with excellent final properties.

Literature Cited

1. Arthur, J. C. Jr.; Blouin, F. A. Am. Dyestuff Reptr. 1962, 51, 1024.
2. Yoshimura, S. Sen-i Gakkaishi 1965, 21, 358.
3. Tsuzuki, M.; Hagiwara, I.; Shiraishi, N.; Yokota, T. J. Appl. Polymer Sci. 1980, 25, 2721.
4. Mizumachi, H.; Kamidozono, M. Holzforschung 1975, 29, 229.
5. Tadokoro, K.; Sadoh, T.; Nakato, K. Mokuzai Gakkaishi 1976, 22, 309.
6. Handa, T.; Yoshizawa, S.; Ikeda, Y.; Saito, M. Kobunshi Ronbunshū 1976, 33, 147.
7. Okumura, M.; Shiraishi, N.; Sadoh, T.; Yokota, T. J. Soc. Materials Sci. Japan 1977, 26, 465.
8. Kawakami, H.; Shiraishi, N.; Yokota, T. Mokuzai Gakkaishi 1977, 23, 143.
9. Okumura, M.; Aso, K.; Shiraishi, N.; Yokota, T. Holzforschung 1980, 34, 23.
10. Tsuzuki, M.; Hagiwara, I.; Shiraishi, N.; Yokota, T. J. Appl. Polymer Sci., 1980, 25, 2909.
11. Shiraishi, N.; Sato, S.; Yokota, T. Mokuzai Gakkaishi 1975, 21, 297.
12. Shiraishi, N.; Matsunaga, T.; Yokota, T. J. Appl. Polymer Sci., 1979, 24, 2361.
13. Funakoshi, H.; Shiraishi, N.; Norimoto, M.; Aoki, T.; Hayashi, S.; Yokota, T. Holzforschung, 1979, 33, 159.

14. Shiraishi, N.; Okumura, M.; Yokota, T. Mokuzai Gakkaishi, 1976, 22, 232.
15. Shiraishi, N.; Tsubouchi, K.; Matsunaga, T.; Yokota, T.; Aoki, T. Abst. Papers Presented at the 30th National Meeting, Japan Wood Res. Soc., Kyoto, 1980, p. 34.
16. Aoki, T.; Shiraishi, N.; Tanahashi, M.; Yokota, T.; Yamada, T. Wood Research and Technical Note, 1980, No. 15, 61.
17. Shiraishi, N.; Fukuhara, K.; Tsubouchi, K.; Yokota, T.; Aoki, T. Abst. Papers Presented at the 31st National Meeting, Japan Wood Res. Soc., Tokyo, 1981, p. 263.

RECEIVED January 28, 1982.

21

UV Degradation and Accelerated Weathering of Chemically Modified Wood

WILLIAM C. FEIST and ROGER M. ROWELL

USDA Forest Service, Forest Products Laboratory, Madison, WI 53705

The roles of cell wall chemical modification, polymer lumen-fill treatments, and a combination of these two treatments in reducing the degradative effects of ultraviolet light on wood revealed widely varying degrees of effectiveness. Under conditions simulating accelerated outdoor weathering (ultraviolet light irradiation alternating with water spray), butylene oxide- or butyl isocyanate-modified southern pine sapwood performed no better than untreated controls and surface degradation was severe. With ultraviolet light exposure only, surface degradation was much less for both modified wood and untreated wood. Southern pine with a dual treatment of chemical modification with butylene oxide or butyl isocyanate followed by lumen-fill treatment with methyl methacrylate, or southern pine impregnated with methyl methacrylate and polymerized _in situ_, resulted in modified woods that were resistant to accelerated weathering and to ultraviolet light alone. Physical, chemical, and microscopic changes occurring as a result of ultraviolet light irradiation are described.

Wood and wood-based products are durable materials that have long been recognized for their versatility and their attractive engineering and structural properties (1). Like other materials, wood is susceptible to environmental deterioration. Wood exposed outdoors undergoes physical changes and chemical reactions in a process usually referred to as weathering (2, 3). These changes and reactions play an important role when wood products are used for exterior siding or cladding.

The weathering of wood is caused primarily by the action of ultraviolet (UV) light, water, heat, and abrasion. The contribution of these elements to the wood weathering process has

0097-6156/82/0187-0349$6.25/0
© 1982 American Chemical Society

received considerable attention (2-5). The outdoor weathering of wood causes rapid surface color changes followed by surface roughening as the grain raises and the wood checks and cracks. As wetting and drying occur, boards may warp. The wood loses its surface coherence, becomes friable, and splinters and fragments come off. Of the elements contributing to the weathering process, UV irradiation from the sun and stresses imposed by wetting and drying are the most important in the majority of climates.

Wood is an excellent light absorber and almost every chemical constituent in wood is sensitive, with consequential deteriorating effect (6, 7). Of the various chemical constituents in wood, lignin appears to be oxidized and degraded by UV light most rapidly (5, 8, 9, 10). Because of lignin's strong UV light-absorbing characteristics, it may also function as a partial shielding agent to protect wood cellulosic components from photodegradation by UV (9). Fortunately, UV light does not penetrate wood deeper than approximately 75 µm (11). Consequently, the interaction of wood components and UV light is essentially a surface reaction in which the free radical intermediates generated play a major role in surface deterioration and discoloration (5, 6, 8, 12, 13).

Many conventional and experimental surface treatments and modifications for wood have been developed to reduce or eliminate the effects of the various weathering elements (2, 3, 5, 14). The chemical changes occurring and the mechanisms of degradation and color formation during weathering have received considerable attention and have been summarized in an earlier paper in this series (3).

Chemical modification of wood cell wall components has been shown to be successful in imparting resistance to various degrading elements (15, 16, 17). Reduced hygroscopicity of chemically modified woods has been demonstrated (18, 19, 20). There are no chemical modifications published in which UV stabilization and control of hygroscopicity (the main weathering elements) were the primary goals of the modification (15).

In an earlier paper (3), it was shown that the springwood of southern pine chemically modified with butylene oxide or methyl isocyanate for protection from biodegradation was not adequately protected against the degradative effects of accelerated weathering (water and UV irradiation). Increasing the dimensional stability of the wood and blocking lignin phenolic hydroxyl groups apparently was not enough to stop the effects of these weathering elements. Filling the wood lumens with methyl methacrylate reduced the rate of springwood erosion by about 50%. The polymer encasing the wood elements probably reduced the rate of water uptake and retarded the leaching of degradation products. It acted as a glue-like material helping to hold the surface cellulosic fibers in place even though the lignin component may have been degraded by UV light. A combination of butylene oxide or methyl isocyanate cell wall modifying treatment followed by methyl methacrylate lumen-filling treatment resulted in a system that protected the wood for 90 days of accelerated weathering. The addition of methyl methacrylate to the cell wall-modifying chemical treatment provided additional water repellency and possible lignin stabilization and had a significant effect on reducing weathering.

The chemical modification of wood could play a very important role in controlling the natural weathering process. The primary purpose of this continuing research was to determine the effects of chemical modification of wood on weatherability and to elucidate the mechanism(s) of UV degradation of modified wood. Chemical modification of wood cell walls with butyl isocyanate (BuIso) or butylene oxide (BuO), lumen-filling modification with methyl methacrylate (MMA), and combined cell wall modification and lumen-filling modifications were compared to unmodified southern pine. Physical, microscopic, and chemical changes occurring on the wood surfaces after UV irradiation in the controlled accelerated weathering environments were evaluated for springwood and summerwood. Both UV light and UV light/water combinations of exposure were included in the studies.

Experimental

Wood Specimens. Southern pine sapwood blocks, 2.4 x 2.4 x 0.6 cm^3 (longitudinal x radial x tangential), were prepared from a single freshly cut log. All specimens were cut and planed to provide a smooth vertical-grain surface and ovendried for 20 hrs at 105°C. The 70 specimens were randomized for use in different portions of the study.

Chemical Modification. Ovendried blocks were reacted in a stainless steel vessel at 120°C, 150 psi nitrogen pressure, 10 blocks with BuIso and 10 with BuO monomer using a technique described earlier (18, 19). BuIso was reacted with wood in the presence of dimethylformamide (65/35, v/v) and BuO with triethylamine (95/5, v/v). Another set of 10 blocks was treated in a glass chamber with MMA monomer containing 5% trimethylol propane trimethacrylate as a cross-linking agent and 0.25% azobisisobutyronitrile as a catalyst (3). Separate blocks in a fourth group were first modified with either the BuIso or BuO system (10 each) and then treated with MMA. The BuIso or BuO systems result in chemically modified (cell wall-bound) material and MMA in lumen-filling modification (3). All specimens were ovendried after treatment and weighed to determine chemical add-on expressed as weight percent gain (WPG) (19) (Table I). The UV light-only specimens were exposed in an accelerated weathering chamber with UV irradiation only; the light/water specimens were exposed to alternating cycles of light and water.

Accelerated Weathering. The radial faces of all test specimens were exposed to a 6500-W xenon arc light source (which

Table I.--Specimen preparation for accelerated weathering, vertical-grained southern pine sapwood

Specimen	Treatment	WPG[1/]	
		Light only	Light/ water
Control	None	--	--
Butyl isocyanate (BuIso)	BuIso + dimethylformamide, 120°C	27.2	26.3
Butylene oxide (BuO)	BuO + triethylamine catalyst, 120°C	28.4	30.3
Methyl methacrylate (MMA)	MMA, catalyst, 70°C	49.6	48.1
BuIso + MMA	BuIso treatment followed by MMA	25.2 43.9	27.3 39.0
BuO + MMA	BuO treatment followed by MMA	30.6 51.1	30.4 57.3

1/ Weight percent gain as a result of treatment, determined from ovendry weight before and after treatment; average of 2 replicates.

closely approximates natural sunlight spectrum in the visible and UV regions) in an enclosed chamber at 45°-50°C and 50% relative humidity (4). One set of specimens was exposed to this radiation for 24 hrs/day. The second set of specimens was exposed to light alternated with a spray of distilled water at ambient temperatures with the light off; each cycle consisted of 4 hrs of distilled water spray followed by 20 hours of light. Exposure time is always expressed as hours of exposure to light. Erosion of springwood and summerwood was measured using a technique described earlier for vertical-grained specimens (4). Erosion was determined after 600, 1,200, and 1,800 hrs of light.

The specimens for erosion measurement had the upper one-half of the exposed face protected with a stainless steel cover.

Weight Loss. Specimens for weight loss determinations were placed in both accelerated weathering chambers (light cycle and light/water cycle) with fully exposed faces. The specimens were ovendried and weighed before and after exposure (1,800 hrs of light) and their weight loss calculated as a function of ovendried weight after chemical modification (Table II).

Chemical Analysis. Three separate specimens were used for chemical analyses for each modification. The first was the unexposed wood. The second was the outer 0.5 mm of wood (removed by slicing with a razor) exposed in the accelerated weathering chamber (referred to as "outer specimen"). The third specimen was the remainder of the exposed specimen after removal of the 0.5 mm of exposed wood surface (referred to as "inner specimen"). All specimens were ground to pass a 40-mesh screen and ovendried for 16 hrs at 105°C before chemical analysis.

Lignin determinations were by a method similar to Technical Association of the Pulp and Paper Industry Standard T13 (21). Samples were treated with 72% sulfuric acid for 1 hr at 30°C and 3% sulfuric acid for 4 hrs at reflux temperature to hydrolyze and solubilize the wood carbohydrate. The insoluble residue was measured gravimetrically as lignin. The hydrolysate from the lignin determination was used for the reducing sugar analyses (22). All values shown are uncorrected for extractives, chemical add-on from modification, and for the small amount of degradation during hydrolysis (Table III).

Scanning Electron Microscopy (SEM). Selected specimens, both unweathered and artificially weathered, were mounted on 9-mm-diameter circular holders with a mixture of silver paint and cellulose acetate cement. The stub holders were then transferred to a high-vacuum evaporating unit and coated with 10-20 nm of gold. The specimens were examined with a scanning electron microscope (Cambridge Stereoscan) at 20 kV. Both radial surfaces (original, exposed, and unexposed) and tangential surfaces (split through springwood with a razor, exposed and unexposed) were examined.

Table II.--Rate of accelerated weathering and overall weight loss of southern pine sapwood after 1,800 hr light exposure

Specimen	Ovendry weight loss, %		Erosion rate, μm/hr[1/]			
			Springwood		Summerwood	
	Light only	Light/ water	Light only	Light/ water	Light only	Light/ water
Control	0.5	5.7	0.008	0.150	0.008	0.042
BuIso	2.0	5.2	0.017	0.150	0.008	0.033
BuO	1.3	6.4	0.033	0.183	0.017	0.067
MMA	0.4	3.5	0.008	0.033	0.008	0.017
BuIso + MMA	1.6	2.7	0.042	0.017	0.025	0.008
BuO + MMA	1.0	2.5	0.017	0.017	0.008	0.008

1/ Erosion rate determined between 1,200 and 1,800 hr, expressed in μm/hr of light exposure.

Selected specimens from both light and light/water exposure studies were photographed using conventional camera systems for comparative purposes.

Results and Discussion

The effectiveness of cell wall chemical modification and lumen-filling modification in protecting wood against the weathering elements of UV light or UV light in combination with alternating water spray was determined on southern pine sapwood modified with BuIso, BuO, or MMA or combinations (Table I). Cell wall modification was in the range of 25 to 31 WPG, the most effective ranges for these treatments (18, 19); lumen-fill modification was in the range of 39 to 57 WPG. Combined treatment specimens were prepared by chemical modification with BuIso or BuO followed by MMA impregnation and reaction.

Weight Loss. Determination of weight loss by weighing specimens before and after accelerated exposure, and measurement of the amount of material eroded from the exposed wood surface are reliable methods of assessing degradation due to UV light. Ovendry weight loss values (Table II) show that some modifications are effective in reducing weight loss when compared to unmodified controls after exposure to UV light and water cycles. Weight loss is low in the UV light-only exposure because there is no leaching of degraded products by water. Weight loss values for cell wall-modified specimens were higher than those observed for unmodified control specimens. This loss may be the result of residual bound monomer being vaporized or of grafted polymer being degraded.

Weight loss values for specimens in the UV light/water exposure study were from 2 to 11 times greater than for specimens in the UV light-only exposure. MMA modification, either alone or in combination with cell wall modification, reduced weight loss by 40-60% as compared to unmodified controls. All these weight loss values (Table II) indicate loss of wood substance only from the surface of the exposed specimen because UV light does not penetrate wood deeply (11) and the weathering process is a surface phenomenon (2, 4).

Erosion of Wood Substance. Wood elements are lost during weathering or accelerated weathering as lignin is degraded by UV light and the adhesive character of this wood component is reduced (2). Degraded lignin and cellulose-rich fibers are washed away by the action of water and the wood gradually erodes away. In softwoods, springwood erodes much faster than does summerwood (1, 2).

The summerwood of specimens chemically modified with BuIso or BuO eroded faster than unmodified controls when exposed to UV light/water cycles (Figure 1). All specimens containing MMA, either alone or in combination with BuIso or BuO, were more

Table III.--Chemical analyses of chemically modified southern pine after accelerated weathering--continued

Specimen[1/]	WPG[2/]		Lig-nin[3/] %	Total sug-ars,[4/] %	Lignin + sugars, %	Sugar ratios				
	Bonded	Filled				Glu-cose	Xy-lose	Man-nose	Arabi-nose	Galac-tose
Control										
Unexposed	--	--	28.6	66.3	94.9	65.0	10.2	16.1	2.8	5.9
Light-outer	--	--	26.9	65.7	92.6	67.0	9.8	15.9	2.6	4.7
-inner	--	--	29.1	66.3	95.4	64.3	11.1	15.0	2.9	6.7
Light/water-outer	--	--	24.8	70.6	95.4	69.3	8.7	14.3	1.8	5.9
-inner	--	--	28.3	67.5	95.8	65.9	10.0	16.9	2.2	5.0
BuIso										
Unexposed	28.0	--	33.1	45.8	78.9	72.8	9.4	13.8	1.5	2.5
Light-outer	27.2	--	32.2	45.8	78.0	72.1	9.1	13.0	1.5	4.3
-inner	27.2	--	33.8	46.4	80.2	72.6	8.2	13.2	1.8	4.2
Light/water-outer	26.3	--	30.7	49.6	80.3	74.7	8.7	12.9	1.1	2.6
-inner	26.3	--	34.9	45.8	80.7	71.3	9.6	14.0	1.6	3.5
BuO										
Unexposed	31.0	--	29.6	34.1	63.6	87.9	6.1	4.5	0.7	0.8
Light-outer	28.4	--	30.1	35.1	65.2	83.9	6.7	6.5	1.4	1.5
-inner	28.4	--	32.5	34.9	67.4	83.9	6.9	6.4	1.3	1.5
Light/water-outer	30.3	--	27.9	36.7	64.6	87.5	5.9	4.7	0.8	1.1
-inner	30.3	--	31.6	33.8	65.4	87.7	6.3	4.6	0.6	0.8

Page 1 of 2

Table III.--Chemical analyses of chemically modified southern pine after accelerated weathering

Specimen[1/]	WPG[2/]		Lignin[3/]	Total sugars,[4/]	Lignin + sugars,	Sugar ratios				
	Bonded	Filled	%	%	%	Glucose	Xylose	Mannose	Arabinose	Galactose
MMA										
Unexposed	--	49.8	48.5	48.4	96.9	67.4	10.0	16.8	2.6	3.2
Light-outer	--	49.6	47.6	46.1	93.7	66.5	10.3	16.3	2.0	4.9
-inner	--	49.6	50.5	45.6	96.1	65.9	10.8	15.5	3.2	4.6
Light/water-outer	--	48.1	46.7	48.8	95.5	67.2	10.7	16.2	1.7	4.2
-inner	--	48.1	48.7	48.0	96.7	65.7	11.8	16.0	2.7	3.8
BuIso + MMA										
Unexposed	28.0	42.3	53.9	31.4	85.4	72.9	8.0	14.3	1.2	3.6
Light-outer	25.2	43.9	47.8	33.3	81.1	75.6	8.7	13.6	1.1	1.0
-inner	25.2	43.9	55.8	28.5	84.3	74.6	8.8	13.6	1.4	1.6
Light/water-outer	27.3	39.0	48.6	34.4	83.0	75.6	8.1	12.5	1.2	2.6
-inner	27.3	39.0	52.7	30.5	83.2	73.0	8.7	13.6	1.7	3.0
BuO + MMA										
Unexposed	31.5	55.9	53.6	20.7	74.3	88.2	5.8	4.5	1.0	0.5
Light-outer	30.6	51.1	50.0	23.3	73.3	84.6	6.9	6.6	0.9	1.0
-inner	30.6	51.1	53.6	22.5	76.1	85.3	6.8	6.5	0.6	0.8
Light/water-outer	30.4	57.3	54.2	20.1	74.3	88.1	6.0	3.7	0.2	2.0
-inner	30.4	57.3	53.8	20.8	74.6	87.4	6.1	4.4	0.5	1.6

1/ Outer = 0.5 mm thickness of exposed surface; Inner = portion under exposed surface.
2/ Bonded = chemically bonded; Filled = lumen-filled modification.
3/ Material insoluble in 72% H_2SO_4, based on O.D. weight.
4/ Reducing sugars dissolved by acid hydrolysis, based on O.D. weight.

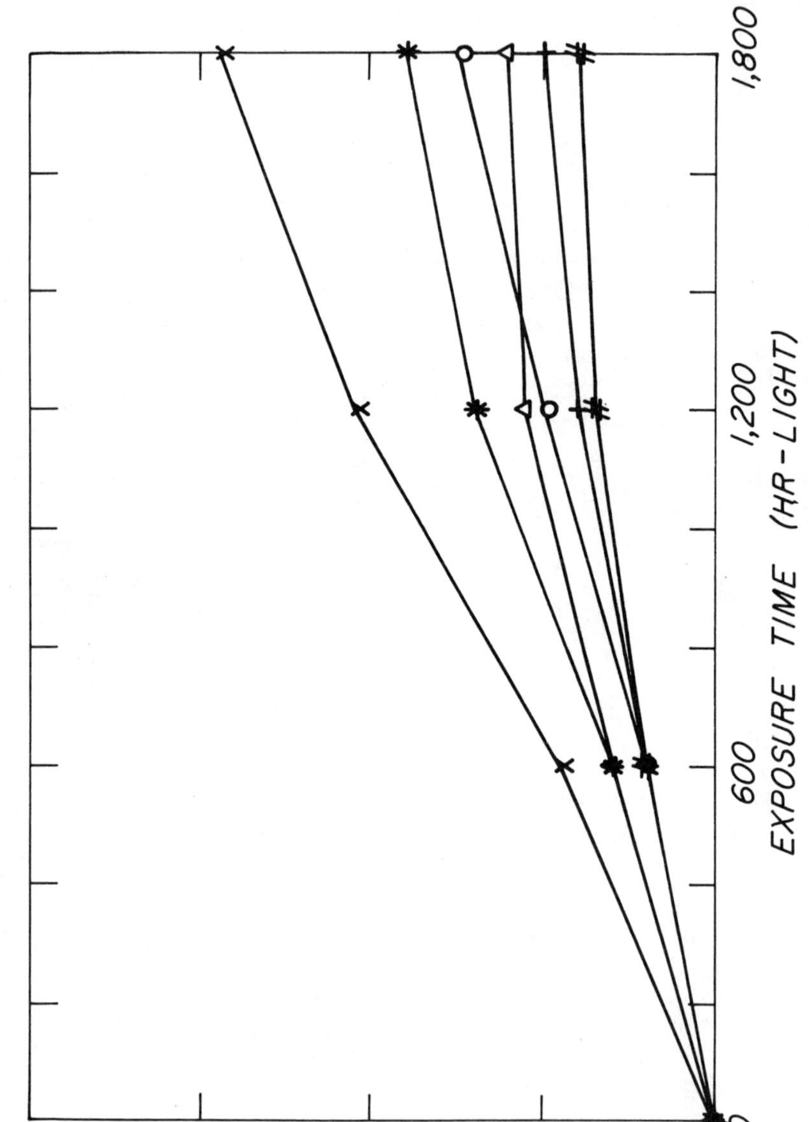

Figure 1. Erosion of Southern pine summerwood in UV-light/water-spray accelerated weathering chamber. Key: ○, *control;* +, *MMA;* *, *BuIso;* ×, *BuO;* △, *BuIso* + *MMA;* #, *BuO* + *MMA* (See *Table I for abbreviations).*

resistant to erosion than unmodified controls. The BuO + MMA specimens showed the greatest resistance to erosion and these specimens also showed the lowest weight loss in the UV light/water exposure (Table II). The decreased summerwood erosion observed for the MMA-filled specimens is consistent with earlier observations for springwood (3) and suggests that the glue-like properties of MMA polymer in the wood are the most important factors operating when these modified woods are exposed to accelerated weathering.

The rate of erosion of southern pine springwood (Table II) and the extent of erosion (Figure 2) is significantly greater than that observed for summerwood, whether the wood is modified or not. Lumen-fill modification with MMA or MMA combinations was effective in controlling loss of springwood cells. Chemical modification with BuIso or BuO was ineffective in controlling erosion. For both springwood and summerwood (Figures 1 and 2), chemical modification of sapwood with BuO yielded a wood in which the overall erosion of wood substance increased during accelerated weathering.

The springwood erosion rates of the MMA-containing specimens, calculated in μm/hr between 1,200 and 1,800 hrs of light exposure, were 1/5 to 1/10 that of unmodified controls after light/water exposure (Table II); summerwood erosion rates were 1/3 to 1/5 those of the controls. BuO or BuIso erosion rates were equal to or greater than those of unmodified controls for both summerwood and springwood in the light/water exposure cycles.

Ovendry wood substance weight loss, springwood and summerwood erosion, and overall erosion rates were all significantly less for specimens exposed to the light-only exposure cycle (Table II). The maximum erosion value for springwood in all specimens, modified or unmodified, after 1,800 hrs of light-only exposure was 40 μm and the minimum was 10 μm; summerwood was approximately 1/2 that of springwood. These compare to values of 60 to 330 μm for specimens in the light/water exposure cycle (Figure 2, 1,800-hr exposure). Without the action of water (leaching, washing), degraded wood substance will still adhere to the wood surface and wood, modified or unmodified, will erode or wear away only slowly.

Physical Changes. Physical changes for the exposed wood can be observed in representative samples of all exposed specimens (Figures 3 and 4). The difference in color between UV light-only and UV light/water exposure is the most obvious change occurring during exposure (2).

The color of exposed wood in the UV light-only exposure is darker than that of the protected wood for all specimens except those chemically modified with BuO (Figure 3). Light-colored woods like southern pine usually darken on exposure to UV light (1, 2). Modification with MMA does not alter the chemical structure of wood, and color changes (darkening) observed are almost identical to those of the control specimens. The BuIso-modified

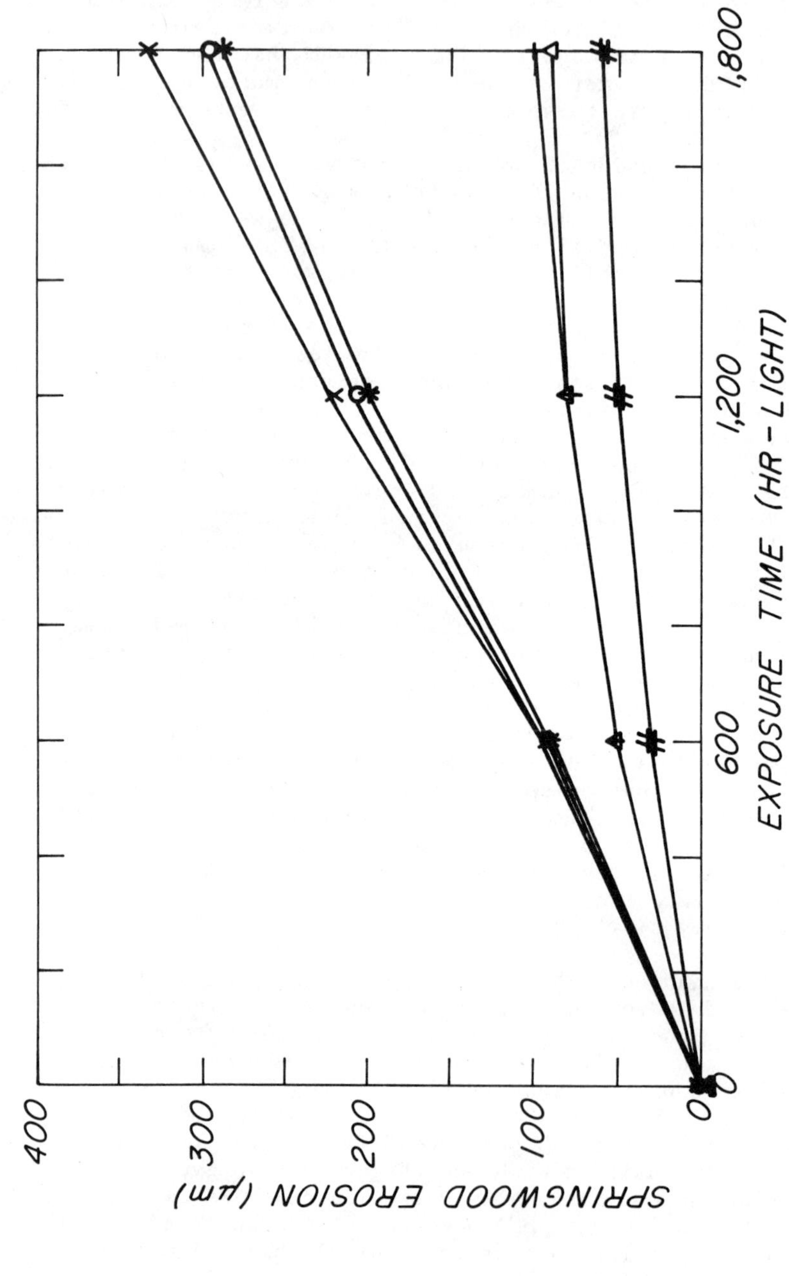

*Figure 2. Erosion of Southern pine springwood in UV-light/water-spray accelerated weathering chamber. Key: ○, Control; +, MMA; *, BuIso; ×, BuO; △, BuIso + MMA; ♯, BuO + MMA (See Table I for abbreviations).*

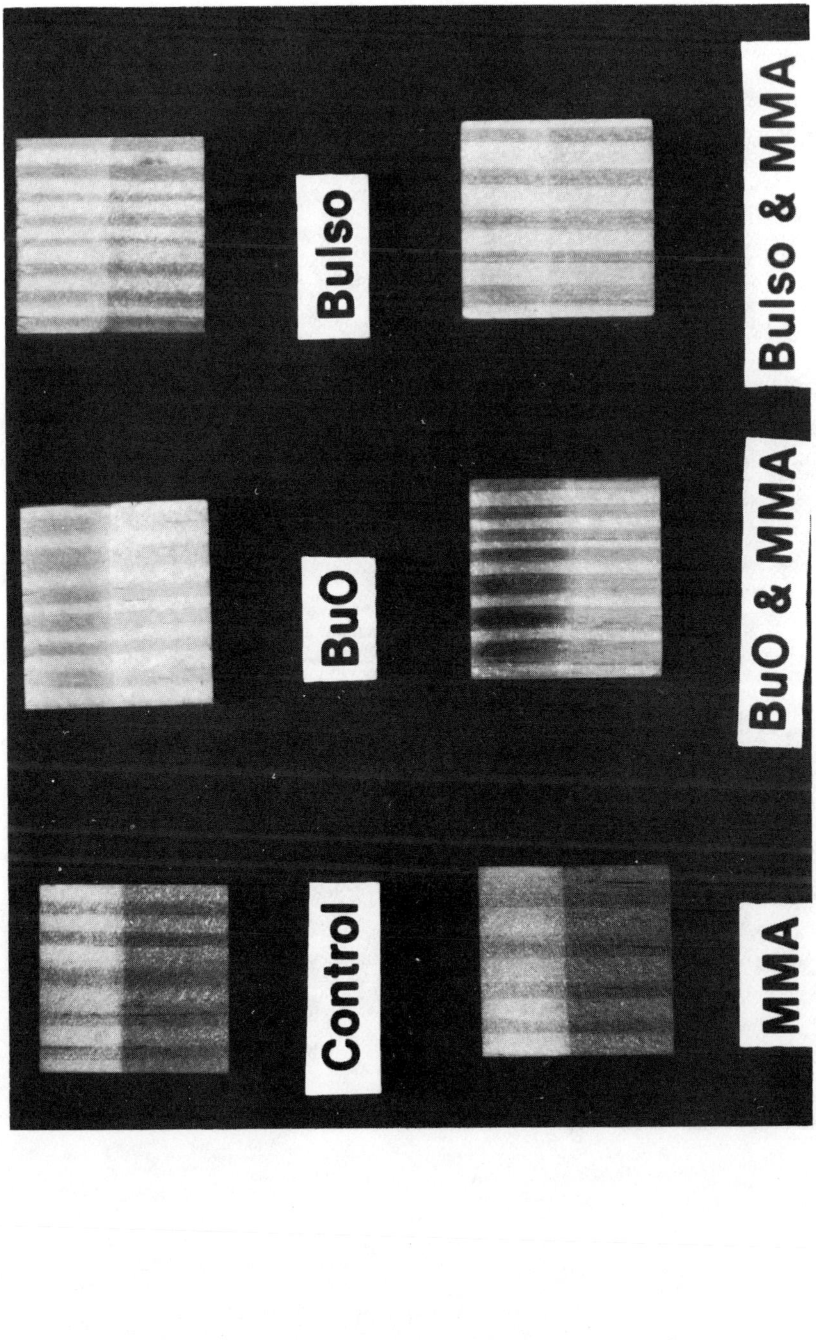

Figure 3. Southern pine specimens after 1800 h exposure to UV light. Upper half of each specimen was protected with a stainless steel plate.

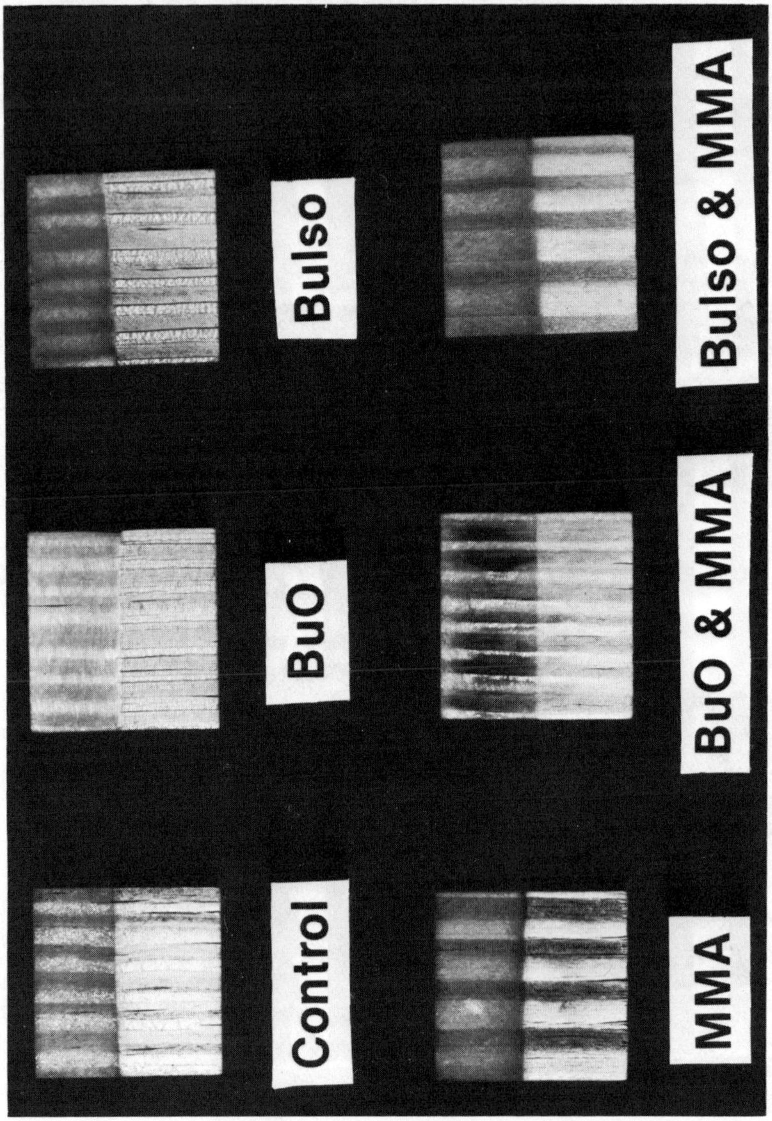

Figure 4. Southern pine specimens after 1800 h exposure to UV light in light/water cycling. Upper half of each specimen was protected with a stainless steel plate.

specimens darkened but to a lesser degree than did the exposed control. Exposed BuIso + MMA-modified wood was somewhat lighter than unexposed wood. BuO-modified wood became lighter in color on exposure to UV light. BuO treatment also resulted in some weight loss after UV light exposure, as did the BuIso treatment (Table II). The bleaching of the BuO-modified wood and differences in weight loss and erosion suggest there may be differences in how these chemicals attach to the hydroxyls as compared to BuIso, and a different degree of sensitivity of chromophoric groups to UV light results.

When UV light-exposed wood is washed with a distilled water spray in the light/water cycles, degraded wood components (presumably primarily lignin) are washed away (2) and shrinking and swelling stresses are developed. Color and other physical changes occurring on such exposure were readily apparent (Figure 4). All woods, modified or unmodified, were bleached and roughened upon exposure. Springwood was eroded preferentially. Original differences in color due to chemical modification and normal anatomical differences were minimized upon exposure. The role of MMA in retaining surface integrity and reducing erosion of wood substance was readily apparent in the exposed portion of the samples (Figure 4).

Chemical Analyses. Chemical analysis of the control sample shows that the sum of the lignin and carbohydrate components (lignin plus total sugars) accounts for 93 to 96% of the wood substance (Table III). In the MMA specimens, the high lignin content shows that the MMA polymer is included in the lignin value. The MMA must be insoluble in the H_2SO_4 procedure and isolated as if it were lignin. Because total sugars are based on a percentage of the total sample, they are low due to the presence of a high percentage of MMA in the specimen. The ratio of sugars is approximately the same for methacrylate-treated wood as for the controls.

The BuIso-modified specimens show a much lower total sugar content, which, in turn, accounts for the lower total mass balance (lignin plus total sugars) (Table III). The reason for this is not known but perhaps part of the explanation is how modified sugars react in the sugar analysis. A substituted sugar unit may not undergo the same colormetric reactions as an unsubstituted sugar. The ratio of sugars is lower in the BuIso specimens as compared to the controls but not as low as BuO-modified wood. The sugars in the hemicellulose fraction are the ones that are lost. It is also possible that these more accessible sugars are more highly substituted and do not respond to the colormetric analysis. Because of the real or apparent loss of sugars in the BuIso- and BuO-modified woods, the total mass balance for these treatments is very low.

Even though there is very little weight loss (after exposure) in specimens dual-treated with either BuIso or BuO combined with MMA, there is a loss in total mass balance and a real or apparent loss of sugars (Table III). The MMA polymer is again showing up in the lignin analysis which, on a total sample % basis, makes the total sugar fraction lower.

It is interesting to note that in specimens modified with BuO (both with and without MMA) the hemicellulose sugar content is greatly reduced. This may be due to the stability of the ether linkage in these modified specimens and the dissolution of substituted sugars in the separation step. The sugars would not be reflected in the sugar analysis and would subsequently appear to have been degraded by the test conditions. The sugars could, in fact, be present in a modified form that did not survive the separation procedure. It is also possible that these stable substituted sugars are less detectable or not detectable by the sugar analysis.

Lignin values were always lower for the exposed outer-surface wood than for unexposed specimens or inner portions of the exposed specimens. This effect has been observed previously (5) and reflects the degradation and solubilization of the lignin component.

The chemical changes occurring when wood or chemically modified wood is exposed to UV light or UV light/water and the significance in using these chemical changes to assess weathering degradation will be addressed in detail in a future publication.

Microscopic Observations. Scanning electron photomicrographs of the planed radial surface of unmodified southern pine before accelerated weathering (Figure 5, upper) and after 1800 hrs UV light exposure (Figure 5, center) show pit degradation and some mild check formation as a result of exposure (2, 10). In a BuO-treated radial surface (Figure 5, lower) slightly greater degradation of pits is visible, although overall wood element degradation is relatively limited.

When unmodified wood is exposed to UV light/water cycles, significantly more surface degradation is apparent than with UV light-only exposure (Figure 6, upper), and both horizontal and longitudinal checking along with loss of surface fibers is apparent. BuIso-modified radial wood surfaces exhibit more severe degradation (Figure 6, center) consistent with the weight loss and erosion data. Severe cracking is apparent on these planed, originally smooth surfaces. The MMA treatment minimizes surface degradation such as lateral checking or cracking--though some large checks occur--and most surface fibers remain intact (Figure 6, lower). This is consistent with the low erosion rate (Table II; Figures 1 and 2) observed for these specimens.

A combination modification of BuO + MMA (Figure 7, upper) after 1,800 hrs exposure (light/water) results in some longitudinal checks on the specimen surface but overall integrity. A good portion of the exposed surface is still reasonably smooth with very little loss of surface wood elements.

A split tangential surface of a BuIso-modified specimen (Figure 7, center) shows the surface degradation (right edge of

Figure 5. SEM photomicrographs of radial Southern pine surfaces showing effects of 1800 h of UV light in an accelerated weathering chamber. Key: upper, unmodified wood before exposure (50×); center, unmodified wood after exposure (100×); lower, BuO-modified wood after exposure (100×).

Figure 6. SEM photomicrographs of radial Southern pine surfaces showing effects of 1800 h of UV light in light/water cycling in an accelerated weathering chamber. Key: upper, unmodified wood after exposure (50×); center, BuIso-modified wood after exposure (100×); lower, MMA-modified wood after exposure (100×).

specimen) upon UV light/water exposure. The deterioration is only 3 or 4 wood fibers deep and fibers below these are essentially unchanged (2, 5). A split tangential surface of a BuIso-modified specimen before exposure is shown (Figure 7, lower) for comparison; the right edge of the specimen is sound and the integrity of this unexposed treated specimen is apparent.

Summary and Conclusions

The springwood and summerwood of southern pine chemically modified with butyl isocyanate or butylene oxide were not resistant to the degradative effects of UV light. Surface deterioration, color changes, and small weight losses occurred during accelerated weathering exposure (UV light and water spray). Accelerated weathering produced little surface erosion until water washed away degraded wood elements. Degradation and loss of summerwood during accelerated weathering was much less than that found for springwood. This trend was observed for modified and unmodified wood. Summerwood erosion was greater for butylene oxide-modified wood than all others. Weight loss increases markedly as lignin degradation products are washed away by the action of the water, and chemical modification does not reduce this weight loss. Increasing the dimensional stability of the wood and blocking lignin phenolic hydroxyl groups apparently is not enough to stop the extreme degradative effects of UV light in the weathering process. UV absorbers or screens chemically bound may be necessary to protect the exposed wood surfaces.

Filling wood lumens and other void volume with methyl methacrylate polymer reduces the extent of erosion. The erosion rate of springwood and summerwood and wood substance loss during accelerated weathering is significantly reduced when compared to chemically modified or unmodified wood. In UV light exposure even with water spray action, degradation is minimal. The MMA polymer, polymerized *in situ* within the wood structure, probably reduces water uptake and retards subsequent leaching of wood degradation products. The polymer can be regarded as a glue-like material holding the surface wood fibers in place even though the natural "glue" (native lignin) has been degraded on the wood surface by the action of the UV light. As the methacrylate polymer holds the cellulose-rich fibers on the wood surface, the fibers may act as partial screens to protect the underlying ungraded wood substance.

Although chemical modification with butyl isocyanate or butylene oxide was not successful in controlling UV light degradation of wood, a combination of either of these chemical modifications with methyl methacrylate lumen-fill treatment resulted in a modified wood that had good accelerated weathering resistance. The combination of the lumen-filling polymer and the cell wall-modifying chemical treatments provides a dimensional stabi-

Figure 7. SEM photomicrographs of radial and tangential Southern pine surfaces after exposure to UV light in light/water cycling. Key: upper, radial surface of BuO + MMA-modified wood (100×); center, tangential surface (opened by splitting sample) of BuIso-modified wood showing degraded fibers on right; lower, tangential surface (split) of BuIso-modified wood before exposure. Integrity of specimen can be seen.

lization and has a significant positive effect on weathering. Weight losses for these combined chemical treatments were at least 50% less than those of the chemically-modified specimens, and wood erosion and erosion rates were low.

Acknowledgments

The authors gratefully acknowledge the laboratory assistance of W. D. Ellis and P. G. Sotos, the analytical assistance and advice of M. J. Effland, and to T. A. Kuster for SEM photographs.

Literature Cited

1. U.S. Forest Products Laboratory. "Wood Handbook: Wood as an engineering material," USDA Agric. Handb. No. 72, Rev. Washington, D.C., 1974.
2. Feist, W. C. Weathering of wood in structural uses, "Proc. Symp.: Structural use of wood in adverse environments." [Vancouver, B.C., 1978] Soc. Wood Sci. Tech., Madison, Wis. 1982 (In Press)
3. Rowell, R. M.; Feist, W. C.; Ellis, W. D. Wood Sci. 1981, 13(4), 202-208.
4. Feist, W. C.; Mraz, E. A. Forest Prod. J. 1978, 28(3), 38-43.
5. Kalnins, M. A. USDA For. Serv. Res. Pap. FPL 57, Part II. 1966, Forest Prod. Lab., Madison, Wis.
6. Hon, D. N.-S.; Glasser, W. Polym. Plast. Technol. Eng. 1979, 12(2), 159-179.
7. Gellerstedt, G.; Petterson, E. L. Svensk Papperstidn. 1977, 80, 15-21.
8. Hon, D. N.-S.; Ifju, G.; Feist, W. C. Wood and Fiber 1980, 12(2), 121-130.
9. Hon, D. N.-S. J. Polym. Sci. Polym. Chem. Ed. 1975, 13, 2641-2652.
10. Chang, S.-T.; Hon, D. N.-S.; Feist, W. C. Wood Sci. (Accepted for publication)
11. Hon, D. N.-S.; Ifju, G. Wood Sci. 1978, 11(2), 118-127.
12. Hon, D. N.-S.; Feist, W. C. Wood Sci. 1981, 14(1), 41-48.
13. Hon, D. N.-S. J. Appl. Poly. Sci. 1975, 19, 2789-2797.
14. Feist, W. C. USDA For. Serv. Res. Pap. FPL 339. 1979, Forest Prod. Lab., Madison, Wis.
15. Rowell, R. M. Amer. Wood-Preserv. Assoc. Proc. 1975, 71, 41-51.
16. Rowell, R. M. Wood Sci. 1980, 13(2), 102-110.
17. Rowell, R. M. Wood Sci. 1978, 10(4), 193-197.
18. Rowell, R. M.; Ellis, W. C. Wood Sci. 1979, 12(1), 52-58.
19. Rowell, R. M.; Gutzmer, D. I. Wood Sci. 1975, 7(3), 240-246.

20. Rowell, R. M.; Youngs, R. L. USDA For. Serv. Res. Note FPL-0243. 1981, Forest Prod. Lab., Madison, Wis.
21. Technical Association of the Pulp and Paper Industry. Standard Method T13. TAPPI, Atlanta, Ga.
22. Saeman, J. F.; Moore, W. E.; Mitchell, R. L.; Millett, M. A. Tappi 1954, 37(8), 336-345.

RECEIVED December 24, 1981.

INDEX

INDEX

B

C

I

J

K

L

M

N

P

Jacket design by Kathleen Schaner.
Production by Florence H. Edwards.

Elements typeset by Service Composition Co., Baltimore, Md.
Printed and bound by Maple Press Co., York, Pa.